Excel

YEAR 9

Mathematics Extension Revision & Exam Workbook

ESSENTIAL skills

Get the Results You Want!

AS Kalra

Reprinted 2001, 2003, 2004, 2006, 2008, 2010, 2011
Updated in 2014 for the Australian Curriculum
Reprinted 2014, 2015, 2017, 2018, 2020, 2021, 2022

Reviewed in 2024 for the NSW Curriculum and Australian Curriculum Version 9.0 changes

ISBN 978 1 74020 034 9

Pascal Press
PO Box 250
Glebe NSW 2037
(02) 9198 1748
www.pascalpress.com.au

Publisher: Vivienne Joannou
Project editor: Michael Cole-King
Edited by Michael Cole-King
Answers checked by Peter Little
Typeset by Precision Typesetting (Barbara Nilsson) and lj Design (Julianne Billington)
Cover and page design by DiZign Pty Ltd
Printed by Vivar Printing/Green Giant Press

Dedication

This book is dedicated to the new generation of young Australians in whose hands lies the future of our nation and who by their hard work, acquired knowledge and intelligence will take Australia successfully through the 21st century.

This book is also in the loving, living and lasting memory of my dear mum, dad and uncle, who will remain a great source of inspiration and encouragement to me for times to come.

Acknowledgements
I would especially like to express my thanks and appreciation to my dear wife and my dear son, who have helped me to find the time to write this book. Without their help and support, achievement of all this work would not have been possible.

Contents

CHAPTER 5 – Linear and non-linear relationships

CHAPTER 6 – Equations

CHAPTER 7 – Area and volume

CHAPTER 8 – Similarity

CHAPTER 9 – Trigonometry

CHAPTER 10 – Probability

CHAPTER 11 – Data representation and analysis

EXAM PAPERS

ANSWERS

Introduction

There are two workbooks in this series for the Year 9 Australian Curriculum Mathematics course:

- ***Excel Essential Skills Year 9 Mathematics Revision & Exam Workbook*** and
- ***Excel Essential Skills Year 9 Mathematics Extension Revision & Exam Workbook*** (this book).

This book should be completed after the first book. It has been written specifically for the Year 9 Australian Curriculum Mathematics course and forms part of a series of eight Revision & Exam Workbooks for Years 7 to 10. Each book in the series has been specifically designed to help students **revise** their work so that they can prepare for success in their **tests** during the school year and in their **half-yearly** and **yearly** exams.

The emphasis in this book is to challenge and extend students through extensive practice. This will ensure that students are fully prepared for the Advanced Mathematics courses in senior years.

The following features will help students achieve this goal:

- This book is a **workbook**. Students write in the book, ensuring that they have all their questions and working in the same place. This is invaluable when revising for exams—no lost notes or missing pages!
- Each page is a **self-contained, carefully graded unit of work**; this means students can plan their revision effectively by completing set pages of work for each section.
- **Topics are covered in depth,** providing students with practice to enable them to focus on areas of weakness or areas for extension.
- A **Topic Test** is provided at the end of each chapter. These tests are designed to help students test their knowledge of each syllabus topic. Practising tests similar to those they will sit at school will build students' confidence and help them perform well in their actual tests.
- **Three Exam Papers** have been included to test students on the complete Year 9 Mathematics course, helping students prepare for their **half-yearly** and **yearly exams**.
- A **marking scheme** is included in both the Topic Tests and Exam Papers to give students an idea of their progress.
- A **Topic Test and Exam Paper Feedback Chart**, found on the inside back cover, enables students to record their scores in all tests and exams.
- **Answers to all questions** are provided at the back of the book.
- There is a **page reference** to the ***Excel Mathematics Study Guide Years 9–10*** or the ***Excel Advanced Mathematics Study Guide Years 9–10*** in the top right-hand corner of all pages, excluding the tests. If students need help with a specific section, they will find relevant explanations and worked examples on these pages of the study guides.

A note from the author

Mathematics is best learned if you have pen and paper with you and do every question in writing. Do not just read through the book—work through it and answer the questions, writing down all working. If this approach is coupled with a menu of motivation, realistic goal-setting and a positive attitude, it will lead to better marks in the examinations.

My best wishes are with you; I believe this book will help you achieve the best possible results. Good luck in your studies!

AS Kalra, MA, MEd, BSc, BEd

CHAPTER 1
Rational numbers, rates and measurements

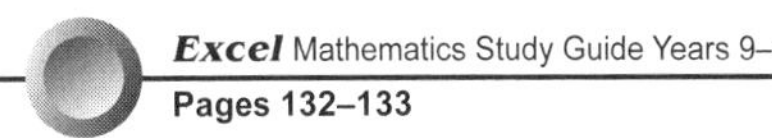

UNIT 1: Rounding

QUESTION 1 What is 2 513 684 rounded to the nearest:

a ten? ______ **b** thousand? ______ **c** million? ______

d ten thousand? ______ **e** hundred? ______ **f** hundred thousand? ______

QUESTION 2 Round off correct to one decimal place.

a 5.78 ______ **b** 6.72 ______ **c** 8.55 ______

d 4.29 ______ **e** 2.97 ______ **f** 0.66 ______

g 0.61 ______ **h** 0.75 ______ **i** 13.95 ______

j 11.66 ______ **k** 0.455 ______ **l** 10.99 ______

QUESTION 3 Round off correct to three decimal places.

a 6.745 2 ______ **b** 8.235 6 ______ **c** 5.738 5 ______

d 0.217 7 ______ **e** 0.006 7 ______ **f** 0.089 9 ______

g 15.0190 ______ **h** 86.007 5 ______ **i** 153.540 6 ______

QUESTION 4 Round off correct to three significant figures.

a 7483 ______ **b** 19 728 ______ **c** 63 584 ______

d 105 674 ______ **e** 3 975 623 ______ **f** 2 029 444 ______

g 0.1678 ______ **h** 13.2595 ______ **i** 7.246 78 ______

j 0.006 529 ______ **k** 0.007 023 77 ______ **l** 0.000 086 235 1 ______

QUESTION 5 Round off correct to four significant figures.

a 19 629 ______ **b** 35 672 919 ______ **c** 856 965 ______

d 30 028.45 ______ **e** 5 912 731 ______ **f** 160 728 ______

g 12.234 5 ______ **h** 3.621 47 ______ **i** 2.907 689 ______

j 0.000 279 15 ______ **k** 0.060 708 55 ______ **l** 0.042 599 ______

QUESTION 6 What is the second significant figure in each of these numbers?

a 29 150 ______ **b** 13.25 ______ **c** 109 200 ______

d 0.00807 ______ **e** 4.795 ______ **f** 0.000 78 ______

UNIT 2: Positive and negative indices

QUESTION 1 Evaluate.

a $2^5 =$ ______ b $3^4 =$ ______ c $5^3 =$ ______

d $10^7 =$ ______ e $1^9 =$ ______ f $2^8 =$ ______

QUESTION 2 Use the index laws to simplify the following, leaving the answers in index form.

a $7^3 \times 7^4 =$ ______ b $5^2 \times 5^5 =$ ______ c $2^4 \times 2^6 =$ ______

d $3^8 \div 3^5 =$ ______ e $11^9 \div 11^2 =$ ______ f $5^{12} \div 5^4 =$ ______

QUESTION 3 Write as fractions in simplest form.

a $\frac{2^2}{2^3} =$ ______ b $\frac{3^5}{3^6} =$ ______ c $\frac{7^8}{7^9} =$ ______

d $\frac{5^6}{5^7} =$ ______ e $\frac{10^4}{10^5} =$ ______ f $\frac{2^4}{2^7} =$ ______

g $\frac{6}{6^3} =$ ______ h $\frac{3^2}{3^4} =$ ______ i $\frac{5^3}{5^{10}} =$ ______

QUESTION 4 Use the index laws to simplify the following, leaving the answers in index form.

a $2^2 \div 2^3 =$ ______ b $3^5 \div 3^6 =$ ______ c $7^8 \div 7^9 =$ ______

d $5^6 \div 5^7 =$ ______ e $10^4 \div 10^5 =$ ______ f $2^4 \div 2^7 =$ ______

g $6 \div 6^3 =$ ______ h $3^2 \div 3^4 =$ ______ i $5^6 \div 5^{10} =$ ______

QUESTION 5 Write as a fraction (without indices).

a $2^{-1} =$ ______ b $3^{-1} =$ ______ c $7^{-1} =$ ______

d $5^{-1} =$ ______ e $10^{-1} =$ ______ f $2^{-3} =$ ______

g $6^{-2} =$ ______ h $3^{-2} =$ ______ i $5^{-4} =$ ______

j $7^{-2} =$ ______ k $2^{-5} =$ ______ l $10^{-6} =$ ______

QUESTION 6 Write in simplest index form (with a negative index).

a $\frac{1}{6} =$ ______ b $\frac{1}{11} =$ ______ c $\frac{1}{13} =$ ______

d $\frac{1}{4} =$ ______ e $\frac{1}{9} =$ ______ f $\frac{1}{8} =$ ______

g $\frac{1}{1000} =$ ______ h $\frac{1}{49} =$ ______ i $\frac{1}{10\,000} =$ ______

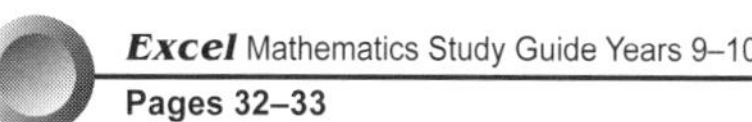

UNIT 3: Powers of 10

QUESTION 1 Write the value of the following.

a $10^3 =$ ______ **b** $10^4 =$ ______ **c** $10^6 =$ ______

d $10^9 =$ ______ **e** $10^7 =$ ______ **f** $10^8 =$ ______

g $10^5 =$ ______ **h** $10^2 =$ ______ **i** $10^{10} =$ ______

j $10^{12} =$ ______ **k** $10^{11} =$ ______ **l** $10^0 =$ ______

QUESTION 2 Write the following as a power of 10.

a 10 = ______ **b** 10 000 = ______ **c** 100 000 = ______

d 100 = ______ **e** 1 = ______ **f** 30 000 000 = ______

g 4000 = ______ **h** 900 000 000 = ______ **i** 1 = ______

j 8 000 000 000 = ______ **k** 70 000 = ______ **l** 500 000 = ______

QUESTION 3 Write the following as a basic numeral.

a $3 \times 10^4 =$ ______ **b** $5 \times 10^6 =$ ______ **c** $8 \times 10^2 =$ ______

d $9 \times 10^3 =$ ______ **e** $6 \times 10^5 =$ ______ **f** $7 \times 10^9 =$ ______

g $2 \times 10^5 =$ ______ **h** $9 \times 10^4 =$ ______ **i** $4 \times 10^5 =$ ______

j $6 \times 10^7 =$ ______ **k** $5 \times 10^8 =$ ______ **l** $3 \times 10^3 =$ ______

QUESTION 4 Complete the following using as a power of 10.

a 3000 = ______ **b** 20 000 = ______ **c** 50 000 = ______

d 600 000 = ______ **e** 600 = ______ **f** 40 000 000 = ______

g 800 000 = ______ **h** 90 000 = ______ **i** 600 000 = ______

j 700 000 = ______ **k** 1 000 000 = ______ **l** 30 000 = ______

QUESTION 5 Write as a fraction.

a $10^{-1} =$ ______ **b** $10^{-4} =$ ______ **c** $10^{-2} =$ ______

d $10^{-3} =$ ______ **e** $10^{-5} =$ ______ **f** $10^{-6} =$ ______

g $10^{-7} =$ ______ **h** $10^{-10} =$ ______ **i** $10^{-8} =$ ______

QUESTION 6 Write as a decimal.

a $10^{-1} =$ ______ **b** $10^{-6} =$ ______ **c** $10^{-9} =$ ______

d $10^{-5} =$ ______ **e** $10^{-2} =$ ______ **f** $10^{-4} =$ ______

g $10^{-3} =$ ______ **h** $10^{-7} =$ ______ **i** $10^{-8} =$ ______

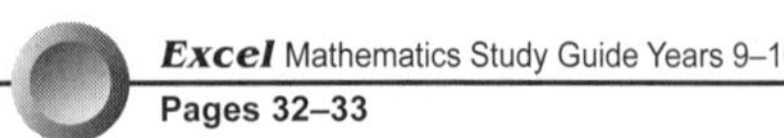

UNIT 4: Scientific notation

Question 1 Express the following in scientific notation.

a 7350 = ______ **b** 52 500 = ______

c 815 000 = ______ **d** 9 386 000 = ______

e 3 500 000 = ______ **f** 6856 = ______

g 69 500 = ______ **h** 43 687 = ______

i 7 864 300 = ______ **j** 853 630 = ______

k 19 643 = ______ **l** 983 000 = ______

Question 2 Express the following in scientific notation.

a 0.0075 = ______ **b** 0.000 982 = ______

c 0.054 = ______ **d** 0.000 095 = ______

e 0.5283 = ______ **f** 0.000 681 3 = ______

g 0.0098 = ______ **h** 0.654 = ______

i 0.6325 = ______ **j** 0.0017 = ______

k 0.000 007 18 = ______ **l** 0.000 835 2 = ______

Question 3 Write the basic numeral for the following.

a $8.7 \times 10^2 =$ ______ **b** $8.0 \times 10^4 =$ ______

c $4.9 \times 10^3 =$ ______ **d** $7.8 \times 10^5 =$ ______

e $2.5 \times 10^5 =$ ______ **f** $1.7 \times 10^4 =$ ______

g $3.7 \times 10^{-2} =$ ______ **h** $4.6 \times 10^{-3} =$ ______

i $9.3 \times 10^{-4} =$ ______ **j** $2.3 \times 10^{-2} =$ ______

Question 4 Use your calculator to answer the following correct to 3 significant figures.

a $(1.2 \times 10^8) \times (2.3 \times 10^4) =$ ______ **b** $(2 \times 10^4) \times (2.5 \times 10^3) =$ ______

c $4.9 \times (1.8 \times 10^8) =$ ______ **d** $(5.9 \times 10^6) \div (2.3 \times 10^3) =$ ______

e $(8.5 \times 10^2) \times (6.3 \times 10^{-4}) =$ ______ **f** $8.1 \times 10^{-2} \times 6.3 \times 10^8 =$ ______

g $(4.5 \times 10^6) \times (3.2 \times 10^3) =$ ______ **h** $(5.6 \times 10^6) \div (2.8 \times 10^2) =$ ______

UNIT 5: Scientific notation calculations

QUESTION **1** Simplify, giving your answers in scientific notation.

a $(3 \times 10^4) \times (2.1 \times 10^5)$ ______ b $(8.5 \times 10^7) \times (2.1 \times 10^2)$ ______
c $(4.5 \times 10^5) \times (1.5 \times 10^{-3})$ ______ d $(8.32 \times 10^2) \times (4.8 \times 10^3)$ ______
e $(1.25 \times 10^7) \times (2.6 \times 10^2)$ ______ f $(3.6 \times 10^4) \times (2.1 \times 10^2)$ ______
g $(9.6 \times 10^3) \times (2.6 \times 10^2)$ ______ h $(6.3 \times 10^2) \times (2.4 \times 10^3)$ ______

QUESTION **2** Give these answers in scientific notation.

a $(4.5 \times 10^7) \div (1.5 \times 10^3)$ ______ b $(8.2 \times 10^9) \div (4.1 \times 10^6)$ ______
c $(2.4 \times 10^6) \div (1.2 \times 10^4)$ ______ d $(9.6 \times 10^7) \div (1.6 \times 10^3)$ ______
e $(9.8 \times 10^3) \div (2.4 \times 10^2)$ ______ f $(2.24 \times 10^{-4}) \div (3.2 \times 10^3)$ ______
g $(8.4 \times 10^5) \div (4.2 \times 10^2)$ ______ h $(8.62 \times 10^5) \div (1.3 \times 10^{-2})$ ______

QUESTION **3** Use a calculator to evaluate these and give your answers in scientific notation.

a $(4.8 \times 10^7) \div (2.4 \times 10^5)$ ______ b $(2.5 \times 10^2) \times (3.6 \times 10^3)$ ______
c $\dfrac{8.5 \times 10^6}{1.7 \times 10^2}$ ______ d $(7.5 \times 10^3) \times (2.1 \times 10^7)$ ______
e $(80)^2 \times (9 \times 10^6)$ ______ f $(2.72 \times 10^6) \times (1.2 \times 10^{-4})$ ______
g $85\,000 \times 9600$ ______ h $337.8 \times (1.25 \times 10^7)$ ______
i $(90 \times 60)^2 \div (3.65 \times 10^{-4})$ ______ j $\sqrt{\dfrac{8.96 \times 10^6}{4.32 \times 10^{-2}}}$ ______

QUESTION **4** Simplify, giving your answers in scientific notation (correct to 4 significant figures where necessary).

a $68\,000 \times 569\,000$ ______ b $5689 \div 1.2567$ ______
c $85\,000 \times 7 \times 21\,000$ ______ d $6.3 \times 8\,169\,000$ ______
e $70\,960 \times 250\,390$ ______ f $5696 \div (698 \times 7653)$ ______
g $\sqrt{6.96 \times 10^{12}}$ ______ h $(6.135 \times 10^{15})^2$ ______
i $(8.9 \times 10^{35}) + (1.5 \times 10^{30})$ ______ j $\sqrt[3]{3.72 \times 10^8}$ ______

QUESTION **5** Use index laws to evaluate the following and give your answers in scientific notation.

a $(3 \times 10^6)^3$ ______ b $(5 \times 10^{-3})^2$ ______
c $\sqrt{7 \times 10^{14}}$ ______ d $(6 \times 10^7) \times (3 \times 10^3)^3$ ______
e $(5.8 \times 10^5) \div (2.9 \times 10^{-5})$ ______ f $(5 \times 10^8) \times (8 \times 10^9)$ ______
g $(6 \times 10^5) \times (8 \times 10^6)$ ______ h $(3.24 \times 10^{-4}) \div (3.21 \times 10^8)$ ______
i $(8.9 \times 10^3) \times (5.2 \times 10^{-8})$ ______ j $(2.8 \times 10^{-8}) \div (1.4 \times 10^{-9})$ ______
k $(1.6 \times 10^{-4}) \div (8 \times 10^{-5})$ ______ l $(6.4 \times 10^5) \div (8.51 \times 10^{-8})$ ______

QUESTION **6** Use index laws to evaluate the following and give your answers in scientific notation.

a $(8.4 \times 10^6) \times (3.2 \times 10^4)$ ______ b $(3.24 \times 10^8) \div (2.4 \times 10^6)$ ______
c $(5.86 \times 10^4) - (3.5 \times 10^3)$ ______ d $(5.2 \times 10^{-2})^2$ ______
e $(7.9 \times 10^5)^3$ ______ f $\dfrac{8.8 \times 10^{-4}}{2.2 \times 10^{-5}}$ ______
g $(6.4 \times 10^5) \times (8 \times 10^{-3})$ ______ h $\sqrt{6.36 \times 10^8}$ ______
i $(8.9 \times 10^{-6}) + (4.5 \times 10^{-4})$ ______ j $(2.4 \times 10^{-3})^3$ ______

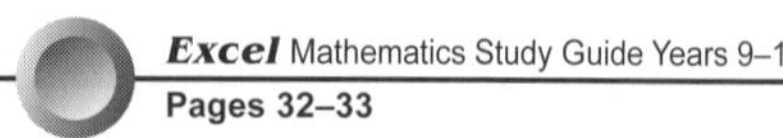
Excel Mathematics Study Guide Years 9–10
Pages 32–33

UNIT 6: Comparing numbers in scientific notation

QUESTION 1 Choose the larger number from each pair.

a 2×10^3 or 5×10^3 ______

b 7×10^9 or 8×10^9 ______

c 5.3×10^4 or 1.5 or 10^6 ______

d 9.5×10^0 or 2.1×10^6 ______

e 3×10^{-2} or 3×10^{-4} ______

f 4.5×10^{-3} or 6.3×10^2 ______

g 6.3×10^{-5} or 3.6×10^{-5} ______

h 8.3×10^{-3} or 5.2×10^{-5} ______

QUESTION 2 Write each group of numbers in ascending order (from the smallest to the largest).

a 3.5×10^8, 3.5×10^5, 3.5×10^{-3} ______

b 8×10^{-4}, 8×10^{-9}, 8×10^{-6} ______

c 3.1×10^4, 2.5×10^{-5}, 1.86×10^3 ______

d 8×10^6, 8×10^4, 8×10^2 ______

e 3.8×10^7, 2.1×10^7, 5.4×10^7 ______

f 6×10^{-2}, 6×10^{-5}, 6×10^{-4} ______

g 3.5×10^{-3}, 3.9×10^{-3}, 5.6×10^{-3} ______

h 8.9×10^0, 3.6×10^5, 5.7×10^{-2} ______

QUESTION 3 Write each group of numbers in descending order (from the largest to the smallest).

a 2.8×10^7, 1.5×10^7, 3.2×10^7 ______

b 8×10^3, 5×10^3, 9×10^3 ______

c 3×10^9, 3.5×10^9, 2.5×10^9 ______

d 4×10^{-5}, 4×10^{-6}, 4×10^{-3} ______

e 5.1×10^{-6}, 2.5×10^{-6}, 3.7×10^{-6} ______

f 3.8×10^2, 4.6×10^3, 3.9×10^{-4} ______

g 2.5×10^{-7}, 3.6×10^{-2}, 4.9×10^{-1} ______

h 5.4×10^4, 3.5×10^3, 8.2×10^6 ______

QUESTION 4 Write in the order indicated.

a 5×10^4, 7×10^3, 8×10^5 (smallest to largest) ______

b 5.3×10^5, 6.7×10^5, 3.2×10^5 (largest to smallest) ______

c 8.5×10^{-3}, 3.7×10^{-2}, 2.5×10^{-4} (smallest to largest) ______

d 6.4×10^{-2}, 5.4×10^{-1}, 6.2×10^{-3} (largest to smallest) ______

e 7.69×10^6, 8.35×10^5, 9.6×10^2 (ascending order) ______

f 9.2×10^3, 8.5×10^3, 7.9×10^3 (descending order) ______

g 3.5×10^{-6}, 5.4×10^{-3}, 6.2×10^{-5} (ascending order) ______

h 5.17×10^{-4}, 3.17×10^{-3}, 8.15×10^{-6} (descending order) ______

QUESTION 5 Select the smaller number from each pair.

a 5.3×10^7 or 8.6×10^5 ______

b 5.04×10^0 or 5.04×10^{-2} ______

c 8.6×10^3 or 1.5×10^7 ______

d 5.8×10^{-4} or 3.2×10^{-6} ______

e 5.79×10^{-6} or 9.57×10^{-6} ______

f 5×10^{-7} or 5×10^{-6} ______

g 3.71×10^{-7} or 9.4×10^3 ______

h 8.6×10^{-6} or $9.7\ \ 10^2$ ______

QUESTION 6 Write in order from smallest to largest.

a 8×10^4, 8×10^3, 8×10^5, 8×10^2 ______

b 5.2×10^5, 3.8×10^5, 8.2×10^5, 7.6×10^5 ______

c 9.1×10^{-2}, 3.8×10^{-3}, 5.4×10^{-4}, 6.3×10^{-5} ______

d 7×10^{-3}, 8.1×10^{-3}, 9.2×10^{-3}, 4.8×10^{-3} ______

e 4.3×10^{-2}, 4.3×10^{-5}, 4.3×10^{-3}, 4.3×10^{-6} ______

f 3.7×10^0, 5.7×10^4, 3.6×10^{-4}, 4.9×10^{-2} ______

g 5.9×10^2, 6.8×10^3, 9.2×10^2, 8.6×10^4 ______

h 3.42×10^3, 4.56×10^{-2}, 8.31×10^{-3}, 5.12×10^2 ______

UNIT 7: Problem solving and scientific notation

QUESTION 1 Write in scientific notation the number of centimetres in 49 km.

QUESTION 2 The distance around Earth's equator is 40 075 km. Express this in scientific notation.

QUESTION 3 The distance between Earth and the sun is 1.521×10^8 km. Express this as a basic numeral.

QUESTION 4 Light travel approximately 9 500 000 000 000 km in one year. Express this in scientific notation.

QUESTION 5 A star has an average distance of 52 800 000 000 000 km from Earth. Express this number in standard notation.

QUESTION 6 The star nearest to Earth is approximately 41 600 000 000 000 km away. Express this distance in scientific notation.

QUESTION 7 The diameter of a star is estimated as 2 773 000 000 km. Write this distance in standard form.

QUESTION 8 The width of a small virus is 1×10^{-4} mm. Write this as an ordinary decimal number.

QUESTION 9 The diameter of an ammonia molecule is 2.97×10^{-8} cm. Write this as an ordinary decimal number.

QUESTION 10 The diameter of Earth is approximately 13 000 km. Write this in centimetres in scientific notation.

QUESTION 11 The sun is approximately 150 000 000 km from Earth. Write this distance in metres standard form.

QUESTION 12 A large molecule has a diameter of 0.000 000 14 mm. Express this in scientific notation.

QUESTION 13 Express in standard notation.

a the number of centimetres in 50 km.

---

b the number of grams in 6 t.

---

c the number of square metres in 120 ha.

---

d the number of millilitres in 380 kL.

QUESTION 14 Express in scientific notation.

a $8235 \div 9\,000\,000$

---

b the product of 8.5×10^7 and 2.6×10^{-5}

---

c the value ab, given that $a = 3.2$ million and $b = 48\,320$

---

d 8.53×10^9 divided by 3.2×10^4

QUESTION 15 The mass of an oxygen atom is approximately 0.000 000 000 000 000 000 026 559 mg. Express this mass in scientific notation.

Rational numbers, rates and measurements

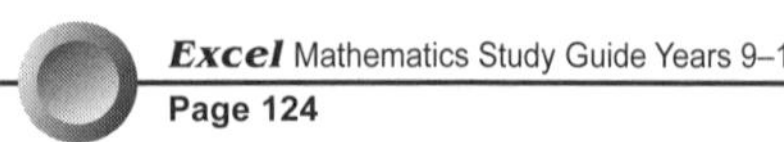

UNIT 8: Units of measurement

QUESTION 1 Complete.

a 5 km = ____________ m **b** 6.2 m = ____________ mm **c** 37 cm = ____________ mm

d 0.7 m = ____________ cm **e** 750 mL = ____________ L **f** 87 250 g = ____________ kg

g 290 000 L = ____________ kL **h** 16 t = ____________ kg **i** 3 mm = ____________ m

j 0.5 km = ____________ m **k** 0.06 g = ____________ mg **l** 0.002 m = ____________ mm

m 1 km = ____________ cm **n** 600 mL = ____________ L **o** 67 500 kg = ____________ t

p 7 mm = ____________ cm **q** 12500 mg = ____________ g **r** 9.5 ha = ____________ m^2

s 15 years = ____________ months **t** 8 min = ____________ s **u** 15120 s = ______ h ______ min

v 1 decade = ____________ years **w** 1 century = ____________ years **x** 1 millennium = ________ years

QUESTION 2 Complete the table.

Prefix	nano	micro	milli	(unit)	kilo	mega	giga	tera
	n	μ	m	-----	k	M	G	T
Meaning				1	1000			
				-----	10^3			

QUESTION 3 Complete.

a 3 ML = ____________ L **b** 7 μm = ____________ m **c** 18 μg = ____________ mg

d 60000 kL = ____________ ML **e** 7000 nm = ____________ μm **f** 8 GL = ____________ L

g 5 Tm = ____________ km **h** 23 000 GL = ____________ ML **i** 60 ng = ____________ g

j 200 000 m = ____________ Gm **k** 0.05 g = ____________ μg **l** 0.008 mm = ____________ μm

QUESTION 4 'As 'kilo' means thousand there are 1000 bytes in a kilobyte.' Is this statement correct? Discuss.

____

____

QUESTION 5 Approximate these conversions using powers of ten.

a 6 MB ≈ ____________ B **b** 92 000 kB ≈ ____________ MB **c** 4 GB ≈ ____________ kB

d 5 TB ≈ ____________ MB **e** 350 GB ≈ ____________ TB **f** 45 000 MB ≈ ____________ GB

QUESTION 6 What fraction of a second is:

a 1 nanosecond? ____________________ **b** 1 microsecond? ____________________

QUESTION 7 How many:

a microseconds are there in 1 hour? **b** nanoseconds are there in 1 day?

____________________________ ____________________________

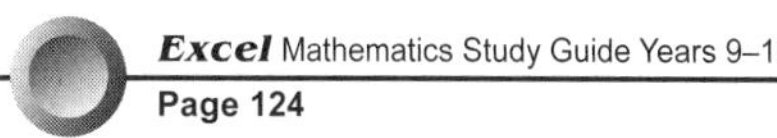

UNIT 9: Accuracy of measurements

QUESTION 1 Billy said: 'This table is exactly two metres long.' Is he correct? Briefly comment.

QUESTION 2

10.152 m

6.237 m

a What is the length of this rectangle to the nearest metre? ____________

b What is the width of the rectangle to the nearest metre? ____________

c What is the area of the rectangle using the rounded measurements from parts a and b?

d Find the area using the measurements given in the diagram.

e Round the answer in part **d** to the nearest square metre. ____________

f Which result is the most accurate for the area of the rectangle to the nearest square metre? Briefly comment.

QUESTION 3 Alex calculated the circumference of the Earth given that the Earth's radius is 6400 km. He gave the answer as 40 212.4 km. Is this a sensible answer? Comment.

QUESTION 4 Each of these measurements is given, correctly, to the nearest 10 m. Write the limits between which the true lengths must lie.

a 40 m ____________ b 360 m ____________

c 1500 m ____________ d 2.3 km ____________

QUESTION 5 Each of these measurements is given correct to one decimal place. Write the limits between which the true lengths must lie.

a 7.8 m ____________ b 3.4 cm ____________

c 21.5 km ____________ d 156.7 m ____________

QUESTION 6 A tape measure is marked in centimetres. To what accuracy can the tape measure be used?

QUESTION 7 A set of scales measures amounts in kilograms. There are four divisional marks between each kg mark on the scales. To what accuracy can the scales be used?

1 kg 2 kg

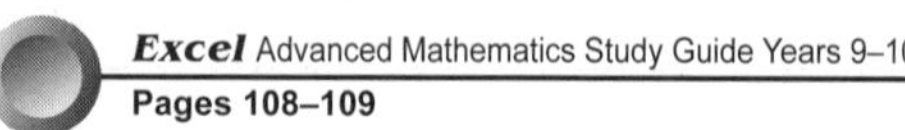

UNIT 10: Rates and proportions

QUESTION 1 A car travels 540 km in $7\frac{1}{2}$ hours.

a What is its average speed?

b How long will it take to travel 450 km at this rate?

QUESTION 2 The speed of sound (at sea level) is 380 m/s.

a Approximately how long does it take sound, at sea level, to travel 1 km?

b If it takes 15 seconds for sound to travel a certain distance, what is that distance?

QUESTION 3 Mary wants to download a file of size 75 MB.

a At a speed of 256 kB/s how long will the file take to download?

b How much faster will the file download at a speed of 1.5 MB/s?

QUESTION 4 A leaky tap can fill a 300 mL cup in $1\frac{1}{2}$ hours.

a How much water is wasted each week?

b What is the cost per week at $2.25/kL?

QUESTION 5 State whether the two measures would be in direct or indirect proportion.

a The amount of petrol used and the distance travelled. ______

b The time for the journey and the speed. ______

c The circumference of a circle and its diameter. ______

d The number of passengers hiring a boat and the cost per person. ______

QUESTION 6 The number, n, of certain trees that can be planted in a particular space is directly proportional to the length, l, of the space, such that $n = kl$ where k is a constant. If 64 trees can be planted in a space of length 200 m, find:

a k

b the number of trees if the length is 350 m

c the length of the space needed for 144 trees

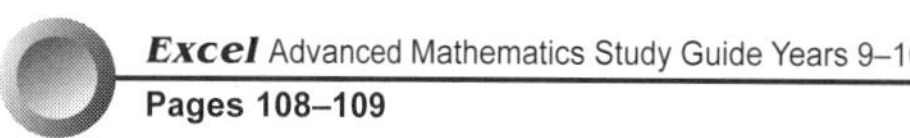

UNIT 11: Rates and conversions

QUESTION 1 Complete.

a \$560 per month = \$__________ per year
b 5 mL per second = __________ mL per minute
c 3.2 t per hour = __________ t per day
d 300 kg per minute = __________ kg per second
e 13 m per minute = __________ m per hour
f 72 m^2 per hour = __________ m^2 per minute

QUESTION 2 A speed of 20 metres per second is how many:

a metres per minute? __________
b metres per hour? __________
c kilometres per hour? __________

QUESTION 3 A speed of 90 km/h is how many:

a metres per hour? __________
b metres per minute? __________
c metres per second? __________

QUESTION 4 Change.

a 15 m/s to km/h
b 126 km/h to m/s
c 6 mL/s to L/h

QUESTION 5 This graph was drawn to convert acres to hectares. Use the graph to answer the following questions.

a How many hectares is 64 acres? __________
b How many hectares is 79 acres? __________
c How many acres is 34 hectares? __________
d How may acres is 7 hectares? __________
e How many hectares is 5000 acres? __________
f A farmer knows that a particular paddock is 40 acres in size. He wants to plant some seed at the rate of 20 kg per ha. How much seed will he need? __________
g The seed comes in 25 kg bags. How many bags will the farmer need? __________

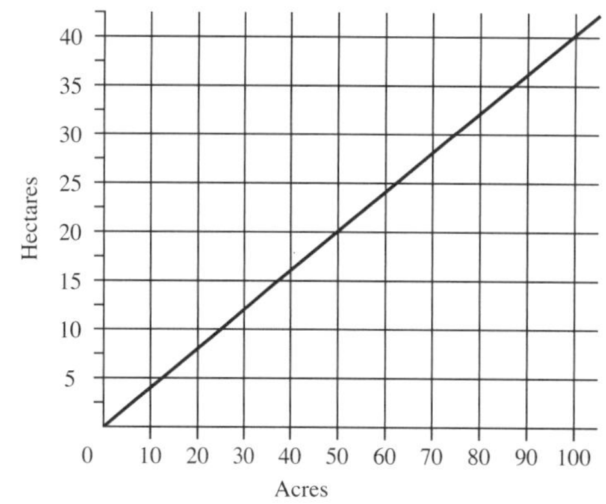

QUESTION 6 When Alice travelled to England she knew that \$10 in Australian currency was equivalent to 4.5 British pounds.

a Use this information to draw a conversion graph.
b How many British pounds would be equivalent to \$400? __________
c How many Australian dollars would be equivalent to 3600 pounds? __________

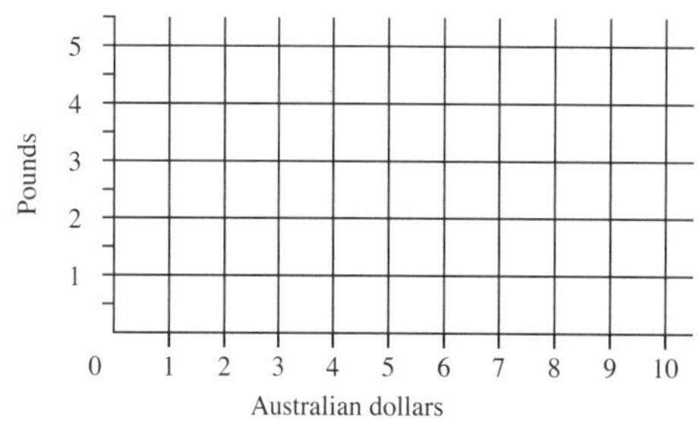

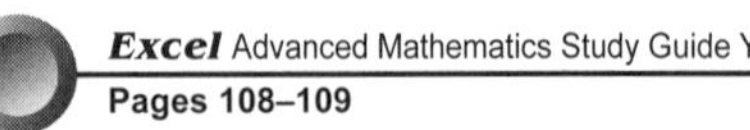

UNIT 12: Graphs and rates of change

QUESTION 1 This graph shows the trip Max made by car.

a What time did Max leave home? ____________

b How long was Max away from home? ____________

c How far from home was Max at 9 am? ____________

d When was Max first 150 km from home?

____

e What do the horizontal lines on the graph represent?

____

f When was Max travelling fastest? Briefly explain how you can tell this.

____

____

g What was the average speed on the return journey? ____________

Distance from home (km): 100, 200
Time: 8 am, 9, 10, 11, 12, 1, 2, 3, 4 pm

QUESTION 2 Briefly explain why this travel graph cannot represent a real journey.

____

____

____

distance
time

QUESTION 3 Choose the graph that matches the description of the rate of change.

a increasing at a constant rate ____________

b decreasing at a decreasing rate ____________

c increasing at an increasing rate ____________

d decreasing at a constant rate ____________

e decreasing at an increasing rate ____________

f increasing at a decreasing rate ____________

A P t B P t C D P t E F P t

QUESTION 4 This container is being filled by water that is being poured into it at a constant rate.

A
B
C
D

a In which section will the height of the water rise fastest? ____________

b In which section will the height of the water rise slowest? ____________

c In which section will the height be changing at a variable rate? ____________

d When the height is increasing at a variable rate will this be at an increasing or decreasing rate? ____________

e Sketch a graph of the height of the water over time.

height
time

Rational numbers, rates and measurements

TOPIC TEST — PART A

Instructions
- This part consists of 10 multiple-choice questions.
- Fill in only ONE CIRCLE for each question.
- Each question is worth 1 mark.

Time allowed: 10 minutes — **Total marks: 10**

Marks

1 Write $\frac{1}{(10^3)^4}$ with a negative index. — 1

(A) 10^{-3} (B) 10^{-4} (C) 10^{-7} (D) 10^{-12}

2 4.05×10^{-6} equals — 1

(A) 0.000 040 5 (B) 0.000 004 05 (C) 0.000 405 (D) 0.000 000 405

3 At an average speed of 80 km/h how far will a car travel in $3\frac{3}{4}$ hours? — 1

(A) 213 km (B) 231 km (C) 276 km (D) 300 km

4 The number 0.0079 has been written correct to a certain number of significant figures. How many? — 1

(A) 2 (B) 3 (C) 4 (D) 5

5 Which fraction is equivalent to 2^{-3}? — 1

(A) $\frac{1}{4}$ (B) $\frac{1}{6}$ (C) $\frac{1}{8}$ (D) $\frac{1}{9}$

6 $(3 \times 10^5) \div (6 \times 10^{-3})$ equals — 1

(A) 5×10^3 (B) 50 000 (C) 5×10^7 (D) 5×10^8

7 Which graph shows P increasing at a decreasing rate? — 1

(A) 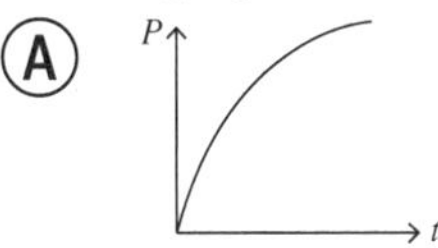

(B)

(C) 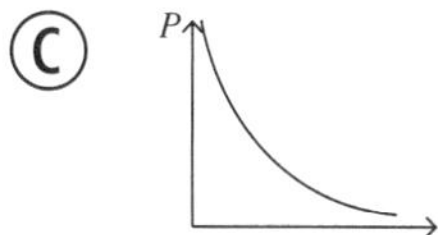

(D) 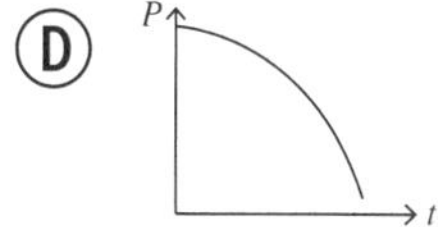

8 The length of a piece of timber was given as 5.73 m correct to two decimal places. Between which measurements does the length lie half way? — 1

(A) 573 cm and 574 cm (B) 5730 mm and 5735 mm
(C) 5.725 m and 5.73 m (D) 5725 mm and 5735 mm

9 Express 0.0059 in standard notation. — 1

(A) 59×10^{-2} (B) 59×10^{-4} (C) 5.9×10^{-3} (D) 5.9×10^{-4}

10 Light travels at the speed 3×10^8 m/s. About how long does it take light to travel 1 metre? — 1

(A) 3 seconds (B) 3 nanoseconds (C) 3 milliseconds (D) 3 microseconds

Total marks achieved for PART A 

Rational numbers, rates and measurements

TOPIC TEST — PART B

Time allowed: 20 minutes — **Total marks: 15**

Marks

1 Find 0.662 170 5 ÷ 215.34 giving the answer:

a correct to 4 decimal places. ______

b to 3 significant figures. ______

c in scientific notation. ______ 3

2 A particular plane can fly at 900 km/h.

a How far will the plane travel in three-quarters of an hour? ______

b How long will the plane take to travel 7425 km? ______

c Change the speed into m/s. ______ 3

3 Complete these conversions.

a 0.7 μg = ______ mg

b 2.5 TB ≈ ______ MB 2

4 a Water flows from a hose at the rate of 150 mL/s. How many litres will flow from the hose in an hour?

______

b A swimming pool holds 90 kL of water. How long, in days, hours and minutes will it take to fill the pool from the hose?

______ 2

5 A particular set of plates is shaped so that the perimeter, P, of any plate is directly proportional to the distance, d, across its centre. So $P = kd$ where k is a constant. One of the plates is 15 cm across its centre and has a perimeter of 54 cm. Find

a the value of k. ______

b the perimeter of a plate that is 24 cm across its centre. ______

c the distance across the centre of a plate with perimeter 126 cm. ______ 3

6 Water is pouring into this container at a constant rate.

a In which section (top, middle or bottom) will the water level rise fastest? ______

b Sketch a graph to show the water level over the time the container takes to fill. 2

height

time

Total marks achieved for PART B /15

Chapter 2
Algebraic techniques

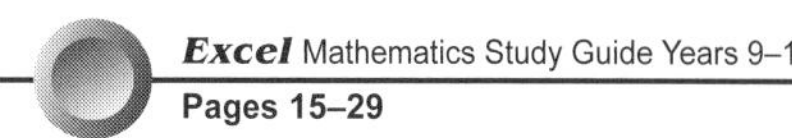
Excel Mathematics Study Guide Years 9–10
Pages 15–29

UNIT 1: Algebraic expressions

Question 1 Write algebraic expressions for the following.

a The sum of a and b = ____________

b The product of x and y = ____________

c The square of m = ____________

d The square root of p = ____________

e The sum of $7x$ and $2y$ = ____________

f The cube of k = ____________

g The square of $5x$ = ____________

h The difference between $8p$ and $3q$ = ____________

i The number $3x$ divided by 7 = ____________

j Nine times the square of a = ____________

Question 2 Write the algebraic expressions for the following.

a The cost of m pens at \$$d$ each = ____________

b The number of minutes in T hours = ____________

c If x is an odd number, the next odd number after x = ____________

d The distance travelled by a person at k km/h in h hours = ____________

e The perimeter of a square of side length l cm = ____________

Question 3 Write an algebraic expression for each of the following, using grouping symbols if necessary.

a Double k and divide the result by 15

____

b Multiply $3a$ and $9b$ and to this result add 7

____

c Eight times the sum of $5x$ and $11y$

____

d Add 14 to $3x$ and multiply the result by 9

____

e The product of a and $2b + 3c$ subtracted from $9x$ ____________

Question 4 Write an algebraic expression for the following.

a $2x$ is divided by $3y$ and z is added to it

____

b The number of metres in k kilometres

____

c The number of km in M metres

____

d The number of grams in Y kilograms

____

e The number of millimetres in x metres

____

f The number of hours in s seconds

____

Question 5 Explain the difference between each pair of algebraic expressions. Then find the value of each when $m = 3$ and $n = 5$

a m^2 and $2m$

____

____

b m^3 and $3m$

____

____

c $2m^2$ and $(2m)^2$

____

____

d $\frac{3}{m^2}$ and $\frac{1}{3m^2}$

____

____

e m^2n and mn^2

____

____

f $m^2 + n^2$ and $(m + n)^2$

____

____

Algebraic techniques

Excel Mathematics Study Guide Years 9–10
Pages 15–29

UNIT 2: Addition and subtraction in algebra

Question 1 Add the following expressions.

a $4x + 12x =$ ______ **b** $7x + 11x =$ ______

c $9x + 8x =$ ______ **d** $20x + 14x + 3x =$ ______

e $15x + 28x =$ ______ **f** $5a + 7a + 9a =$ ______

g $5ab + 10ab + 2ab =$ ______ **h** $7mn + 2mn + mn =$ ______

i $6p + 3p + 9p =$ ______ **j** $8x^2 + 9x^2 + 3x^2 =$ ______

k $15a^2 + 6a^2 + 2a^2 =$ ______ **l** $5n + 8n + 10n + n =$ ______

Question 2 Subtract the following expressions.

a $18a - 3a =$ ______ **b** $9x - 8x =$ ______

c $17y - 12y =$ ______ **d** $10m - 3m - m =$ ______

e $15x - 4x - 5x =$ ______ **f** $7xy - 2xy - xy =$ ______

g $14x^2 - 5x^2 - 3x^2 - x^2 =$ ______ **h** $16n - 4n - n - 2n =$ ______

i $12p - 3p - p - 2p =$ ______ **j** $10a^2 - 2a^2 - a^2 - 4a^2 =$ ______

k $8y - 3y - y - y =$ ______ **l** $9x - 7x - x - 2x - 3x =$ ______

Question 3 Simplify the following expressions by adding or subtracting.

a $10a + 5a - 4a =$ ______ **b** $7x + 8x - 3x - 6x =$ ______

c $16a - 4a + 12a - 8a =$ ______ **d** $8mn - 3mn + 2mn =$ ______

e $5p^2 + 7p^2 - p^2 - 2p^2 =$ ______ **f** $16ab + 8ab - 7ab - ab =$ ______

g $9t + 7t + 6t - 15t =$ ______ **h** $15a + 7a - a - 2a =$ ______

i $4m^2 - 3m^2 + 8m^2 - m^2 =$ ______ **j** $16t + 8t - 7t - t =$ ______

k $9x - 3x + 2x - x =$ ______ **l** $8mn + 6nm - 5mn =$ ______

Question 4 Simplify the following.

a $8a + 3b - 5a + b =$ ______ **b** $16x + 4x - 5y + 7y =$ ______

c $18a^2 + 9a^2 - 5b^2 - b^2 =$ ______ **d** $14m + 5n - 3m - 2n =$ ______

e $6a + 9b + 3b - 5a =$ ______ **f** $8m + 3n - 2n - 6m =$ ______

g $14x + 5x - 6y - 3y =$ ______ **h** $9p + 7q - p - q =$ ______

i $16ab^2 + 3a^2b - ab^2 - 2a^2b =$ ______ **j** $9x + 3x - 2y - 6y =$ ______

Question 5 Simplify the following expressions.

a $25 - 12x + 8x - 7 =$ ______ **b** $8x^2 + 7x^2 - 9y^2 =$ ______

c $15a + 7b - 8a =$ ______ **d** $9m + 4n - 6m - n =$ ______

e $20x + 4y - 6x - 2y =$ ______ **f** $16xy + 4yx - 7yz - yz =$ ______

g $15p + 8p - 9q =$ ______ **h** $5ab + 3ba + 9ab - ab =$ ______

i $14t + 10 - 6t - 12 =$ ______ **j** $4xy + 9yz - 3yx - 8zy =$ ______

Algebraic techniques

UNIT 3: Multiplication of pronumerals

QUESTION 1 Simplify the following.

a $5 \times 3a =$ ______ **b** $7 \times 5y =$ ______

c $9 \times 8x =$ ______ **d** $15 \times 2b =$ ______

e $\frac{1}{2} \times 6x =$ ______ **f** $\frac{1}{3} \times 15y =$ ______

g $\frac{2}{5} \times 25a =$ ______ **h** $\frac{1}{50} \times 100xy=$ ______

i $6a \times 7b =$ ______ **j** $9x \times 15y =$ ______

QUESTION 2 Simplify the following expressions.

a $4 \times 8xy =$ ______ **b** $3 \times 16ab =$ ______

c $10 \times (-8a) =$ ______ **d** $8 \times (-5a) =$ ______

e $-5 \times 6x =$ ______ **f** $-3 \times 15y =$ ______

g $-2 \times 9xy =$ ______ **h** $-7 \times 3ab =$ ______

i $-6 \times 15abc =$ ______ **j** $-2 \times 18ab =$ ______

QUESTION 3 Simplify the following expressions.

a $5m \times 8m =$ ______ **b** $9a \times 7ab =$ ______

c $6 \times 3 \times 4b =$ ______ **d** $6xy \times -3x \times -y =$ ______

e $5a \times 6a =$ ______ **f** $3ab \times 2a \times b =$ ______

g $mn \times 5n =$ ______ **h** $-4 \times (-7a) =$ ______

i $7abc \times -3a =$ ______ **j** $-4 \times (-3ab) =$ ______

QUESTION 4 Simplify the following.

a $2a \times -3a =$ ______ **b** $5a \times -2a \times -a =$ ______

c $5x \times 2x \times 3x =$ ______ **d** $3mn \times 2m \times -4 =$ ______

e $3a \times 2a \times -a =$ ______ **f** $8xy \times 15 =$ ______

g $5xy \times -2x \times -3xy =$ ______ **h** $(-3a) \times (-5a) =$ ______

i $8a \times 2a \times -a =$ ______ **j** $-2x \times (-x) \times (-3x) =$ ______

QUESTION 5 Simplify the following expressions.

a $-4x \times -3y =$ ______ **b** $6ab \times -2ab \times 0 =$ ______

c $(-5a) \times (-3) \times (-2y) =$ ______ **d** $-x \times 3x \times 5y =$ ______

e $5mn \times -7m =$ ______ **f** $(-2x) \times (-5) \times (-3y) =$ ______

g $-6a \times (-9a) =$ ______ **h** $(-8ab) \times (-7a) =$ ______

Algebraic techniques

UNIT 4: Division of pronumerals

QUESTION 1 Divide the following.

a $27ab \div 9 =$ ______

b $16x^2y \div 2xy =$ ______

c $15pq \div 3p =$ ______

d $36ab \div 9a^2 =$ ______

e $8xy \div y =$ ______

f $-8x \div -4 =$ ______

g $12x^2 \div x =$ ______

h $-15abc \div -3a =$ ______

i $15a^2 \div 5 =$ ______

j $18a^2 \div 3a =$ ______

QUESTION 2 Simplify the following divisions.

a $-12ab \div a^2 =$ ______

b $-8a \div -4 =$ ______

c $-36xy \div x =$ ______

d $-36mn \div -12m =$ ______

e $-15x \div 3x =$ ______

f $-a^2 \div 2a =$ ______

g $-9xy \div -3x =$ ______

h $-20mn \div -2m =$ ______

i $-15ab \div -5a =$ ______

j $-36xy \div -4x =$ ______

QUESTION 3 Simplify the following divisions.

a $28xy \div 7x =$ ______

b $-32x \div -8 =$ ______

c $15ab \div -3a =$ ______

d $-42y \div -21 =$ ______

e $-60m \div -10m =$ ______

f $-15x^2 \div 5x =$ ______

g $8a \div (-4) =$ ______

h $-21x^2y^2 \div 6x =$ ______

i $-45xy \div -5x =$ ______

j $-12xyz \div -4x =$ ______

QUESTION 4 Simplify the following.

a $18pq \div 6pq =$ ______

b $10xy \div (-10xy) =$ ______

c $24xyz \div -5xy =$ ______

d $15ab \div 3b \div 5a =$ ______

e $6mn \div -3n =$ ______

f $35a^2 \div 7a \div 5 =$ ______

g $abc \div ab =$ ______

h $mnp \div mp \div n =$ ______

i $15a^2bc \div 6a =$ ______

j $16x^2y \div 8xy =$ ______

QUESTION 5 Simplify the following expressions.

a $28x \div 7x \div 4 =$ ______

b $3abc \div 3a \div b =$ ______

c $21pq \div 7p \div 3q =$ ______

d $(5b)^2 \div 25b =$ ______

e $20x^2 \div 10x \div x =$ ______

f $18xy \div 3x \div 6 =$ ______

g $xyz \div xy \div z =$ ______

h $56a^2 \div 9a \div -3 =$ ______

Algebraic techniques

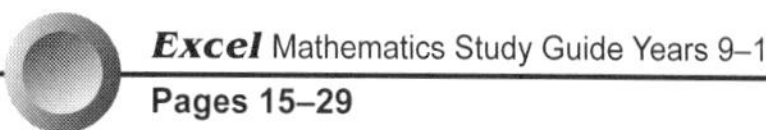

UNIT 5: Multiplication and division in algebra

QUESTION 1 Multiply the following.

a $5 \times 4x =$ ____
b $6x \times 4y =$ ____
c $4xy \times 5x =$ ____
d $9y \times y =$ ____
e $-6m \times 3n =$ ____
f $2p \times 3q =$ ____
g $-6a^2 \times 5a =$ ____
h $9x^3 \times (-2x) =$ ____
i $5ab \times (-4ba) =$ ____
j $-2 \times 4x \times -3y =$ ____
k $(-xy) \times (-yz) =$ ____
l $-20xy \times (-\frac{1}{5}yx) =$ ____

QUESTION 2 Divide the following expressions.

a $14x \div 7 =$ ____
b $18y \div 6y =$ ____
c $10x^2 \div 10 =$ ____
d $36mn \div 9m =$ ____
e $12ab \div ab =$ ____
f $27 \div 9x =$ ____
g $-3xy \div x =$ ____
h $-48a \div 6a =$ ____
i $2xyz \div xy =$ ____
j $15xy \div -3x =$ ____
k $-28ab \div -7a =$ ____
l $-64abc \div 16b =$ ____

QUESTION 3 Work out the following divisions.

a $36ab \div (-4a) =$ ____
b $12xy \div (-12xy) =$ ____
c $m^2n^2 \div mn \div m =$ ____
d $18xy \div 2x =$ ____
e $14a^2b^2 \div 7abc =$ ____
f $24x \div 8x \div x =$ ____
g $40x^2y^2 \div 10xy \div 2y =$ ____
h $26abc \div ac \div 26 =$ ____
i $21 \div 14ab =$ ____
j $27a^2 \div (-9a) =$ ____

QUESTION 4 Simplify the following expressions.

a $5 \times 3k \times 2ky =$ ____
b $4x \times 2y \times 3z =$ ____
c $14x \times 3 \times 2x =$ ____
d $10x \div 5 \times 3x =$ ____
e $18xy \div xy \div 18 =$ ____
f $9m \times 7n \div 3n =$ ____
g $8x \times 9y \div 3x =$ ____
h $42a^2b^2c^2 \div 7abc \div 2 =$ ____
i $a^2b \times ab \div 3a =$ ____
j $16xy \times 5x \div 8y =$ ____

QUESTION 5 Simplify the following.

a $xy \times 8yz \div 4xz =$ ____
b $15am \div 5m \div 3a =$ ____
c $9x - 3 \times 2x =$ ____
d $14xy \div 2x \times 4y =$ ____
e $14a^2 - 4a \times 2a =$ ____
f $4 \times 6xy \div xy =$ ____
g $\frac{10x \times 5y}{25xy} =$ ____
h $\frac{14a \times 5b}{7a^2b} =$ ____
i $\frac{(4a)^2 \times (5b)^2}{40ab} =$ ____
j $\frac{3a^2 \times 4b}{6a \times 2b^2} =$ ____

Algebraic techniques

UNIT 6: The index laws (1)

QUESTION 1 Simplify the following, writing your answer in index form.

a $x^2 \times x^3 =$ ______ **b** $y^3 \times y^3 =$ ______ **c** $a^3 \times a^4 =$ ______

d $m^3 \times m^4 =$ ______ **e** $p^3 \times p^7 =$ ______ **f** $n^{10} \times n^3 =$ ______

g $a^2 \times a^3 \times a^4 =$ ______ **h** $x^5 \times x^2 \times x =$ ______ **i** $y^5 \times y^9 =$ ______

QUESTION 2 Simplify the following.

a $x^9 \div x^3 =$ ______ **b** $x^{12} \div x^4 =$ ______ **c** $x^7 \div x =$ ______

d $y^7 \div y^4 =$ ______ **e** $y^9 \div y^2 =$ ______ **f** $a^9 \div a =$ ______

g $m^{25} \div m^{14} =$ ______ **h** $m^{12} \div m^{10} =$ ______ **i** $m^{15} \div m^{11} =$ ______

QUESTION 3 Simplify the following.

a $5x^4 \times x^5 =$ ______ **b** $9x^5 \times x^2 =$ ______ **c** $a^{11} \times 3a^5 =$ ______

d $m^7 \times 5m^2 =$ ______ **e** $8k^3 \times 7k^5 =$ ______ **f** $5a^7 \times 8a^6 =$ ______

g $m^3n^2 \times m^5n^7 =$ ______ **h** $x^2y^4 \times x^5y^7 =$ ______ **i** $x^2y^2 \times x^4y^2 =$ ______

QUESTION 4 Simplify, giving answers in index form.

a $x^9 \div x^6 =$ ______ **b** $y^8 \div y^5 =$ ______ **c** $a^7 \div a^5 =$ ______

d $15m^5 \div 5m^3 =$ ______ **e** $16n^7 \div 8n^4 =$ ______ **f** $36a^8 \div 9a^6 =$ ______

g $24y^6 \div 8y =$ ______ **h** $x^6y^3 \div x^4y =$ ______ **i** $a^7b^6 \div a^4b^3 =$ ______

QUESTION 5 Simplify the following products.

a $2a^2 \times 3a^3 =$ ______ **b** $6p^2 \times p^3 =$ ______ **c** $9y^3 \times 9y^8 =$ ______

d $7m^3 \times 5m^4 =$ ______ **e** $8a^3 \times 4a^5 =$ ______ **f** $2x^7 \times x^5 \times x^3 =$ ______

g $6x^5 \times 4x^7 =$ ______ **h** $9a^2 \times 8a^7 =$ ______ **i** $10p^5 \times 8p^6 =$ ______

j $8x^8 \times 3x^{12} =$ ______ **k** $a^4b^4 \times a^3b^3 =$ ______ **l** $9x^2 \times 4x^5y^5 =$ ______

QUESTION 6 Simplify these divisions.

a $6a^{12} \div 3a^8 =$ ______ **b** $36m^6 \div 9m^4 =$ ______ **c** $20a^8b^7 \div 5a^7b^6 =$ ______

d $15a^{15} \div 5a^5 =$ ______ **e** $12k^{15} \div 3k^{10} =$ ______ **f** $12a^6b^9 \div 6a^4b^4 =$ ______

g $48a^{48} \div 12a^{36} =$ ______ **h** $48a^7 \div 12a^5 =$ ______ **i** $36x^{12}y^8 \div (-4x^2y^7) =$ ______

j $16m^{16} \div 4m^9 =$ ______ **k** $18m^6n^8 \div 9m^4n^6$ ______ **l** $64a^8b^9 \div 8a^6b^7 =$ ______

Algebraic techniques

UNIT 7: The index laws (2)

QUESTION 1 Simplify the following.

a $(a^2)^3 =$ ______ b $(b^5)^4 =$ ______ c $(a^5)^6 =$ ______

d $(x^3)^7 =$ ______ e $(b^2)^7 =$ ______ f $(x^7)^8 =$ ______

g $(3x^3)^2 =$ ______ h $(3x^2)^4 =$ ______ i $(3b^4)^3 =$ ______

QUESTION 2 Simplify these expressions.

a $2(a^3)^2 =$ ______ b $3(y^5)^5 =$ ______ c $(8x^2y^3)^2 =$ ______

d $5(m^5)^2 =$ ______ e $6(x^3)^7 =$ ______ f $(9ab^2)^3 =$ ______

g $a(x^6)^4 =$ ______ h $(4a^2)^3 =$ ______ i $(5m^4)^3 =$ ______

j $x^2(y^7)^6 =$ ______ k $(9p^2)^2 =$ ______ l $(10a^3b^3)^3 =$ ______

QUESTION 3 Use index laws to simplify the following.

a $a^0 =$ ______ b $m^0 =$ ______ c $x^0 =$ ______

d $(2x)^0 =$ ______ e $(3x^2)^0 =$ ______ f $(4mn)^0 =$ ______

g $9x^0 =$ ______ h $2a^0 =$ ______ i $(5a)^0 =$ ______

QUESTION 4 Simplify the following.

a $4m^0 =$ ______ b $(6m)^0 =$ ______ c $8n^0 =$ ______

d $5m^0 \times (8m)^0 =$ ______ e $a^6b^0 =$ ______ f $(x^2)^0 =$ ______

g $(5y^3)^0 =$ ______ h $a^7 \div a^7 =$ ______ i $9x^5 \div 9x^5 =$ ______

j $a^0 + b^0 =$ ______ k $2x^0 + 3y^0 =$ ______ l $7p^0 + 4^0 =$ ______

m $p^0 - q^0 =$ ______ n $3x^0 + (3x)^0 =$ ______ o $4(2k)^0 =$ ______

QUESTION 5 Give the answers in simplest index form.

a $(3^2)^3 =$ ______ b $(2^3)^2 =$ ______ c $(x^3)^2 =$ ______

d $(m^3)^4 =$ ______ e $4(x^5)^6 =$ ______ f $(2y^3)^4 =$ ______

g $(a^3)^2 \times a^4 =$ ______ h $(y^7)^2 \div y^6 =$ ______ i $(a^4)^3 \times (a^5)^2 =$ ______

Algebraic techniques

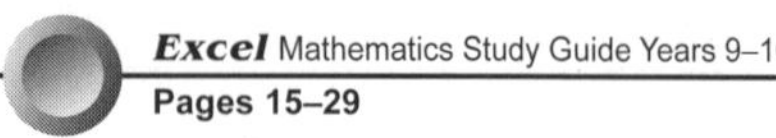

UNIT 8: Combinations of the index laws

QUESTION 1 Simplify.

a $\left(\frac{x^3}{y}\right)^2 =$ ______ b $\left(\frac{x^2}{y^3}\right)^3 =$ ______ c $\left(\frac{a^4}{b^3}\right)^5 =$ ______

d $\left(\frac{m^8}{m^3}\right)^2 =$ ______ e $\left(\frac{a^5}{b^7}\right)^2 =$ ______ f $\left(\frac{m^8}{n^7}\right)^2 =$ ______

g $\left(\frac{m^6}{2}\right)^4 =$ ______ h $\left(\frac{x^2}{y^3}\right)^4 =$ ______ i $\left(\frac{a^9}{b^2}\right)^3 =$ ______

QUESTION 2 Simplify.

a $(4a^3)^2 \times a^5 =$ ______

b $(2x^2)^3 \times x^9 =$ ______

c $(a^3)^4 \times (2a^5)^2 =$ ______

d $(x^4y^5)^2 \times (xy)^5 =$ ______

e $(p^2q)^5 \times (p^3q^4)^2 =$ ______

f $(2a)^2 \times (6ab)^2 =$ ______

g $(5x^2y^3)^2 \times (2xy)^4 =$ ______

h $(2p^4q^5)^7 \div (4p^6q^4)^4$ ______

i $(3x^8y)^4 \div 9(x^9)^3$ ______

j $(m^2n^2p)^4 \times m^2n =$ ______

k $96m^{12} \div 8m^4 \div 6m^2 =$ ______

l $12x^3 \times 5x^5 \div 20x^7$ ______

m $9a \times 6a^2 \times 4a^3 =$ ______

n $32x^5 \div 8x^4 \times 5x =$ ______

o $(5x)^4 \div 5x^2 =$ ______

QUESTION 3 Simplify.

a $5a^0 \times (4a)^0 =$ ______ b $8x^0 \times (8x)^0 =$ ______ c $(9y)^0 \div 9y^0 =$ ______

d $(5x)^0 =$ ______ e $(7p)^0 \times (5q)^0 =$ ______ f $(6m)^0 \div 6 =$ ______

g $8x^0 \div (4x)^0 \times 5y^0 =$ ______ h $8a^0 =$ ______ i $9x^0 \div (4x)^0 \times 7x^0 =$ ______

QUESTION 4 Simplify.

a $\frac{(7a^2)^3}{(7a^3)^2} =$ ______ b $\frac{12m^4 \times 18m^3}{9m^2 \times 4m^5} =$ ______ c $\left(\frac{6a^2}{3a}\right)^3 \times (8a)^2 =$ ______

d $\frac{(8c^2)^3}{4c^2 \times 6c^3} =$ ______ e $\frac{9m^6 \times 8m^9}{24m^8} =$ ______ f $\left(\frac{a^4}{3}\right)^2 \times a^5 =$ ______

g $\left(\frac{y^3}{4}\right)^3 \times y^8 =$ ______ h $\left(\frac{6y^2}{8y}\right)^3 \times (4y)^2 =$ ______ i $\frac{48x^8y^6}{8x^4y^4} =$ ______

j $\frac{m^6n^8 \times (m^2n)^3}{(mn)^4} =$ ______ k $\frac{(6a^4)^2}{(6a^2)^4} =$ ______ l $12x^3 \times 5x^5 \div 20x^7 =$ ______

m $\frac{20x^6 \times (2x^3)^2}{10x^8} =$ ______ n $\left(\frac{9k^2}{3k}\right)^3 \times (4k^2)^2 =$ ______ o $\frac{(2y^5)^3}{2y^2 \times (2y)^2} =$ ______

UNIT 9: Negative indices with variables

QUESTION 1 Write in fractional form.

a $x^{-2} =$ ______ **b** $a^{-1} =$ ______ **c** $e^{-3} =$ ______ **d** $p^{-7} =$ ______

e $2m^{-2} =$ ______ **f** $5n^{-4} =$ ______ **g** $6x^{-6} =$ ______ **h** $\frac{1}{4}a^{-3} =$ ______

i $(3x)^{-3} =$ ______ **j** $(7y)^{-2} =$ ______ **k** $(2a)^{-5} =$ ______ **l** $(xy)^{-4} =$ ______

QUESTION 2 Write in index form (with a negative index).

a $\frac{1}{x^4} =$ ______ **b** $\frac{1}{a^6} =$ ______ **c** $\frac{1}{e^{10}} =$ ______ **d** $\frac{1}{x^9} =$ ______

e $\frac{3}{n^3} =$ ______ **f** $\frac{4}{m^8} =$ ______ **g** $\frac{a}{b^5} =$ ______ **h** $\frac{7}{a^{12}} =$ ______

i $\frac{1}{4a^2} =$ ______ **j** $\frac{1}{8x^3} =$ ______ **k** $\frac{1}{2x^7} =$ ______ **l** $\frac{1}{81a^4} =$ ______

QUESTION 3 Simplify, giving the answer in index form.

a $x^7 \times x^{-2} =$ ______ **b** $a^{-5} \times a^8 =$ ______ **c** $m^{-2} \times m^{-3} =$ ______

d $5p^6 \times 2p^{-3} =$ ______ **e** $6k^{-2} \times 4k^{-5} =$ ______ **f** $8x^{-7} \times 2x^4 =$ ______

g $x^7 \div x^{-3} =$ ______ **h** $a^4 \div a^{-8} =$ ______ **i** $b \div b^{-6} =$ ______

j $9m^{-5} \div 3m^2 =$ ______ **k** $10n^{-5} \div 2n^5 =$ ______ **l** $12a^3 \div 2a^{-4} =$ ______

m $6a^{-2} \times 3a^{-2} =$ ______ **n** $35x^7 \div 5x^{-3} =$ ______ **o** $4x^6 \times 2x^{-6} =$ ______

p $a^2b^{-3} \times a^{-2}b^4 =$ ______ **q** $p^3q^4 \times p^{-2}q^{-2} =$ ______ **r** $m^5n^{-2} \div m^8n^{-3} =$ ______

QUESTION 4 Simplify, giving each answer in fractional form.

a $x^{-3} \div x$ = ______ = ______ **b** $a^{-4} \times a^{-2}$ = ______ = ______ **c** $x^{-5} \div x^{-3}$ = ______ = ______

d $5p^6 \times 2p^{-7}$ = ______ = ______ **e** $8a^{-8} \div 2a^{-2}$ = ______ = ______ **f** $x^2y^{-4} \div x^{-3}y$ = ______ = ______

UNIT 10: Grouping symbols in algebra

QUESTION 1 Expand the following expressions.

a $3(x+2) =$ ______
b $2(a+5) =$ ______
c $4(2y-1) =$ ______
d $3(6a+7) =$ ______
e $5(8-a) =$ ______
f $6(2k-3) =$ ______
g $5n(n-1) =$ ______
h $3(4-3a) =$ ______
i $7(2n+7) =$ ______
j $y(2y+7) =$ ______
k $m(m+10) =$ ______
l $2a(3a-7) =$ ______

QUESTION 2 Remove the grouping symbols.

a $-2(2a+3) =$ ______
b $-3(5n-4) =$ ______
c $-(y+8) =$ ______
d $-5(7+2t) =$ ______
e $-3(5x+18) =$ ______
f $-4(3x-2) =$ ______
g $-(6x+11) =$ ______
h $-2(4x-9) =$ ______
i $-5(4x-5) =$ ______
j $-3(a-14) =$ ______
k $-8(x-10) =$ ______
l $-(2-5x) =$ ______

QUESTION 3 Expand the following expressions.

a $\frac{1}{3}(9x-15) =$ ______
b $\frac{1}{2}(8x-4) =$ ______
c $-\frac{1}{4}(24y-8) =$ ______
d $a^3(2a+3) =$ ______
e $a^2(3a+4b) =$ ______
f $-2y(3y+7) =$ ______
g $-y^2(3y-6) =$ ______
h $4t^2(5t-8) =$ ______
i $-m(3m^2+5m) =$ ______
j $-6p(3p^2+5) =$ ______
k $-4x(8x-1) =$ ______
l $3n(8n^2+7n) =$ ______

QUESTION 4 Expand.

a $-2(5x+y-z) =$ ______
b $-3(2a+3b-4c) =$ ______
c $4(a^2-3a+7) =$ ______
d $-(5t^2-3t+4) =$ ______
e $3(2xy+3xy^2-8x) =$ ______
f $2ab(4a^2b-6ab+3ab^2) =$ ______
g $-5a(3a-2b+4c) =$ ______
h $3p(8p-2q+3r) =$ ______
i $4a(a^2+2ab-3ac) =$ ______
j $-a(2a+3b-9c) =$ ______
k $-t(2t^3+3t^2-5t) =$ ______
l $8(9x-7y+2z) =$ ______

QUESTION 5 Expand.

a $3t(t^4-5t^3+2t^2-8t-7) =$ ______
b $m(5m^4-3m^3+2m^2-m-1) =$ ______
c $x^2(4y^2-3xy+4x-7y) =$ ______
d $ab(a^4-a^3+4ab-2a^2+3ab^2) =$ ______
e $-4a(5a^3-4a^2+3a-2) =$ ______
f $-2y(8y^2+7y-xy+6) =$ ______
g $-ab(a^3+b^2-2ab+c) =$ ______
h $-4x(x^3+y^2-2xy-x) =$ ______

UNIT 11: Expanding and simplifying algebraic expressions (1)

QUESTION **1** Expand and simplify.

a $5(x + 3) + 2x - 5 =$

b $3(a + 2) + 2a - 7 =$

c $7(2m - 1) + 10m - 3 =$

d $6a + 7 - 2(2a + 4) =$

e $8y - 3 - 2(y + 5) =$

f $9x + 2(3x - 1) + 6 =$

g $5t + 6 + 3(t^2 + 5) =$

h $5x - (2x - 1) + 3x =$

i $18 - 2(x - 2) + 4x =$

j $7(2m - 5) - 4m + 1 =$

k $8a + 7 - 2(4a - 1) =$

l $7x + 11 - 2(x - 3) =$

QUESTION **2** Remove the grouping symbols and simplify.

a $5(2a + 4) + 3 =$

b $7(2t - 7) + 5t =$

c $6m + 3(2m - 5) =$

d $9p + 2(8 - 3p) =$

e $10y + 3(8y - 1) =$

f $6(3x - 10) + 5x =$

g $7(3 - n) - 9n =$

h $9y(y + 3) - 4 =$

i $6a - 4(2a - 3) =$

j $25 - 2(4x - 5) =$

k $9x - (3x - 2y + z) =$

l $5t + 3(9 - 2t) - 8 =$

QUESTION **3** Expand and simplify.

a $2(x + 3) + 4(x - 1) =$

b $5a(a^2 - 2a - 3) - a(a + 9) =$

c $3xy(x^2 - y - 7) - x^2(x + 3) =$

d $5(m + 3n) - 3(2m - 6n) - 2(m + 8) =$

e $2t(t^2 - 3t + 3) - 5t(3t^2 - 2t - 1) =$

f $7a^4 - 5a^3 + 2a^2 - 3a - 2(10 - 5a + 3a^2) =$

QUESTION **4** Write in simplest form.

a Add $2a + 3b$ to $7a - 5b$

b Add $5x - 3y + z$ to $8x + 5y - 3z$

c Find the sum of $2m + 3$, $9 - 5m$ and $m - 10$

d Subtract $5a - 7$ from $18a - 10$

e Subtract $y^2 - 4y + 6$ from $4y^2 - 10y + 9$

f From $8t^2 - 5t - 9$ take $5t^2 + 4t - 3$

UNIT 12: Expanding and simplifying algebraic expressions (2)

QUESTION 1 Expand and simplify.

a $5(x + 3) + 3(x + 5)$

b $7(a + 4) + 6(a + 1)$

c $3(m - 7) + 2(m - 1)$

d $-5(n + 2) + 3(n - 2)$

e $2(3x - 1) + 8(x + 2)$

f $4(2x - 3) + 3(2x + 3)$

g $-4(5x - 2) + 5(4x - 1)$

h $-2(6x + 7) - 3(3x + 4)$

i $6(-2x + 3) - 5(x + 4)$

j $x(x + 2) - 3(x + 2)$

k $x(3x + 1) - (x - 3)$

l $a(3a - 1) - 2(3a - 1)$

m $2x(7x - 3) - 5(7x - 3)$

n $8x(x - 2) - x(x - 7)$

o $6x(5x - 3) - 5x(6x - 3)$

p $8(a + b) - 2(4a + b)$

q $9m^2(2m - 1) - 4m(4m - 3)$

r $3a^2(8a - b) - 2a(5a^2 - b)$

QUESTION 2 Expand, and simplify where possible.

a $2x^2y(x + y) + 3xy^2(x - y)$

b $6ab(a - b) - 3(2a + b)$

c $5xy(5x + y) - 6(4x^2 - y)$

d $3(5 - 4x) - 7(2y + 1)$

e $-2mn(m + 2) + 3mn(n - 4)$

f $-(x + 5) - x(x - 6)$

g $4x(y - z) + 3y(3x - z)$

h $7x^2y(3x^2 - y) + 2xy^2(x - y^2)$

i $4pq(p^2 - q) - p^2(p^2 - q)$

j $4x^6(3x^4 - 2y^2) - 8y^2(2xy - 5)$

k $-3a^2(a^3 - 4) - 2a(a^5 + 3)$

l $2(xy + x + y) - 3(2x - y)$

Excel Mathematics Study Guide Years 9–10
Pages 15–29

UNIT 13: Substitution

QUESTION 1 Calculate the value of each expression given that $a = -2$, $b = 3$ and $c = 4$

a $a + b =$ ____________ **b** $a + b + c =$ ____________

c $b + c =$ ____________ **d** $c + a =$ ____________

e $a + b - c =$ ____________ **f** $a - b + c =$ ____________

g $3a + 2b =$ ____________ **h** $4b - 5c =$ ____________

i $a + 2b + 3c =$ ____________ **j** $a^2 + b^2 =$ ____________

k $a^2b + b^2a =$ ____________ **l** $\frac{a+b}{ab} =$ ____________

QUESTION 2 If $x = 3$, calculate the value of the following expressions.

a $4x^2 =$ ____________ **b** $(4x)^2 =$ ____________

c $30 - 5x =$ ____________ **d** $(6x - 7)^2 =$ ____________

e $(x - 1)(x - 8) =$ ____________ **f** $\sqrt{x^2 - 5} =$ ____________

g $(x - 2)^3 =$ ____________ **h** $4x^2 \times 5x =$ ____________

i $(x + 2)(x - 2) =$ ____________ **j** $20 - x^2 =$ ____________

k $5x^2 - 8x =$ ____________ **l** $x^2 + 4x - 6 =$ ____________

QUESTION 3 If $x = \frac{1}{2}$ and $y = \frac{1}{3}$, find the value of:

a $x + y =$ ____________ **b** $x - y =$ ____________

c $\frac{x+y}{x-y} =$ ____________ **d** $\frac{x-y}{x+y} =$ ____________

e $\frac{x+y}{x-y} + \frac{x-y}{x+y} =$ ____________ **f** $x^2 + y^2$ ____________

g $x^2 - y^2 =$ ____________ **h** $\frac{xy}{x+y} =$ ____________

i $\frac{x-y}{xy} =$ ____________ **j** $(x + y)^2 =$ ____________

k $(x - y)^2 =$ ____________ **l** $\frac{x}{y} + \frac{y}{x} =$ ____________

QUESTION 4 Given that $x = 8.5$, $y = 5.2$ and $z = 6.4$, find, correct to one decimal place, the value of:

a $xy^2 =$ ____________ **b** $x^2y =$ ____________

c $xy + yz =$ ____________ **d** $(x + y)^2 =$ ____________

e $(x + y)(x - y) =$ ____________ **f** $\sqrt{x + y + z} =$ ____________

g $xyz \div 3 =$ ____________ **h** $\frac{x+y}{y+z} =$ ____________

i $\frac{x}{y} + \frac{y}{z} =$ ____________ **j** $(2x + 3y)^2 =$ ____________

Excel Mathematics Study Guide Years 9–10
Pages 15–29

UNIT 14: Factorisation using common factors

QUESTION 1 Factorise the following by taking the common factor out.

a $4x + 16 =$ __________ b $9a - 27 =$ __________

c $5x - 25x^2 =$ __________ d $7a + 21a^2b =$ __________

e $5ab + 25a^2b^2 =$ __________ f $7m - 21m^2n =$ __________

g $a^3b^2 - a^2b^3 =$ __________ h $14x^3y^3 - 28x^2y^2 =$ __________

i $15ab - 25bc =$ __________ j $12ab + 15a^2 =$ __________

k $x^2y^2 - 7xy =$ __________ l $abc - 6bcd =$ __________

QUESTION 2 Factorise the following by taking the negative common factor out.

a $-4a - 28 =$ __________ b $-3a - 15 =$ __________

c $-8x - 32 =$ __________ d $-10xy - 15y =$ __________

e $-8y + 40 =$ __________ f $-m^3 - m^2 =$ __________

g $-x^3 - 10x^2y^2 =$ __________ h $-6x^2 + 12x =$ __________

i $-10y^2 + 12y =$ __________ j $-5x - 9x^2 =$ __________

k $-3m - 18m^3 =$ __________ l $-9m + 36m^4 =$ __________

QUESTION 3 Factorise the following.

a $a(a + 2) + b(a + 2) =$ __________ b $3(x + y) - a(x + y) =$ __________

c $9(x - y) + 2a(x - y) =$ __________ d $5(2a + 3b) - c(2a + 3b) =$ __________

e $x^2(5 - y) - 3(5 - y)=$ __________ f $x(2x - 9) + 5(2x - 9) =$ __________

g $m(a - b) - n(a - b) =$ __________ h $12(x^2 + 7) - y(x^2 + 7) =$ __________

i $5(x + 8) + y(x + 8) =$ __________ j $4a(3b - 5c) + 2(3b - 5c) =$ __________

k $m(2n - p) - q(2n - p) =$ __________ l $3x^2(2a - 5b) + y^2(2a - 5b) =$ __________

QUESTION 4 Factorise each of the following.

a $mx + my + mz =$ __________ b $ac + bc + cd =$ __________

c $5m - mn + 6mp =$ __________ d $10a + 25b + 35c =$ __________

e $20xy - 8x^2 + 36 =$ __________ f $n^2 - 8mn + 10n =$ __________

g $5a^2 + 15abc - 10a =$ __________ h $xy^2 - 2xy + x^2y =$ __________

i $3a - 9ab - 15a^2 =$ __________ j $5m - 10mn + 20m^2n =$ __________

k $x^3y^2 - 2x^2y^2 + 3x^2y^3 =$ __________ l $12x^2y^2z^2 - x^3y^2 + x^2y^3 =$ __________

Algebraic techniques

TOPIC TEST — PART A

Instructions
- This part consists of 10 multiple-choice questions.
- Fill in only ONE CIRCLE for each question.
- Each question is worth 1 mark.

Time allowed: 10 minutes — **Total marks: 10**

Marks

1 $a^3 \times a^3$ equals

Ⓐ $2a^3$ Ⓑ $2a^6$ Ⓒ a^6 Ⓓ a^9 — 1

2 $15x^{10} \div 5x^5$ equals

Ⓐ $3x^5$ Ⓑ $3x^2$ Ⓒ $10x^5$ Ⓓ $10x^2$ — 1

3 $(4m^3)^2$ equals

Ⓐ $8m^5$ Ⓑ $8m^6$ Ⓒ $16m^9$ Ⓓ $16m^6$ — 1

4 $(5y^3)^0$ equals

Ⓐ $5y^3$ Ⓑ 5 Ⓒ 0 Ⓓ 1 — 1

5 $4x^{-2}$ equals

Ⓐ $\frac{1}{4x^2}$ Ⓑ $\frac{4}{x^2}$ Ⓒ $\frac{-1}{4x^2}$ Ⓓ $\frac{-4}{x^2}$ — 1

6 $5x^2y^3 \times 3xy^4 =$

Ⓐ $15x^2y^7$ Ⓑ $15x^3y^7$ Ⓒ $15x^2y^{12}$ Ⓓ $5x^3y^{12}$ — 1

7 $2(a^4)^3 =$

Ⓐ $2a^7$ Ⓑ $2a^{12}$ Ⓒ $8a^7$ Ⓓ $8a^{12}$ — 1

8 $5a - (2 - a)$ equals

Ⓐ $4a - 2$ Ⓑ $6a - 2$ Ⓒ $5a - 2$ Ⓓ $4a + 2$ — 1

9 If x is an integer, which of the following will always produce an odd number?

Ⓐ x^2 Ⓑ $3x^2$ Ⓒ $2x^2 + 1$ Ⓓ $3x^2 + 2x$ 

10 The correct factorisation of $3xy - x$ is

Ⓐ $3x(y - 1)$ Ⓑ $3x(y - x)$ Ⓒ $x(3y - 1)$ Ⓓ $x(3y - x)$ — 1

Total marks achieved for PART A 

Algebraic techniques

TOPIC TEST — PART B

Time allowed: 20 minutes — **Total marks: 20**

Marks

1 Simplify.

a $(5x^3)^2$

b $52m^8 \div 13m^6$

c $2x^3y^2 \times 5xy^3$

d $8y^0 + (8y)^0 + (8y^8)^0$

e $5p + 3q - 4p + 2q$

f $9x^2 - 2x + 3x^2 - 5x$

g $3x^3y^2 \times 4x^5y^7$

h $14a^4 \div 7a^6$

i $3a^{-4} \times 5a^{-2}$

j $7a^7b^3 \times 7a^6b^4$

k $(7x^2y^3)^\circ$ 11

2 Expand.

a $\frac{3}{4}(16xy + 32x^2 - 12y^2)$

b $-2x^2(3x^2 + 4xy^2)$ 2

3 Expand and simplify.

a $2a + 2b - 2c - (2a + 2b + 2c)$

b $4(5x - 3) - 2(3x + 8)$

c $5(3a - 7) - 4(2 - 8a)$

d $-2x^2(xy - 3) - 3y(x^3 - 4y^2)$

4

4 Simplify. $\dfrac{6a^5 \times 2a^3}{4a^2 \times 3a^4}$ 1

5 Given that $a = \frac{1}{2}$, $b = \frac{1}{3}$, find the value of $(ab)^2 + (a + b)^2$

1

6 Factorise $8a^2 + 24ab - 16a$ 1

Total marks achieved for PART B

/20

CHAPTER 3
Pythagoras' theorem

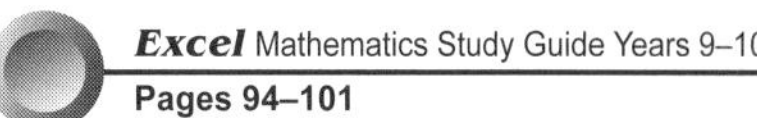

UNIT 1: Naming the hypotenuse of a right-angled triangle

QUESTION 1 Name the hypotenuse of each right-angled triangle.

a
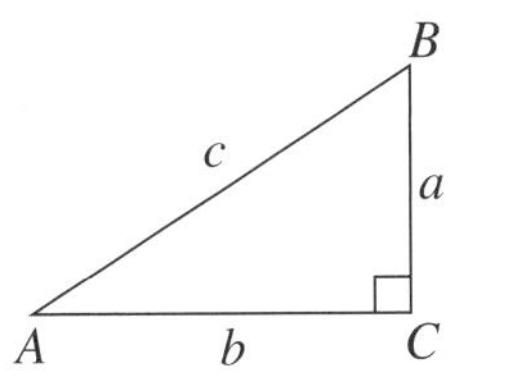

____

b
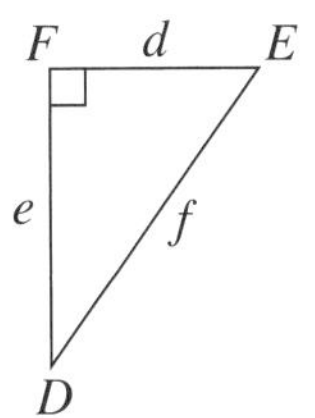

____

c
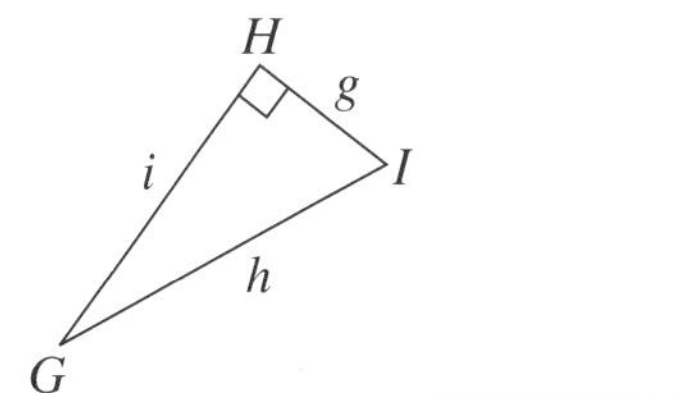

____

d
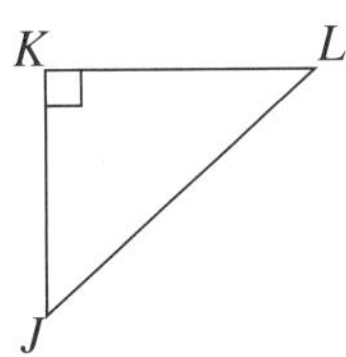

____

e
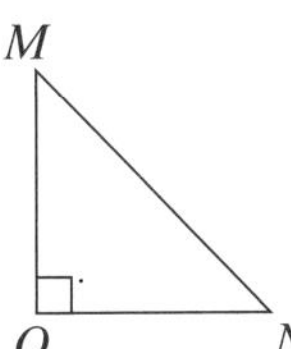

____

f
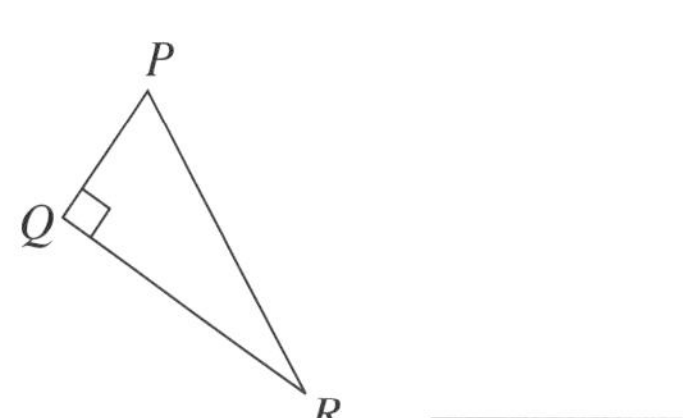

____

QUESTION 2 Name the hypotenuse of each named triangle.

a
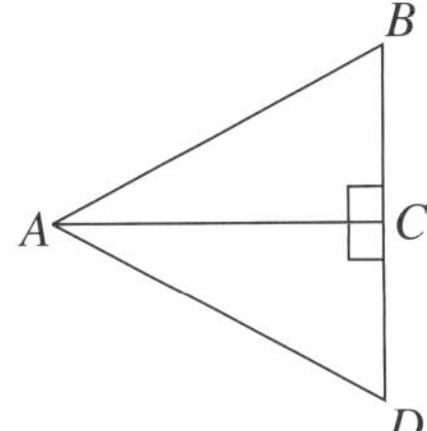

ΔABC __________

b
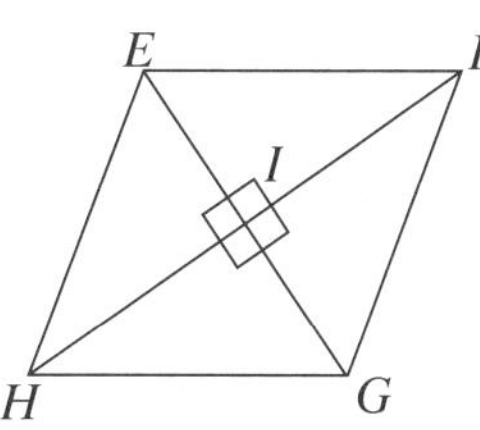

ΔEFI __________

c
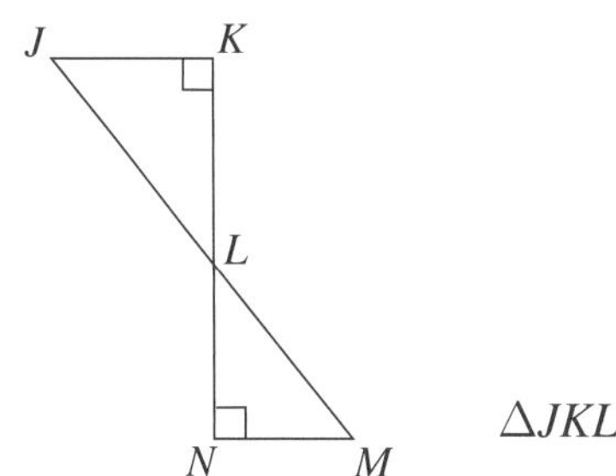

ΔJKL __________

d
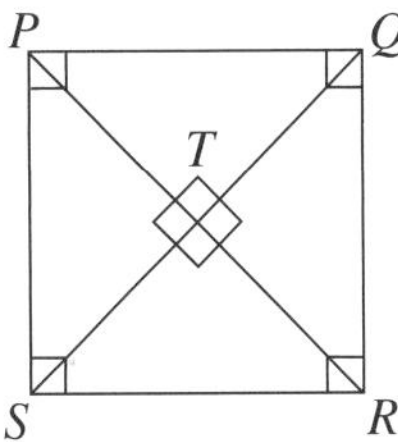

ΔPTQ __________

e
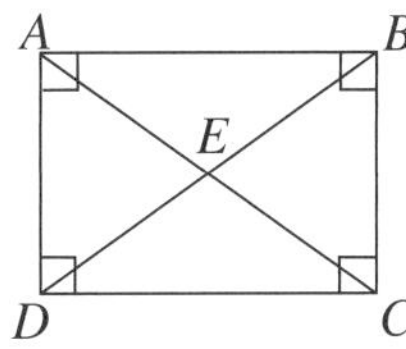

ΔABC __________

f
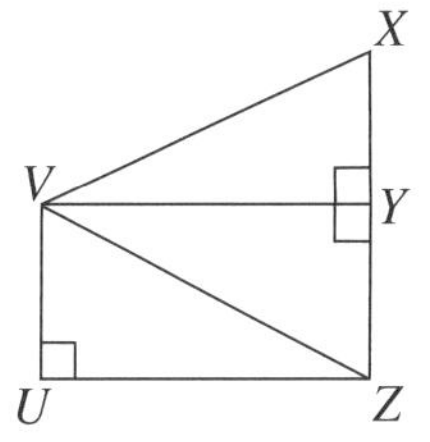

ΔVUZ __________

QUESTION 3 Complete the following statements.

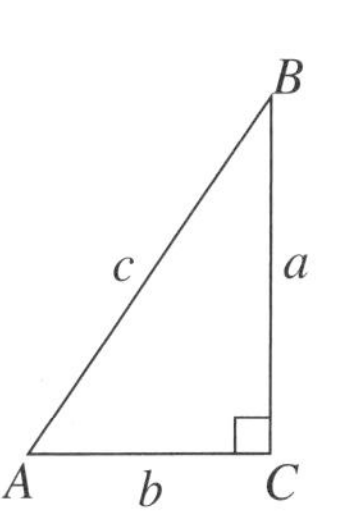

a ____________________ is the length of the hypotenuse.

b ____________________ is the length of the side opposite $\angle A$

c ____________________ is the length of the side opposite $\angle B$

d ____________________ is the length of the side opposite $\angle C$

e ____________________ is the area of the square on the side opposite $\angle A$

f ____________________ is the area of the square on the side opposite $\angle B$

g ____________________ is the area of the square on the side opposite $\angle C$

h ____________________ is the name given to the longest side of ΔABC

Pythagoras' theorem

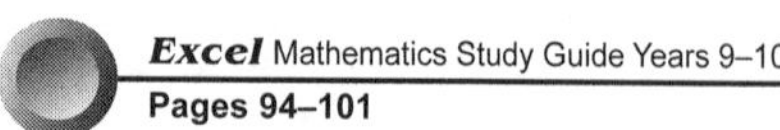

UNIT 2: Naming the sides of a right-angled triangle

QUESTION **1** Complete the table below for each of the following triangles and verify that the square of the hyotenuse is equal to the sum of the squares of the other two sides.

a 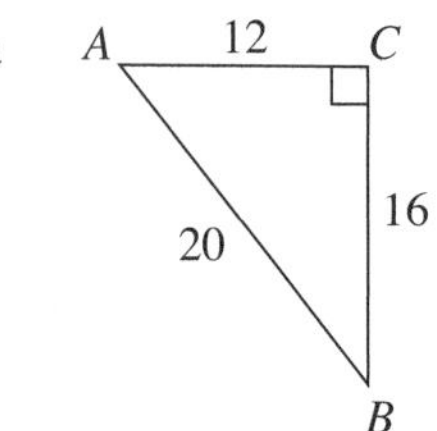

b 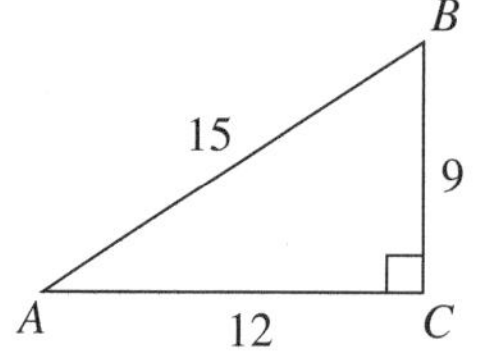

c 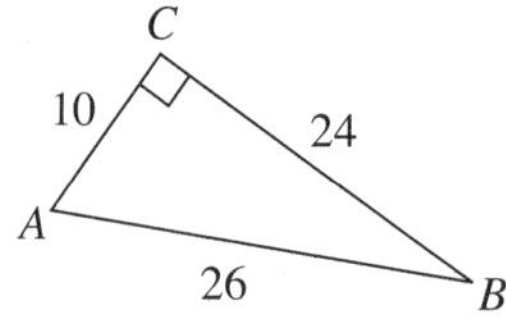

d 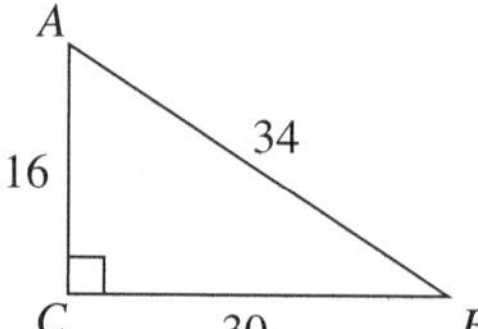

e 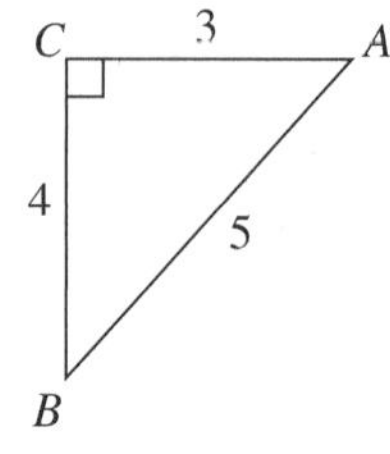

f 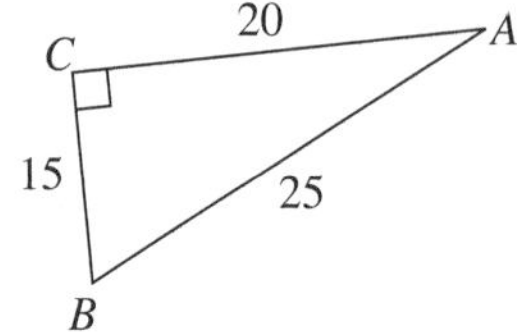

g 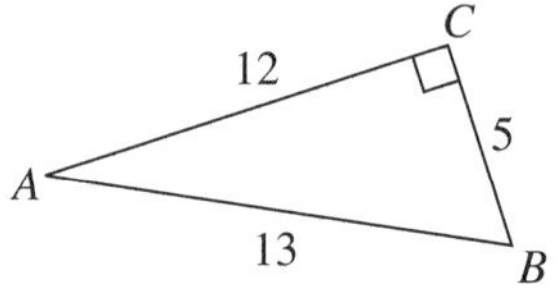

h 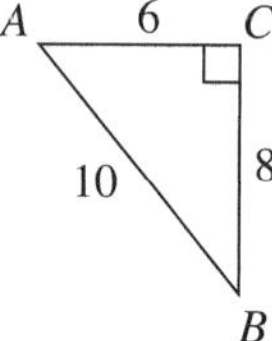

i 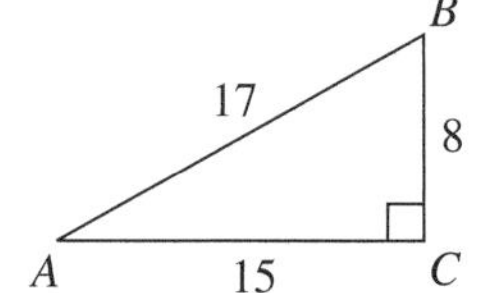

j 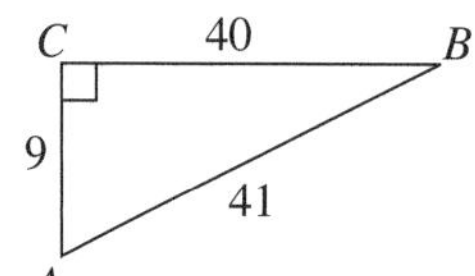

k 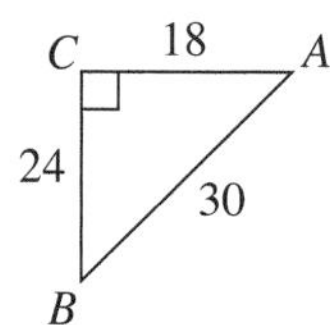

l 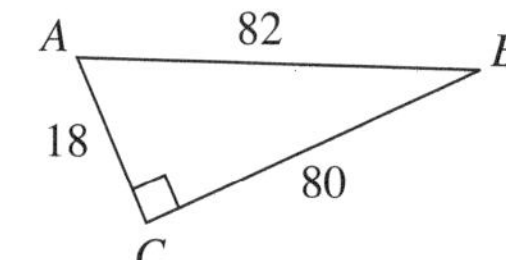

	a	b	c	a^2	b^2	c^2	$a^2 + b^2$
a							
b							
c							
d							
e							
f							
g							
h							
i							
j							
k							
l							

Pythagoras' theorem

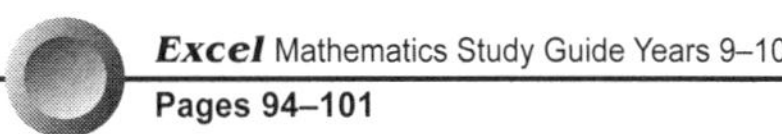

UNIT 3: Selecting the correct Pythagoras' rule

QUESTION 1 Choose the correct expression of Pythagoras' theorem for each triangle.

a **A** $a^2 = b^2 + c^2$

B $b^2 = a^2 + c^2$

C $c^2 = a^2 + b^2$

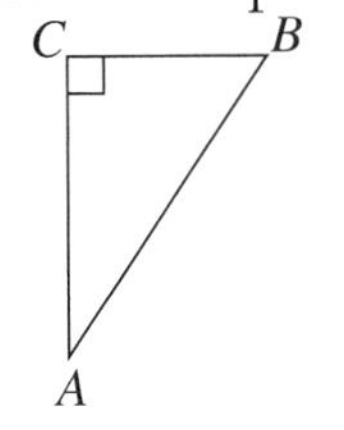

b **A** $y^2 = a^2 + z^2$

B $z^2 = a^2 + y^2$

C $a^2 = y^2 + z^2$

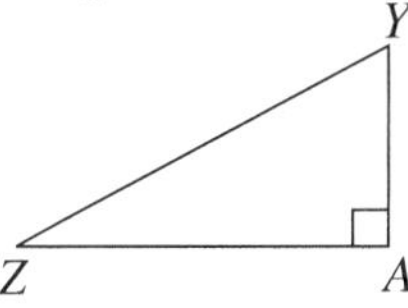

c **A** $d^2 = e^2 + f^2$

B $e^2 = d^2 + f^2$

C $f^2 = d^2 + e^2$

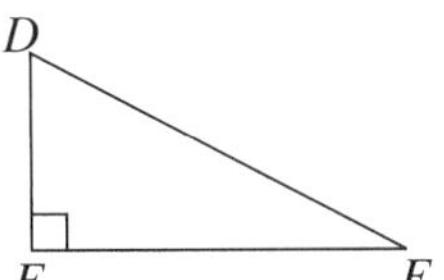

d **A** $b^2 = c^2 + d^2$

B $c^2 = b^2 + d^2$

C $d^2 = b^2 + c^2$

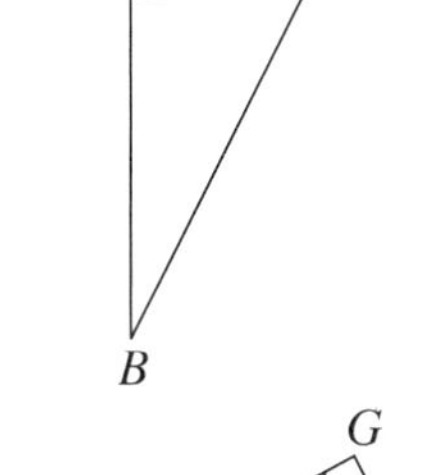

e **A** $g^2 = h^2 + i^2$

B $h^2 = g^2 + i^2$

C $i^2 = g^2 + h^2$

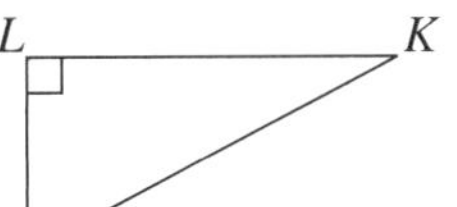

f **A** $e^2 = f^2 + g^2$

B $f^2 = e^2 + g^2$

C $g^2 = e^2 + f^2$

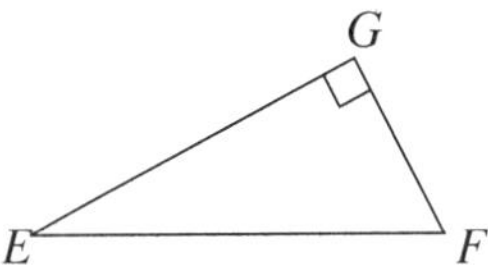

g **A** $j^2 = k^2 + l^2$

B $k^2 = j^2 + l^2$

C $l^2 = j^2 + k^2$

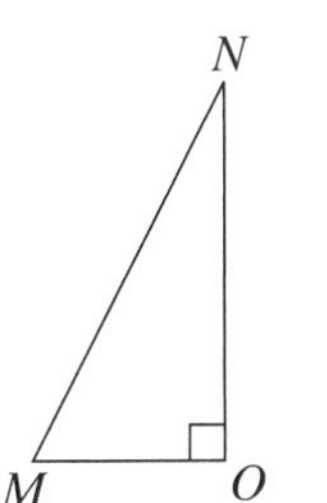

h **A** $h^2 = i^2 + j^2$

B $i^2 = h^2 + j^2$

C $j^2 = h^2 + i^2$

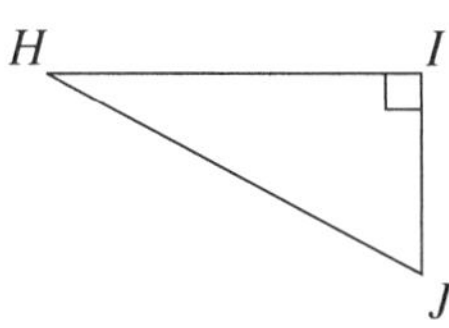

i **A** $m^2 = n^2 + o^2$

B $n^2 = m^2 + o^2$

C $o^2 = m^2 + n^2$

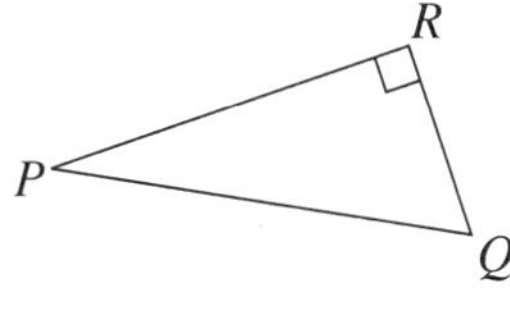

j **A** $k^2 = l^2 + m^2$

B $l^2 = k^2 + m^2$

C $m^2 = k^2 + l^2$

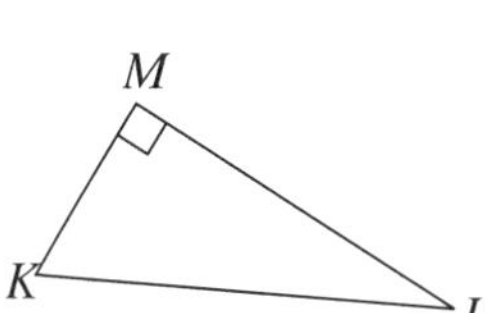

k **A** $p^2 = q^2 + r^2$

B $q^2 = p^2 + r^2$

C $r^2 = p^2 + q^2$

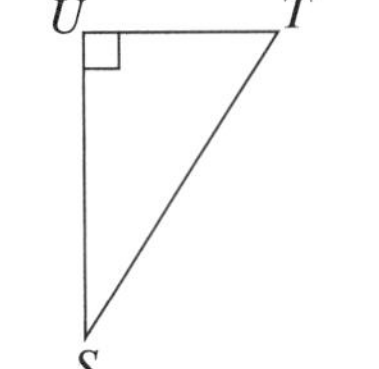

l **A** $n^2 = o^2 + p^2$

B $o^2 = n^2 + p^2$

C $p^2 = n^2 + o^2$

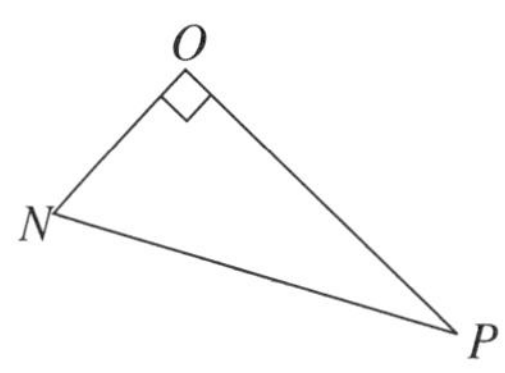

m **A** $s^2 = t^2 + u^2$

B $t^2 = s^2 + u^2$

C $u^2 = s^2 + t^2$

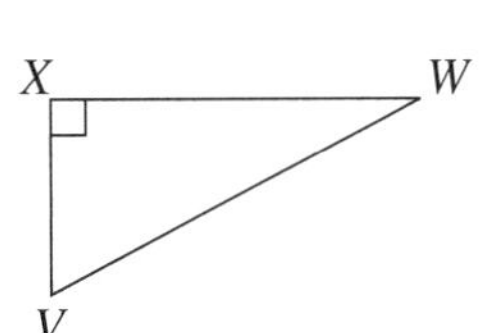

n **A** $q^2 = r^2 + s^2$

B $r^2 = q^2 + s^2$

C $s^2 = q^2 + r^2$

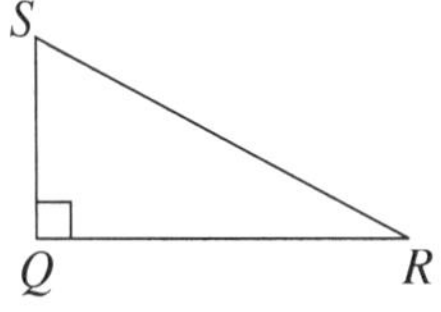

o **A** $v^2 = w^2 + x^2$

B $w^2 = v^2 + x^2$

C $x^2 = v^2 + w^2$

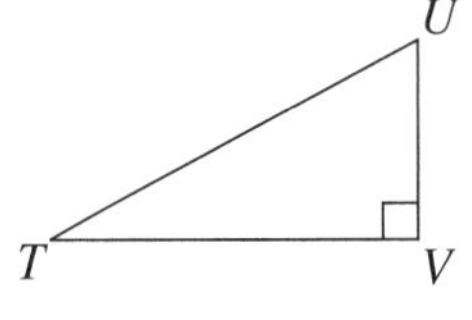

p **A** $u^2 = v^2 + t^2$

B $v^2 = u^2 + t^2$

C $t^2 = u^2 + v^2$

Pythagoras' theorem

Excel Mathematics Study Guide Years 9–10
Pages 94–101

UNIT 4: Squares and square roots

QUESTION 1 Use your calculator, if necessary, to find the following squares.

a $5^2 =$ ________ **b** $15^2 =$ ________ **c** $28^2 =$ ________

d $31^2 =$ ________ **e** $92^2 =$ ________ **f** $9^2 =$ ________

g $56^2 =$ ________ **h** $7^2 =$ ________ **i** $61^2 =$ ________

j $32^2 =$ ________ **k** $85^2 =$ ________ **l** $78^2 =$ ________

QUESTION 2 Find the following.

a $\sqrt{4} =$ ________ **b** $\sqrt{1} =$ ________ **c** $\sqrt{9} =$ ________

d $\sqrt{16} =$ ________ **e** $\sqrt{49} =$ ________ **f** $\sqrt{64} =$ ________

g $\sqrt{25} =$ ________ **h** $\sqrt{81} =$ ________ **i** $\sqrt{100} =$ ________

j $\sqrt{144} =$ ________ **k** $\sqrt{36} =$ ________ **l** $\sqrt{121} =$ ________

QUESTION 3 Use your calculator to find the value of x given that $x > 0$.

a $x^2 = 784$ ________ **b** $x^2 = 289$ ________ **c** $x^2 = 1369$ ________

d $x^2 = 169$ ________ **e** $x^2 = 196$ ________ **f** $x^2 = 2401$ ________

g $x^2 = 441$ ________ **h** $x^2 = 1156$ ________ **i** $x^2 = 324$ ________

j $x^2 = 256$ ________ **k** $x^2 = 225$ ________ **l** $x^2 = 3969$ ________

QUESTION 4 Calculate the following.

a $(1.3)^2 =$ ________ **b** $(5.6)^2 =$ ________ **c** $(7.9)^2 =$ ________

d $(5.2)^2 =$ ________ **e** $(6.7)^2 =$ ________ **f** $(8.35)^2 =$ ________

g $(8.3)^2 =$ ________ **h** $(8.32)^2 =$ ________ **i** $(11.25)^2 =$ ________

j $(9.7)^2 =$ ________ **k** $(5.41)^2 =$ ________ **l** $(22.2)^2 =$ ________

QUESTION 5 Use the calculator square key to find the following squares.

a $(5.61)^2 =$ ________ **b** $(3.2)^2 =$ ________ **c** $(6.31)^2 =$ ________

d $(7.8)^2 =$ ________ **e** $(5.3)^2 =$ ________ **f** $(13.5)^2 =$ ________

g $(5.9)^2 =$ ________ **h** $(6.8)^2 =$ ________ **i** $(15.2)^2 =$ ________

j $(6.7)^2 =$ ________ **k** $(9.2)^2 =$ ________ **l** $(8.95)^2 =$ ________

QUESTION 6 Find these square roots correct to 1 decimal place.

a $\sqrt{5.4} =$ ________ **b** $\sqrt{6.58} =$ ________ **c** $\sqrt{52.7} =$ ________

d $\sqrt{8.1} =$ ________ **e** $\sqrt{3.25} =$ ________ **f** $\sqrt{93.8} =$ ________

g $\sqrt{7.69} =$ ________ **h** $\sqrt{6.75} =$ ________ **i** $\sqrt{62.1} =$ ________

j $\sqrt{8.23} =$ ________ **k** $\sqrt{8.123} =$ ________ **l** $\sqrt{73.8} =$ ________

Pythagoras' theorem

UNIT 5: Finding the length of the hypotenuse

QUESTION 1 Find the length of the hypotenuse in each of the following triangles. All measurements are in centimetres.

a

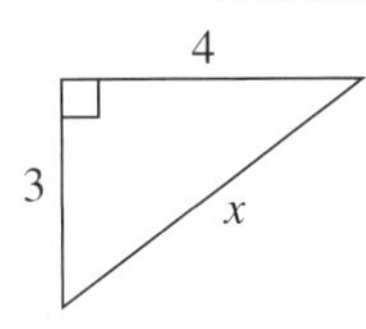

b

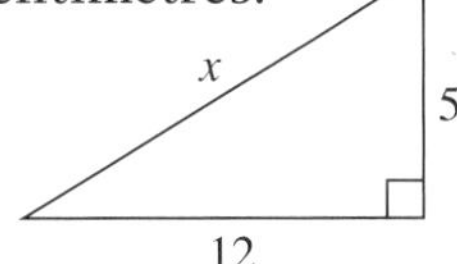

c

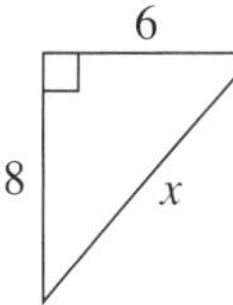

d

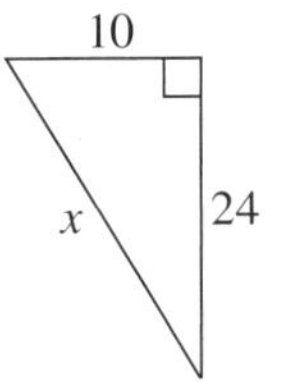

e

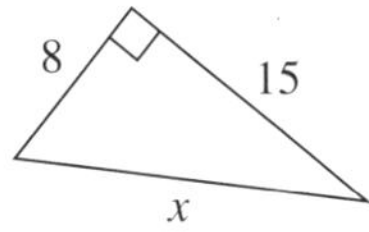

f

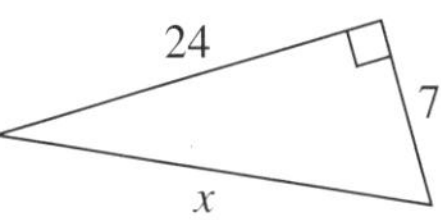

QUESTION 2 Find the length of the hypotenuse correct to 1 decimal place. All measurements are in centimetres.

a

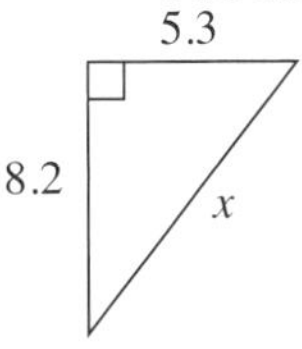

b

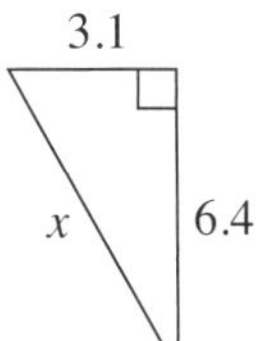

c

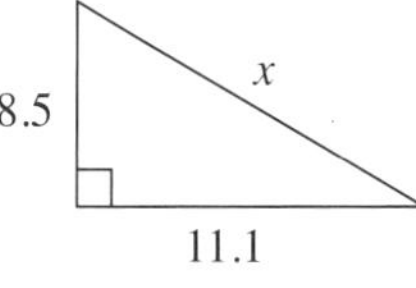

d

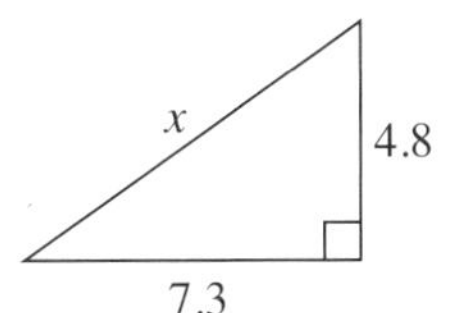

e

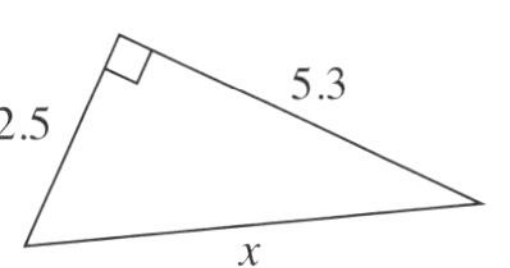

f

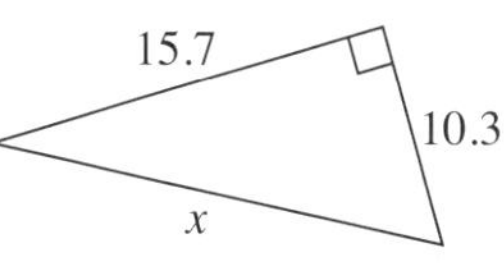

g

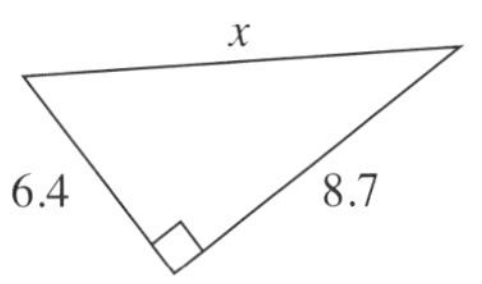

h

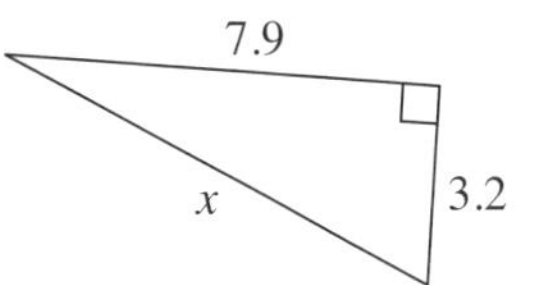

i

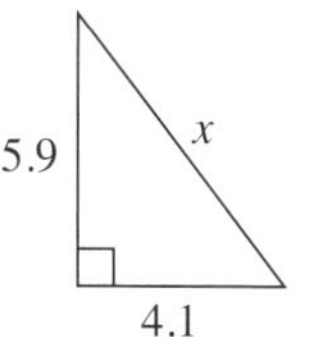

Pythagoras' theorem

UNIT 6: Finding the length of a side

QUESTION 1 Find the length of the unknown side in each of the following triangles. All measurements are in centimetres.

a

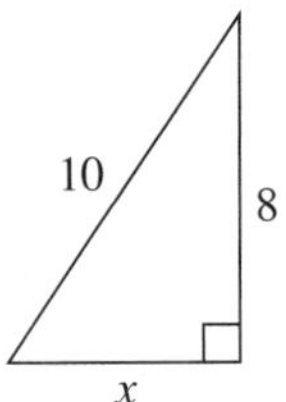

b

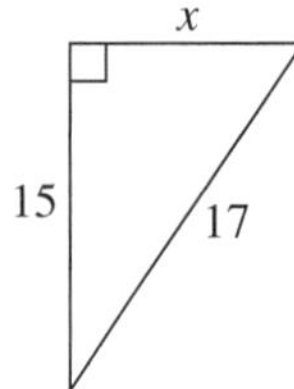

c

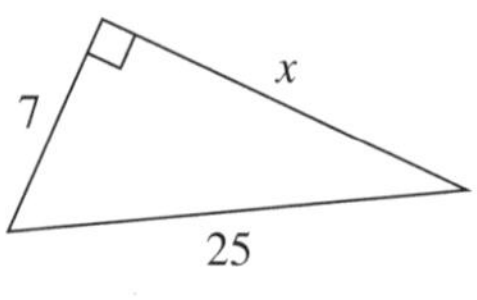

d

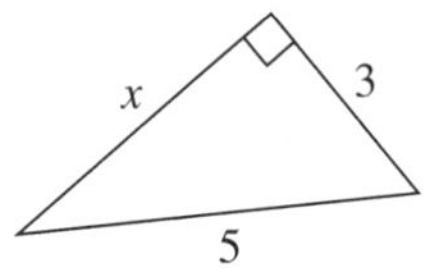

e

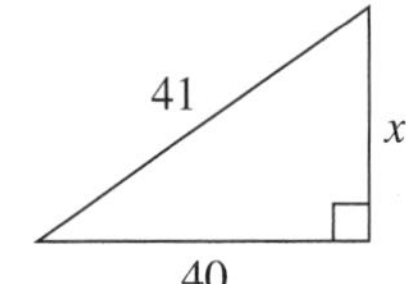

f

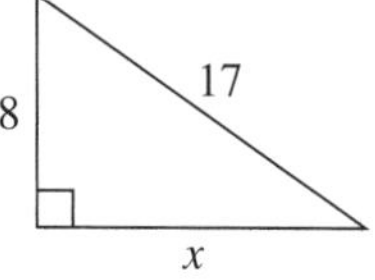

QUESTION 2 Find the length of the unknown side correct to 2 decimal places. All measurements are in centimetres.

a

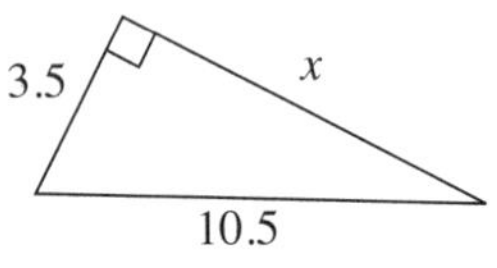

b

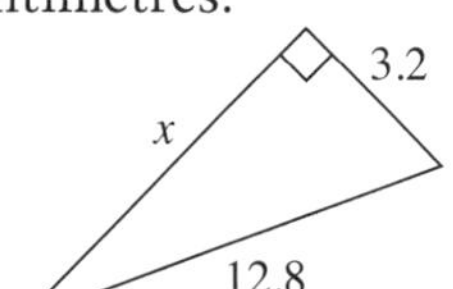

c

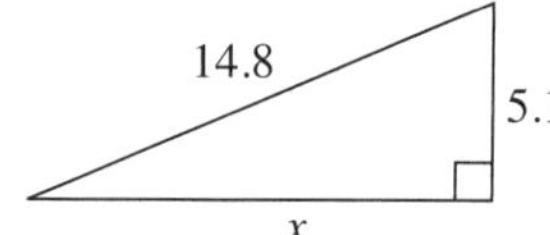

d

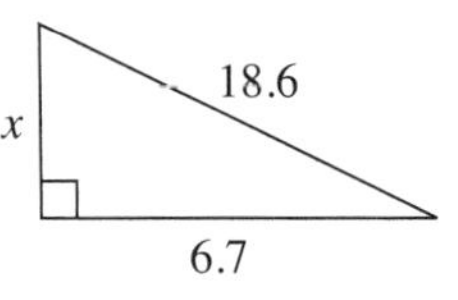

e

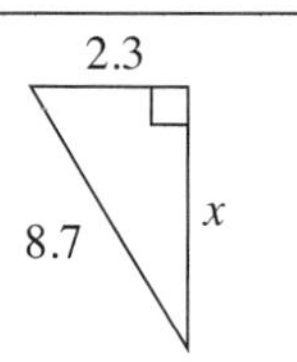

f

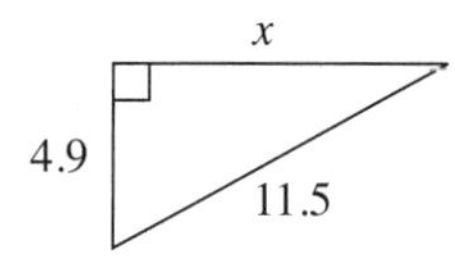

g

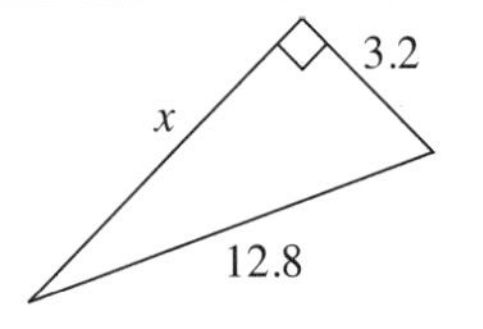

h

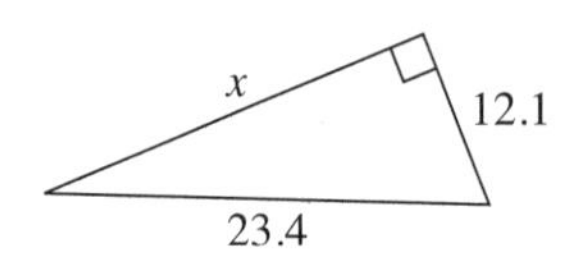

i

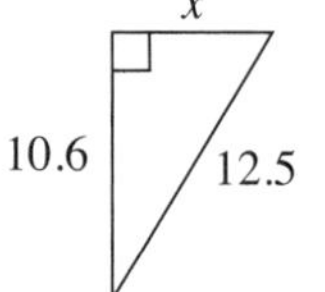

Pythagoras' theorem

UNIT 7: Miscellaneous questions

QUESTION 1 Find the length of the hypotenuse.

a

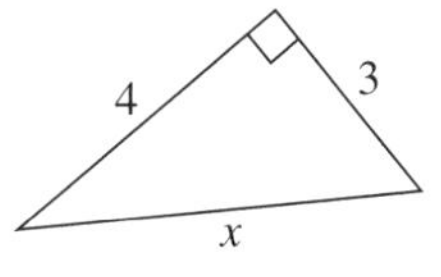

b

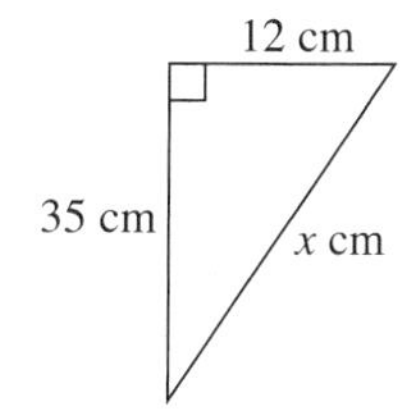

c

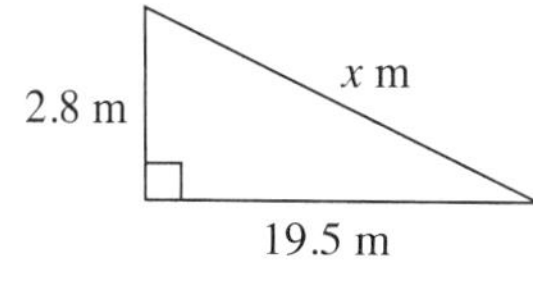

QUESTION 2 Find the length of the side x.

a

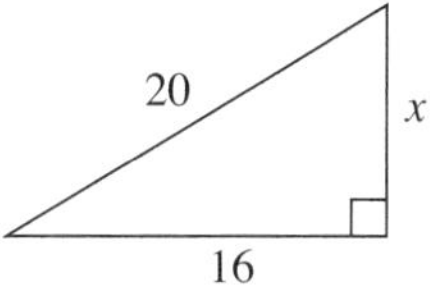

b

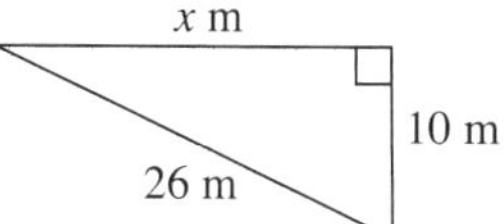

c

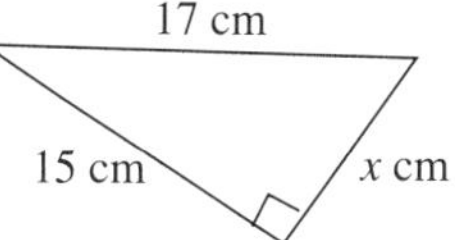

QUESTION 3 Find the length of the unknown side, giving the answer correct to one decimal place.

a

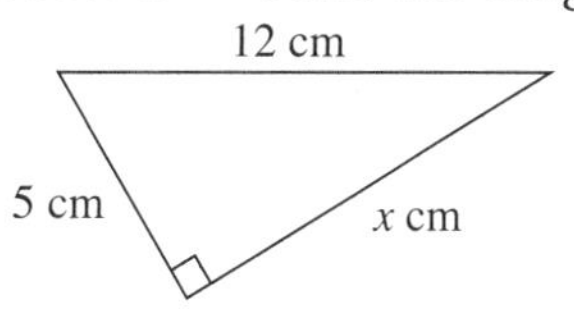

b

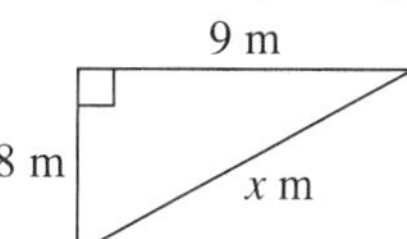

c

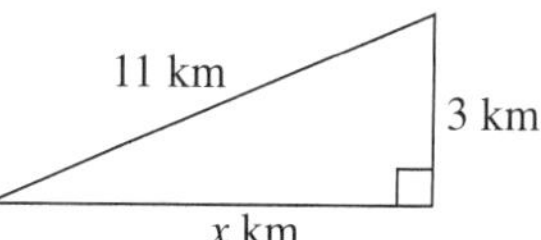

d

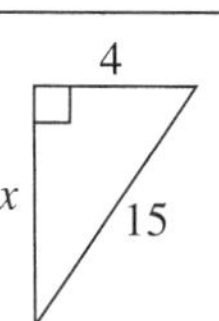

e

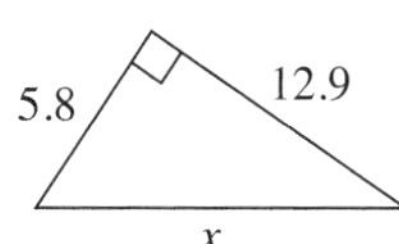

f

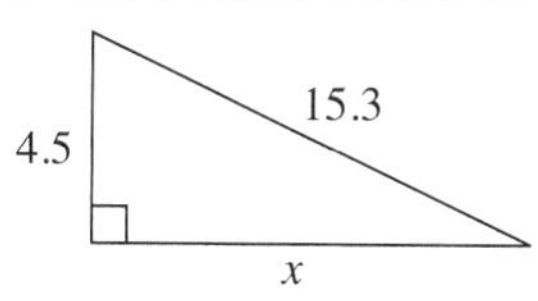

UNIT 8: Mixed questions on Pythagoras' theorem

QUESTION 1 In each of the following triangles find the length of the unknown side. All measurements are in centimetres.

a

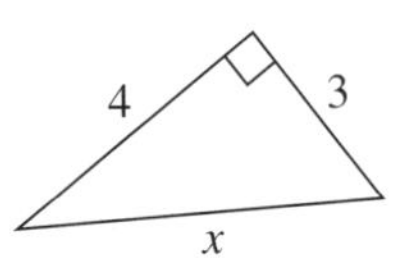

b

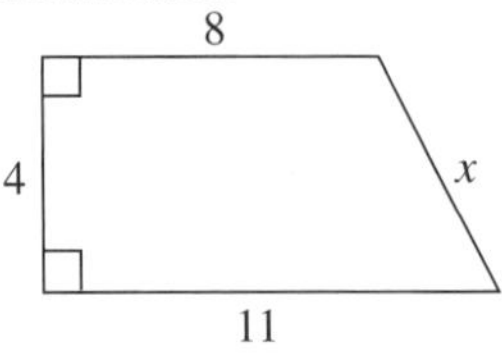

c

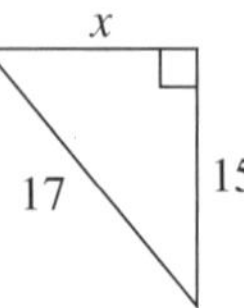

d

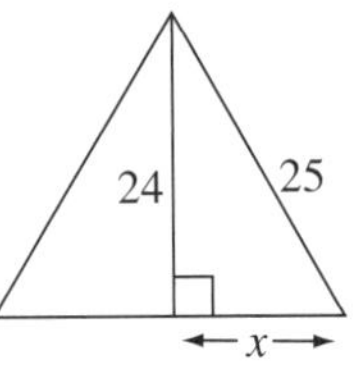

e

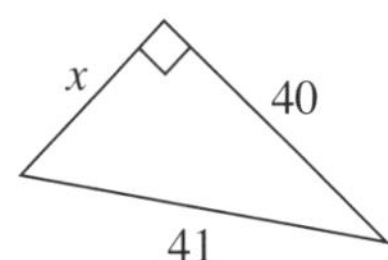

f

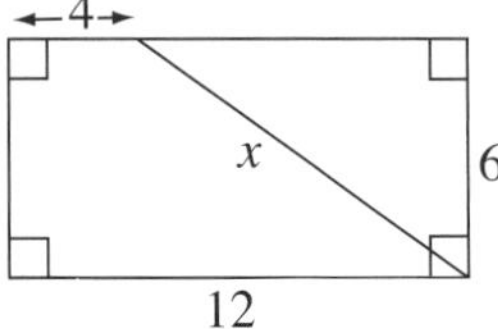

QUESTION 2 Find the length of the unknown sides correct to one decimal place. All measurements are in centimetres.

a

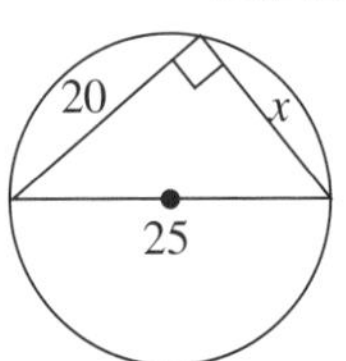

b

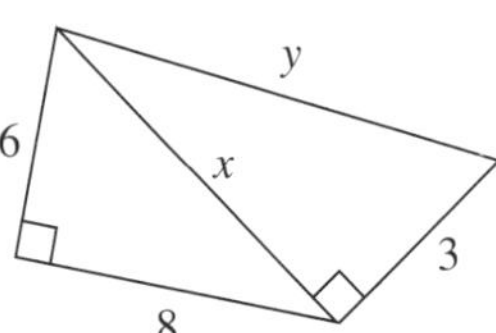

c

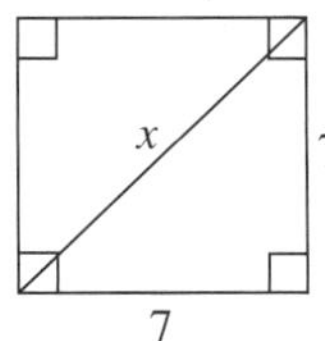

d

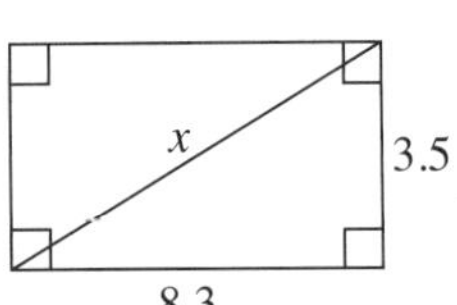

e

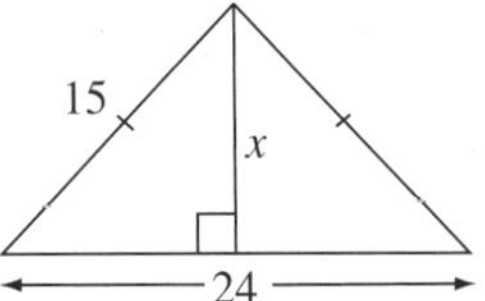

f

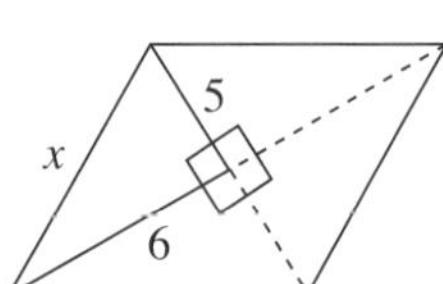

Pythagoras' theorem

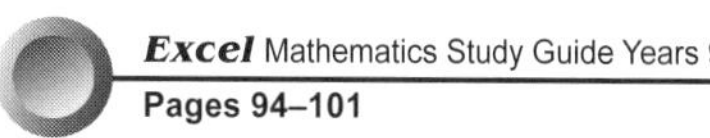

UNIT 9: Pythagorean triads

QUESTION 1 Which of the following are Pythagorean triads?

a {2, 4, 5} ______ **b** {7, 12, 13} ______ **c** {4, 12, 13} ______

d {8, 10, 12} ______ **e** {3, 4, 5} ______ **f** {6, 8, 10} ______

g {5, 12, 13} ______ **h** {8, 13, 17} ______ **i** {8, 15, 17} ______

j {7, 24, 25} ______ **k** {9, 40, 41} ______ **l** {16, 30, 34} ______

QUESTION 2 Prove that the following triangles are right-angled triangles.

a

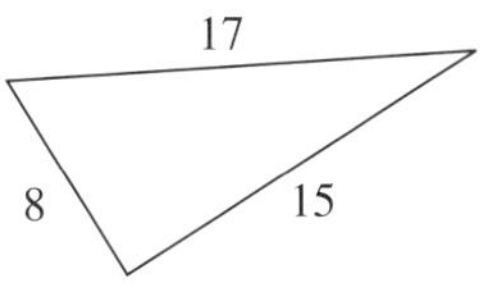

b

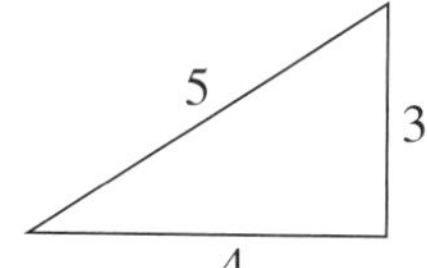

c

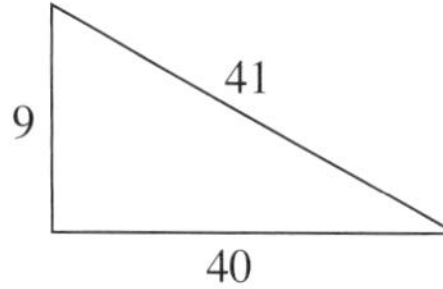

d

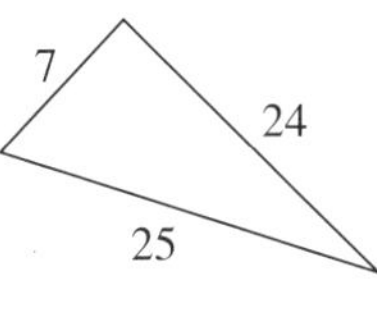

e

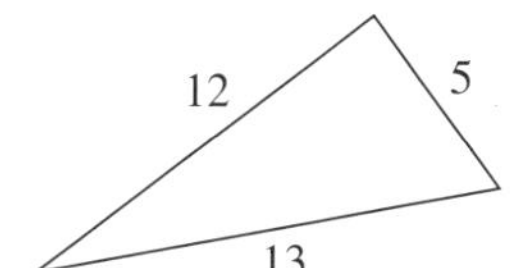

f

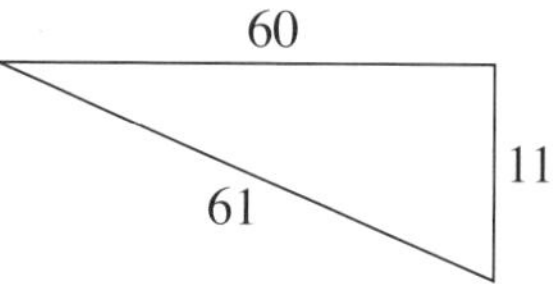

QUESTION 3 Determine whether the triangle is right-angled or not.

a

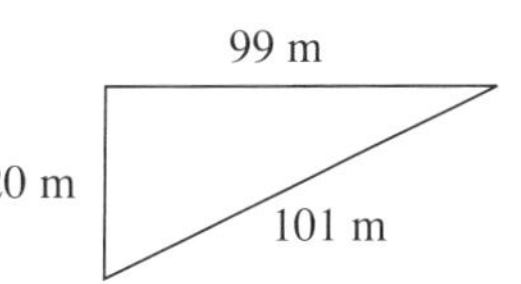

b

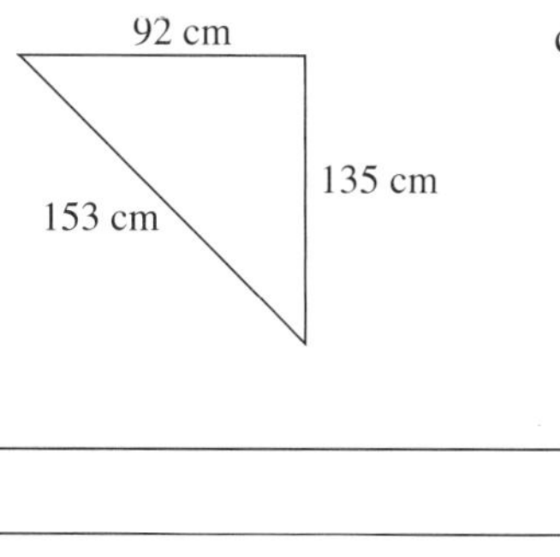

c

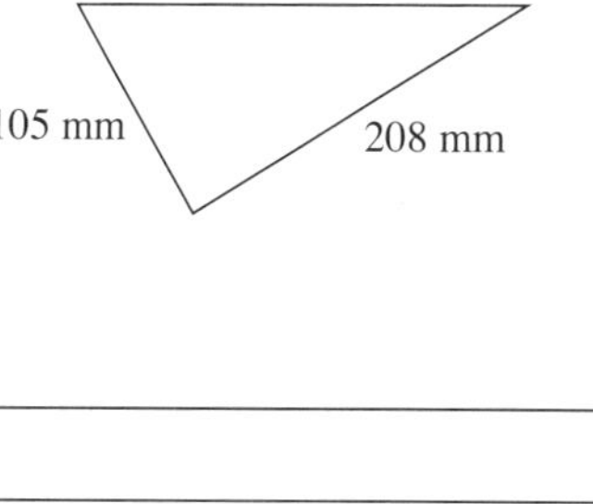

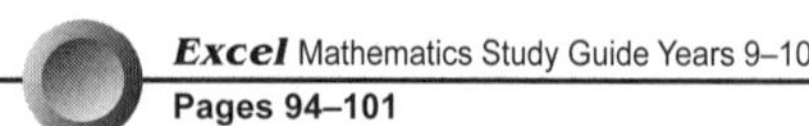

UNIT 10: Applications of Pythagoras' theorem

QUESTION 1 Find the length of the diagonal (to 1 decimal place) of:

a a square of side length 5 cm

b a rectangle 28 cm long and 9 cm wide

QUESTION 2 The radius of the base of a cone is 4.2 m and its slant height is 8 m. Find the height of the cone correct to one decimal place.

h 8 m 4.2 m

QUESTION 3 Find the length of the third side of a right-angled triangle, (to one decimal place), if the:

a longest side is 49 cm, other side is 12 cm

b hypotenuse is 45 cm, other side is 16 cm

QUESTION 4 What is the altitude of an equilateral triangle where sides are each 12 cm (answer correct to 2 decimal places).

QUESTION 5 Find the perimeter of this block of land.

60 m 45 m 32 m 54 m 36 m

Pythagoras' theorem

UNIT 11: Problem solving

QUESTION 1 A 5 metre ladder has its foot 2 metres from the foot of a wall. How far up the wall does the ladder reach (give the answer to the nearest cm)?

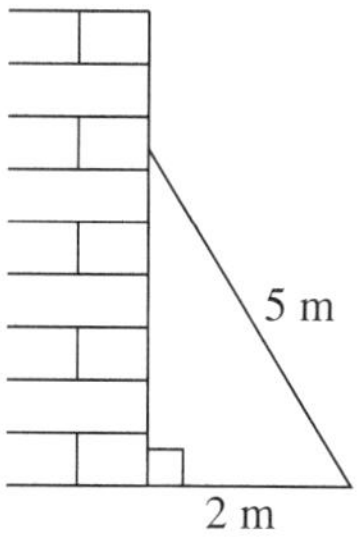

5 m
2 m

QUESTION 2 Two roads are at right angles to each other. Person A walks 8 km on one road and person B walks 15 km on the other road. How far apart are A and B?

QUESTION 3 A 6 metre ladder rests against a wall and its foot is 3 metres away from the base of the wall. How high does the ladder reach up the wall (answer correct to 2 decimal places)?

QUESTION 4 Carlo is building a rectangular gate from steel pipe. The gate is 4.2 m long and 1.2 m high. In order to brace the gate, Carlo wants to add a centre brace and two diagonal braces as shown in the diagram. He has 6 m of pipe left. It this enough for the bracing he wants to do? Justify your answer.

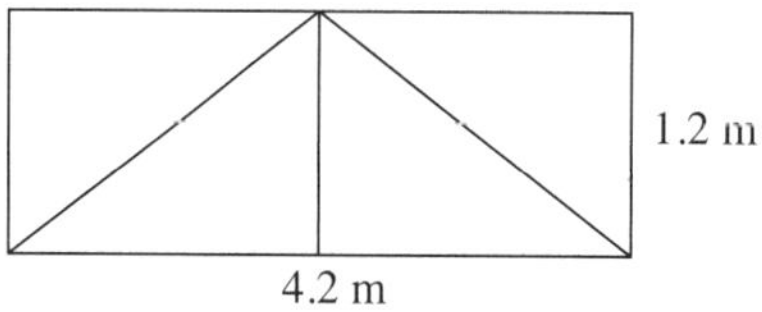

1.2 m
4.2 m

QUESTION 5 Two flag posts are 9 m and 12.5 m long and 24 m apart. Find the length of the string needed to join the tops of the two posts.

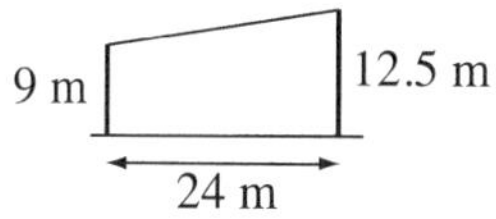

9 m
12.5 m
24 m

Pythagoras' theorem

TOPIC TEST **PART A**

Instructions
- This part consists of 10 multiple-choice questions.
- Fill in only ONE CIRCLE for each question.
- Each question is worth 1 mark.

Time allowed: 10 minutes **Total marks: 10**

Marks

1 $\sqrt{5}$ is closest to

Ⓐ 2 Ⓑ 2.2 Ⓒ 2.23 Ⓓ 2.24 — 1

2 Which of the following is not a Pythagorean triad?

Ⓐ {12, 35, 37} Ⓑ {11, 60, 61} Ⓒ {9, 40, 41} Ⓓ {13, 44, 45} — 1

3 The Pythagorean result for a triangle *ABC* with hypotenuse *BC* is

Ⓐ $a^2 = b^2 + c^2$ Ⓑ $b^2 = a^2 + c^2$ Ⓒ $a^2 = c^2 - b^2$ Ⓓ $c^2 = b^2 + a^2$ — 1

4 If two sides of a right-angled triangle are 7 cm and 24 cm, then the hypotenuse is

Ⓐ 23 cm Ⓑ 24 cm Ⓒ 25 cm Ⓓ 31 cm — 1

5 Which one of the following triads determines a right-angled triangle?

Ⓐ {8, 9, 12} Ⓑ {11, 10, 15} Ⓒ {9, 11, 20} Ⓓ {16, 30, 34} — 1

6 Find the area of a rectangle which has a diagonal 10 cm long and one side 6 cm long.

Ⓐ 40 cm^2 Ⓑ 48 cm^2 Ⓒ 60 cm^2 Ⓓ 80 cm^2 — 1

7 Given that $c^2 = a^2 + b^2$ and $a = 10$ and $b = 24$ and $c > 0$, what is the value of c?

Ⓐ 26 Ⓑ 28 Ⓒ 576 Ⓓ 676 — 1

8 The hypotenuse of a right-angled triangle is 17 cm. If one side is 8 cm, the third side is

Ⓐ 9 cm Ⓑ 11 cm Ⓒ 13 cm Ⓓ 15 cm — 1

9 Which of the following is a Pythagorean triad?

Ⓐ {48, 53, 71} Ⓑ {48, 55, 73} Ⓒ {48, 57, 77} Ⓓ {48, 59, 75} — 1

10 The two shorter sides of a right-angled triangle have lengths 12 cm and 5 cm. What is the square of the length of the hypotenuse?

Ⓐ 13 Ⓑ 119 Ⓒ 169 Ⓓ 289 — 1

Total marks achieved for PART A ___ / 10

Pythagoras' theorem

TOPIC TEST **PART B**

Time allowed: 20 minutes **Total marks: 15**

Marks

1 If $a^2 = 4761$ and $a > 0$, find the value of a ______________ 1

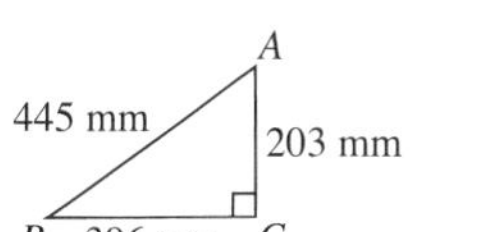

2 Is $\triangle ABC$ a right-angled triangle? ______________ 1

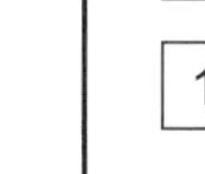

3 Find the value of x, to one decimal place if necessary.

a

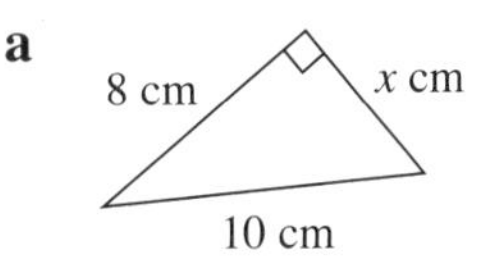

b

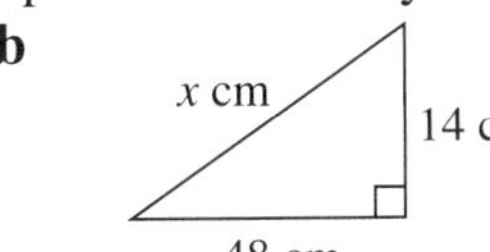

c

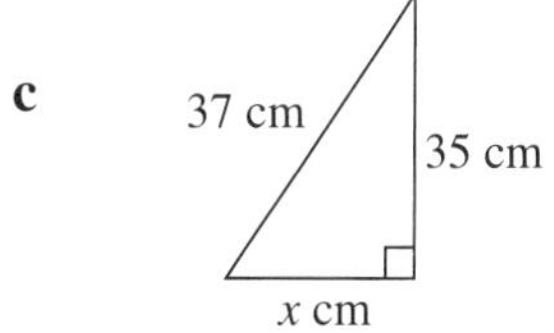

3

d

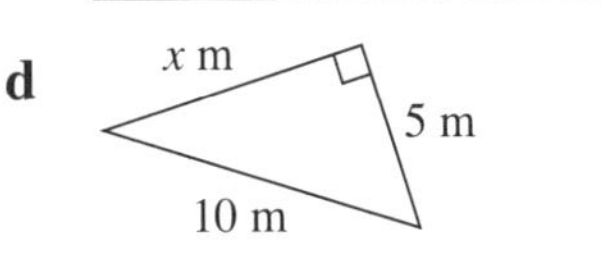

e

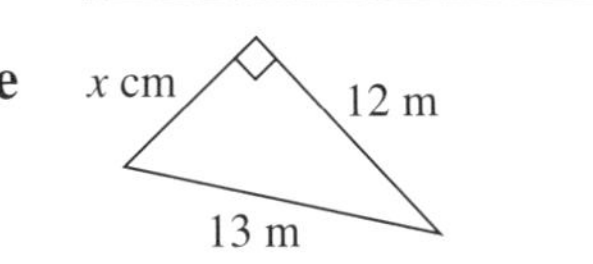

f

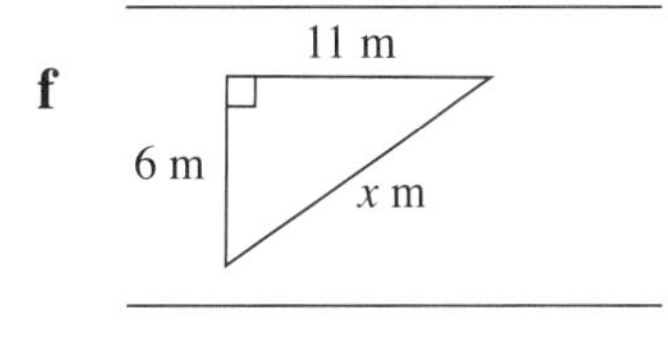

3

g

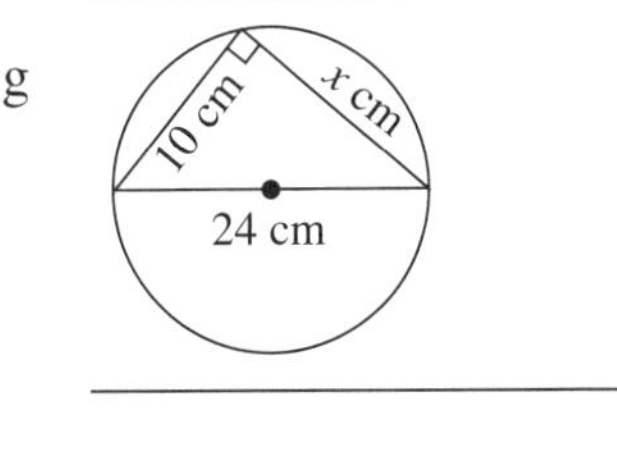

h

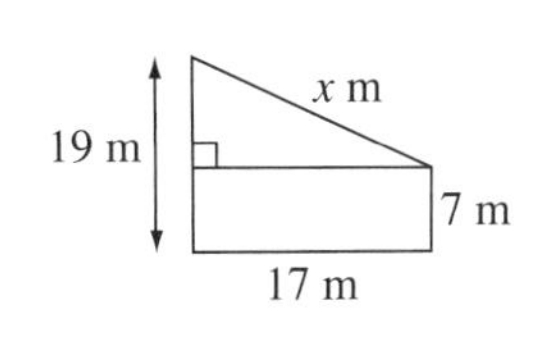

i

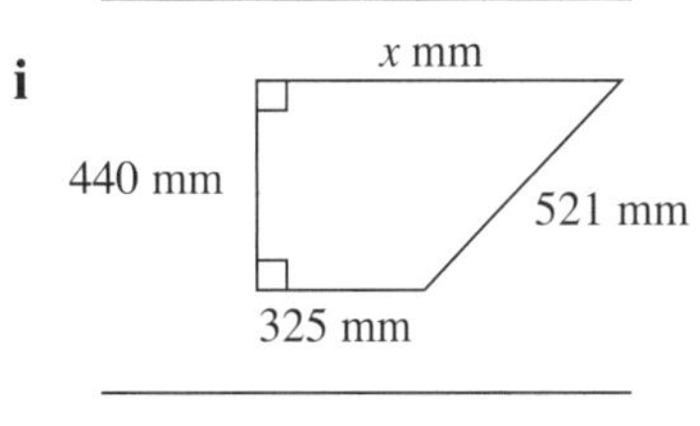

3

4

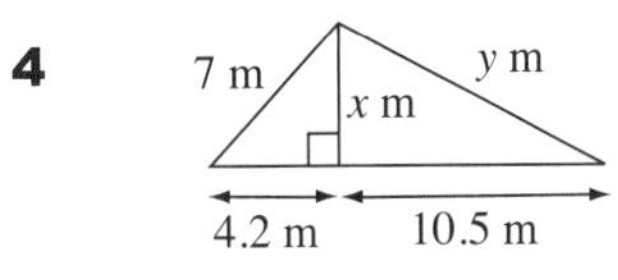

a Find x ______________ 2

b Find y ______________ 2

5

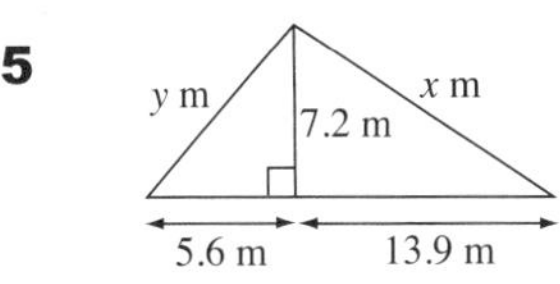

a Find x ______________

b Find y ______________

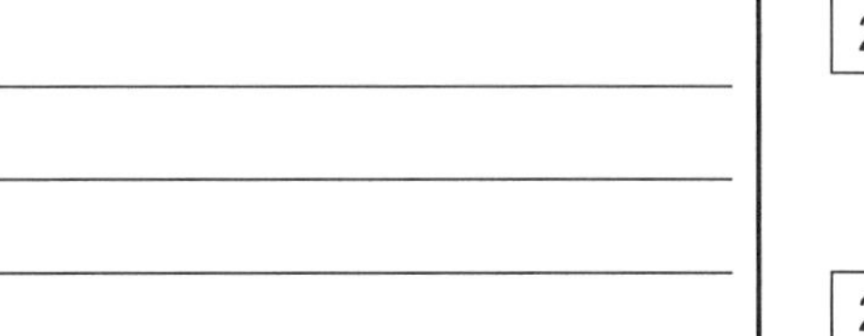

Total marks achieved for PART B /15

CHAPTER 4
Financial mathematics

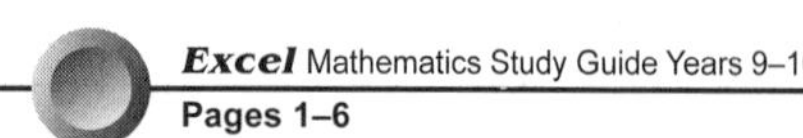
Excel Mathematics Study Guide Years 9–10
Pages 1–6

UNIT 1: Wages and salaries

QUESTION 1 Angela works a basic week of 40 hours and her hourly rate of pay is $12.50. Calculate her weekly wage.

QUESTION 2 Michael works 35 hours per week and his weekly wage is $756. Find his hourly rate of pay.

QUESTION 3 Cleve works 8 hours a day and a nine-day fortnight. If his pay rate is $23.15 per hour, what is his fortnightly pay?

QUESTION 4 Reno works 6 hours on Monday, 8 hours on Tuesday, 7 hours on Wednesday, 9 hours on Thursday and 6 hours on Friday. If he paid $18.20 per hour, what is his weekly pay?

QUESTION 5 John's annual salary is $43 550. How much is he paid each week?

QUESTION 6 Amie receives a salary of $72 852 p.a. What is her gross fornightly pay?

QUESTION 7 Jenny earns $659 per week. What is her annual salary?

QUESTION 8 Daniel receives $3240 per month. Find his:

a annual salary

b weekly pay

QUESTION 9 Yousef is paid $163.50 for working $7\frac{1}{2}$ hours. What will he be paid for working 5 hours at the same rate of pay?

QUESTION 10 Mladdin is on a salary of $67 440 p.a. paid monthly.

a How much does he receive each month?

b Mladdin works 200 hours each month. How much does he receive per hour?

QUESTION 11 Last year the chief executive of a bank received a total remuneration package of $7 774 624.

a How much is this per week?

b A newspaper headline read: 'Bank boss paid $21 300 a day'. Is this correct? Justify your answer.

Financial mathematics

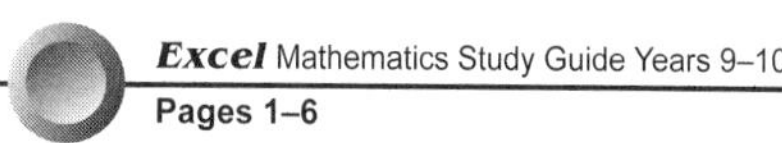

UNIT 2: Overtime and holiday pay

QUESTION 1 A man is paid a basic rate of \$14.70 per hour. Calculate his hourly overtime rate of pay when this is paid at:

a time-and-a-half

b double-time

QUESTION 2 Kelli's normal pay rate is \$16.80 per hour. What will she earn for working:

a 5 hours at time-and-a-half?

b 3 hours at double-time-and-a-half?

QUESTION 3 James is paid \$860 for a 40-hour week. He works 6 hours overtime at time-and-a-half. What is his total income for the week?

QUESTION 4 Michelle gets an annual salary of \$48 630.40. If she receives $17\frac{1}{2}\%$ holiday loading on the 4-week holiday pay period, calculate:

a her normal pay for 4 weeks.

b her holiday loading.

c her holiday pay for 4 weeks.

QUESTION 5

a John receives a gross pay of \$850 for a 40-hour week. Calculate John's hourly rate of pay.

b In one busy week, in addition to his normal 40 hours, John works the following overtime; 6 hours on Saturday at time-and-a-half and 5 hours on Sunday at double-time. Find John's gross pay for that week.

QUESTION 6 Ronnie is an electrician and gets paid \$1200 for a 40-hour week. In one week she works 12 hours overtime, of which 8 hours is at time-and-a-half and 4 hours is at double-time. What are her earnings that week?

QUESTION 7 Jeremy is paid \$38.60 per hour. He works 35 hours every week. Calculate Jeremy's holiday pay if he receives $17\frac{1}{2}\%$ loading on his 4 weeks of vacation time.

QUESTION 8 Brent's normal wage is \$672 for a 40 hour week. He worked overtime and earned \$873.60 in one week.

a Find his normal hourly rate

b How much extra did he earn for overtime?

c How many hours of overtime did he work if he was paid double time for it?

Financial mathematics

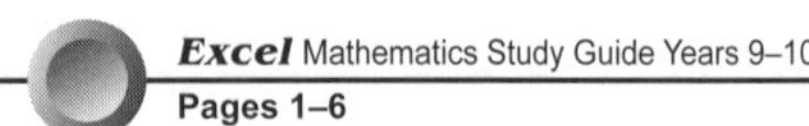

UNIT 3: Commission and piecework

QUESTION 1 Yasmin receives a commission of 5% on sales. How much commission will she receive in a week in which her sales total $11 000?

QUESTION 2 Meena is a sales person and earns $250 a week plus 3.5% commission on sales. Her weekly sales total $60 000. Find:

a her commission

b her total earnings for the week

QUESTION 3 James makes leather belts and is paid $2.55 per belt. How much does he earn for making 230 belts?

QUESTION 4 Dominic is a fruit-picker and is paid $2.30 for every full bag of fruit he picks. How much will he earn in a day if he picks 83 bags of fruit?

QUESTION 5 David sells cars. He is paid a retainer (basic wage) of $350 per week and a commission of 3% on sales made. Find his weekly income in a week in which he sells cars to the value of:

a $45 000

b $70 000

QUESTION 6 Joshua is a real estate agent and receives 2% commission on the first $200 000, $1\frac{1}{2}$% on the next $100 000, $1\frac{1}{4}$% on the next $100 000 and 1% on the value therafter. Find his commission for the selling a property worth $650 000.

QUESTION 7 Sebastian works as a packer on a fruit plantation and is paid $2.00 per box with a bonus of 90 cents for each box packed in excess of 100 boxes per day. Find his income per day in which he packs 135 boxes.

QUESTION 8 Damien works in a factory on a basic wage of $250 a week. In addition to this, he is paid a bonus of 50 cents per article for every article in excess of the weekly quota of 2000. How much will he earn in a week in which 4300 articles are made?

Financial mathematics

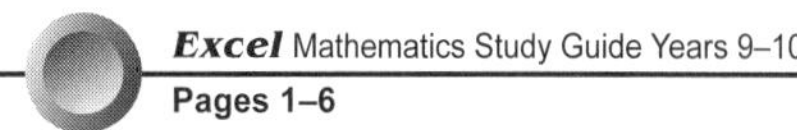

UNIT 4: Deductions from income and taxation

QUESTION 1 John's yearly salary is $66 900. His fortnightly deductions include income tax $870, medicare levy $52 and union fees $8.90. Calculate his fortnightly take-home pay (net pay).

QUESTION 2 Angela received a gross wage of $1230.60 per week. The payments deducted from her weekly wage are tax, 33% of gross weekly wage; health insurance, $35.40 per week; superannuation, 31 units at $2.75 per unit. Calculate her net pay for the week.

QUESTION 3

Taxable income	Tax on this income
0–$18 200	Nil
$18 201–$37 000	19 cents for each $1 over $18 200
$37 001–$80 000	$3572 plus 32.5 cents for each $1 over $37 000
$80 001–$180 000	$17 547 plus 37 cents for each $1 over $80 000
$180 001 and over	$54 547 plus 45 cents for each $1 over $180 000

Mark's gross income is $78 670. His total deductions are $4630. Use this made up table to work out the following.

a Find his taxable income.

b Calculate the amount of tax due.

c If he pays $485 per week in tax, how much refund should he receive for the year?

QUESTION 4 Use the table given above to work out the following.

Jo's taxable income was $48 000. What tax was payable on her income?

Financial mathematics

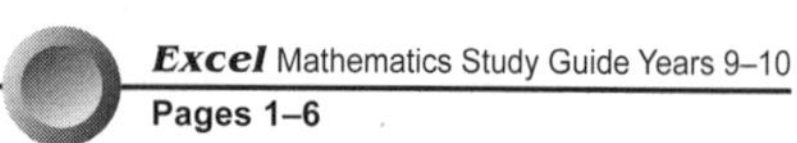

UNIT 5: Simple interest (1)

Question 1 Find the simple interest on:

a \$4500 at 8% p.a. for 2 years.

b \$8000 at 7% p.a. for 6 years.

c \$20 000 at 9% p.a. for 8 years.

d \$7800 at 12% for 3 years.

e \$6500 at 4% for 2 years.

f \$5000 for 5 months at 0.8% per month.

g \$36 000 at 10.25% p.a. for 4 years.

h \$65 000 for 5 years at 6.5% p.a.

i \$82 000 for 2 years at 8.25% p.a.

j \$5900 at 12% p.a. for 6 months.

k \$12 500 at 15% p.a. for 6 months.

l \$13 000 at 16% p.a. for 7 months.

m \$20 500 at $7\frac{1}{2}$% p.a. for 3 months.

n \$20 000 for 25 days at 15% p.a.

Question 2 Find the length of time for:

a \$500 to be the interest on \$1800 at 6% p.a.

b \$850 to be the interest on \$2400 at 8% p.a.

Question 3 Find the percentage rate per annum if:

a \$1500 is the interest on \$5400 after 5 years.

b \$900 is the interest on \$2700 after 2 years.

Financial mathematics

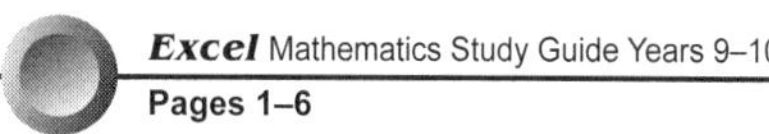

UNIT 6: Simple interest (2)

QUESTION **1** Find the principal required for the simple interest to be:

a $900 on a sum invested for 2 years at 10%

b $250 on a sum invested for 1 year at 9% p.a.

QUESTION **2** $3000 is invested at 5% p.a. simple interest for 4 years. Find the total:

a amount of interest earned.

b value of the investment.

QUESTION **3**

a Find the simple interest rate if a principal of $2500 yields interest of $625 in 2 years.

b An investment yielded $4500 flat rate of interest in 4 years at 9% p.a. Find the principal invested.

c $8500 was invested at 15% p.a. flat rate. Find the number of years the money was invested if the total interest earned was $3825.

d Find the principal required for the simple interest to be $900 on an amount invested for 2 years at 10% p.a.

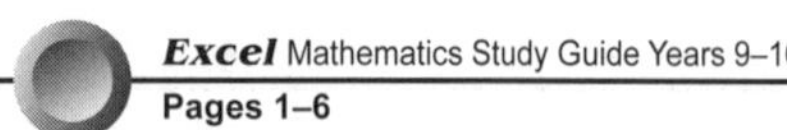

UNIT 7: Simple interest (3)

QUESTION **1** Find the length of time for:

a \$800 to be the simple interest earned on \$2700 invested at 5% p.a.

b \$1250 to be the simple interest earned on \$4500 invested at 7% p.a.

QUESTION **2** Find the percentage rate per annum, if:

a \$1800 is the simple interest earned on \$6900 invested for 4 years.

b \$3000 is the simple interest earned on \$10 000 invested for 3 years.

QUESTION **3** Find the principal required for:

a the simple interest earned to be \$800 on an amount invested for 3 years at 6% p.a.

b the simple interest earned to be \$1220 on an amount invested for 2 years at 8% p.a.

QUESTION **4** Jill borrows \$15 000.

a Find the simple interest she will pay if she takes the loan over 4 years at 7% p.a.

b How much extra will Jill pay if she takes the loan over 4 years at $7\frac{1}{2}\%$ p.a.?

c How much less would Jill pay if she takes the loan over 3 years at $7\frac{1}{2}\%$ p.a. instead of 4 years (at $7\frac{1}{2}\%$ p.a.)?

Financial mathematics

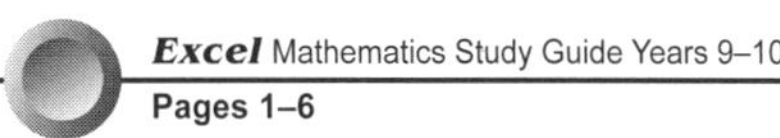

UNIT 8: Borrowing money

QUESTION 1 Nelly wanted to buy a car and approached a bank for a personal loan of $20 000. The loan was approved at an interest rate of 8% pa. She has to repay the loan in 5 years.

a How much interest will Nelly pay?

b What will be her monthly repayment?

QUESTION 2 Michael decided to buy a TV marked at $3000. He pays 20% deposit and the balance over 3 years, with interest charged at 15% on the balance p.a.

a Find the deposit paid.

b Calculate the balance owing.

c Calculate the interest paid.

d Find the total amount to be repaid.

e What is the monthly repayment?

QUESTION 3 Mai buys some furniture priced at $10 500. She pays $1500 deposit and agrees to pay $251.25 per month for four years.

a How much does Mai pay in total?

b How much interest does Mai pay?

c What rate of interest is Mai charged?

QUESTION 4 Chris wants to buy a boat and takes out a loan of $10 000 on which the interest rate charged is 9.5% p.a.. There is also a loan protection fee of 30 cents for each $100 borrowed. The loan is repaid over 5 years in equal monthly instalments.

a Calculate the total amount repaid.

b Find the amount of each monthly repayment.

Financial mathematics

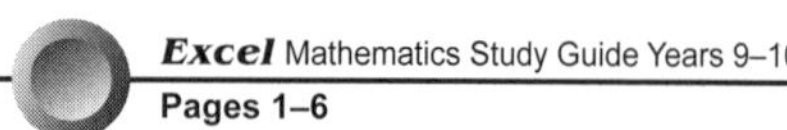

UNIT 9: Home loans

QUESTION 1 Andrew buys a house for $300 000. He borrows 80% of the purchase price from a building society which charges 13.5% p.a. on the amount owing.

a Find the deposit paid.

b What is the amount of interest charged per year on the balance owing?

c If $2900 is paid per month, how much of the balance is paid during the first year (assume simple interest)?

QUESTION 2 Yvette buys a house for $190 000, pays a deposit of $50 000, and then pays off the interest and balance at $850 per month for 25 years. Find:

a the total cost of the house.

b the yearly interest paid.

QUESTION 3 Kate buys a block of land for $150 000. She pays a deposit of $30 000 and borrows the remainder from a bank. The repayments are $1350 per calendar month. The loan is repaid after 10 years. In addition, she paid in cash the following charges; government and other charges = $3000, solicitor's fees = $1345.

a Calculate the amount repaid to the bank.

b Find the total cost of purchasing the land.

c Find the amount paid in excess of $150 000. Express this amount as a percentage of the purchase price.

Financial mathematics

TOPIC TEST **PART A**

Instructions
- This part consists of 10 multiple-choice questions.
- Fill in only ONE CIRCLE for each question.
- Each question is worth 1 mark.

Time allowed: 10 minutes **Total marks: 10**

Marks

1 \$500 invested for 2 years at 10% simple interest p.a. becomes

(A) \$550 (B) \$600 (C) \$625 (D) \$650 1

2 A debt of \$542.40 is to be paid in how many equal installments of \$45.20?

(A) 8 (B) 10 (C) 12 (D) 14 1

3 Melissa's hourly rate of pay is \$15.20 for the first 36 hours and time-and-a-half for every extra hour. How much is she paid for 45 hours?

(A) \$684 (B) \$752.40 (C) \$820.80 (D) \$1026 1

4 Alex is paid \$25.20 per hour and works 38 hours per week. Find his holiday pay for 4 weeks including a $17\frac{1}{2}\%$ holiday loading.

(A) \$3830.40 (B) \$4500.72 (C) \$670.32 (D) none of these 1

5 The tax on a salary of \$58 485, paid at \$11 772 plus 42 cents for each \$1 over \$52 000 is

(A) \$2723.70 (B) \$14 495.70 (C) \$24 563.70 (D) none of these 1

6 Mark receives a retainer of \$300 per week and 15% commission on all sales. How much does he earn in a week in which he sells \$10 000 worth of goods?

(A) \$1500 (B) \$1800 (C) \$1200 (D) \$545 1

7 The simple interest on \$5600 invested at 0.5% per month for 3 years is

(A) \$8400 (B) \$84 (C) \$1008 (D) \$10 080 1

8 An amount of \$12500, when invested for 4 years, earns a total of \$3500 in simple interest. What interest rate is paid?

(A) 5% pa (B) 6% pa (C) 7% pa (D) 8% pa 1

9 Sam paid a total of \$2730 simple interest on a loan. The loan was taken over 5 years and the interest rate was 6.5% pa. What amount did Sam borrow?

(A) \$8400 (B) \$8872.50 (C) \$3549 (D) \$21 000 1

10 An investment of \$5500 grew to \$6160. Simple interest of 4% pa was paid on the investment. For how many years was the money invested?

(A) 2 (B) 3 (C) 7 (D) 28 1

Total marks achieved for PART A /10

Financial mathematics

TOPIC TEST — PART B

Time allowed: 20 minutes — **Total marks: 15**

Marks

1 **a** What is the simple interest on \$2500 at 6% p.a. for 7 months?

____

b A credit card company charges 0.057 53% interest per day. Find the interest charged in 4 weeks on a balance of \$900.

________________ [2]

2 Nathan receives a salary of \$66 900 per annum.

a Calculate the amount he will receive each fortnight.

____

b He pays 5% of his gross salary in superannuation. Calculate his fortnightly superannuation contribution.

________________ [2]

3 Jaani is paid a wage of \$32.85 per hour.

a If Jaani works a normal 38-hour week, calculate his weekly wage.

____

b What will Jaani's wage be in a week when, in addition to his normal hours, he works 5 hours at time-and-a-half and 3 hours at double time?

____

c Calculate the total amount Jaani will receive for his 4 weeks' annual leave if he is paid an annual leave loading of $17\frac{1}{2}\%$ on 4 weeks' normal wages.

________________ [3]

4 Kate buys a boat. The cash price is \$4500. Kate pays no deposit and makes payments of \$225 every month for 2 years. Find the:

a total amount of interest Kate pays.

____

b annual rate of simple interest.

________________ [2]

5 A car is priced at \$16 000. Jed pays 10% deposit and borrows the rest at 9% p.a. simple interest over 3 years. Find the following.

a deposit

____

b amount borrowed

____

c total interest

____

d total amount that must be repaid

____

e amount of each monthly repayment

____

f total amount paid for the car

________________ [6]

Total marks achieved for PART B

CHAPTER 5
Linear and non-linear relationships

Excel Mathematics Study Guide Years 9–10
Pages 52–69

UNIT 1: Horizontal and vertical distances

QUESTION 1 Plot each pair of points on the number plane and find the distance between them.

a $A(1, 2)$ and $B(4, 2)$, AB ____________

b $C(1, 4)$ and $D(3, 4)$, CD ____________

c $E(3, 1)$ and $F(3, 5)$, EF ____________

d $G(2,1)$ and $H(2, 2)$, GH ____________

e $I(4, 0)$ and $J(4, 5)$, IJ ____________

f $K(1, 3)$ and $L(5,3)$, KL ____________

g $M(1,5)$ and $N(5, 5)$, MN ____________

h $Q(0, 1)$ and $P(4, 1)$, QP ____________

QUESTION 2 Plot each pair of points and find the distance between them.

a $A(2, 3)$ and $B(4, 3)$ ____________

b $P(-3, 1)$ and $Q(-3, 4)$ ____________

c $L(-2, -4)$ and $M(3, -4)$ ____________

d $C(1, 4)$ and $D(1, -2)$ ____________

e $E(2, 0)$ and $F(2, 3)$ ____________

f $G(3, 5)$ and $H(3, -1)$ ____________

g $I(6, 2)$ and $J(2, 2)$ ____________

h $S(5, 1)$ and $T(-2, 1)$ ____________

i $U(3, -2)$ and $V(3, 4)$ ____________

QUESTION 3 What is the distance between each pair of points?

a $A(1, 3)$ and $B(5, 3)$

b $C(2, 1)$ and $D(2, 6)$

c $E(-3, 7)$ and $F(-3, 2)$

d $G(1, 2)$ and $H(5, 2)$

e $I(3, -2)$ and $J(-2, -2)$

f $K(-5, 0)$ and $L(2, 0)$

g $M(0, 0)$ and $N(0, 5)$

h $Q(-3, -1)$ and $R(2, -1)$

i $S(2, 4)$ and $T(-4, 4)$

UNIT 2: Using Pythagoras' theorem to find distances

QUESTION 1 Use Pythagoras' theorem to find the distance AB in each diagram. Leave your answers in surd (square root) form where necessary.

a

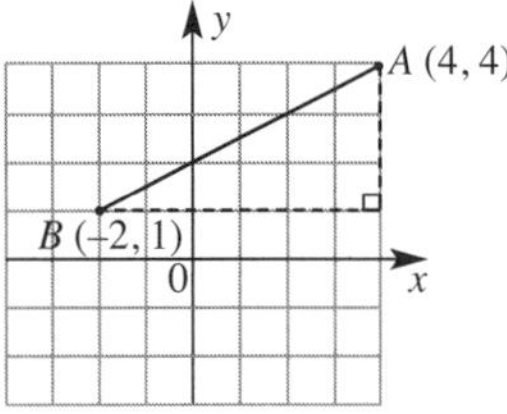

b

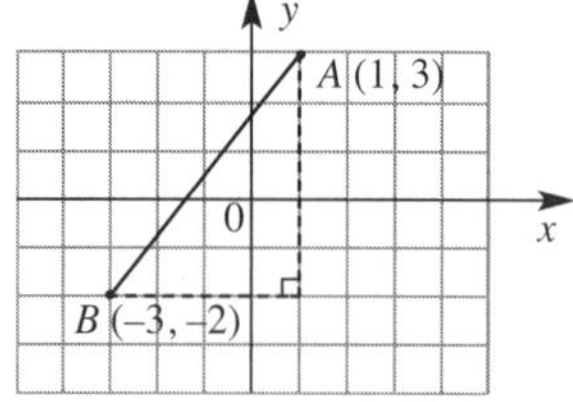

c

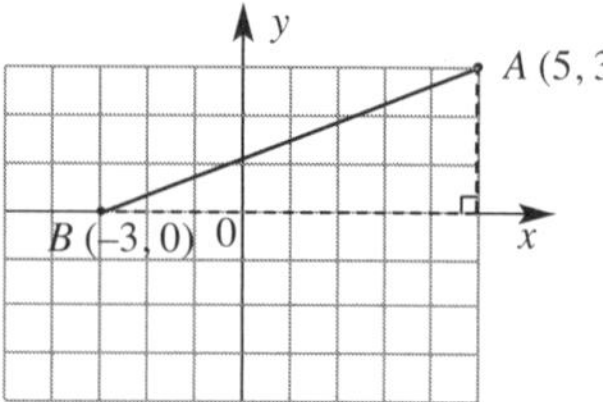

d

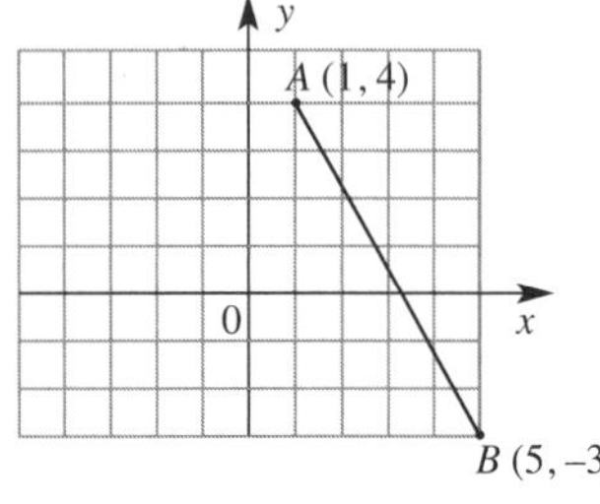

e

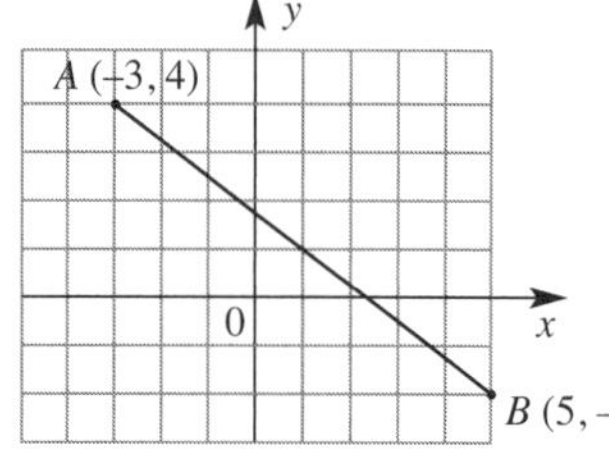

f

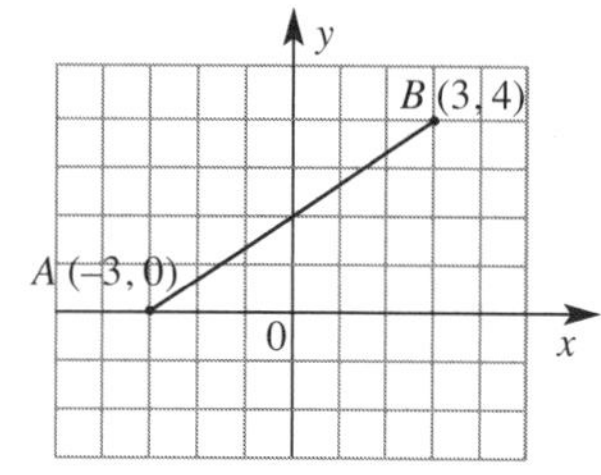

QUESTION 2 Use Pythagoras' theorem to find the distance AB in each diagram. Leave your answers in surd (square root) form where necessary.

a

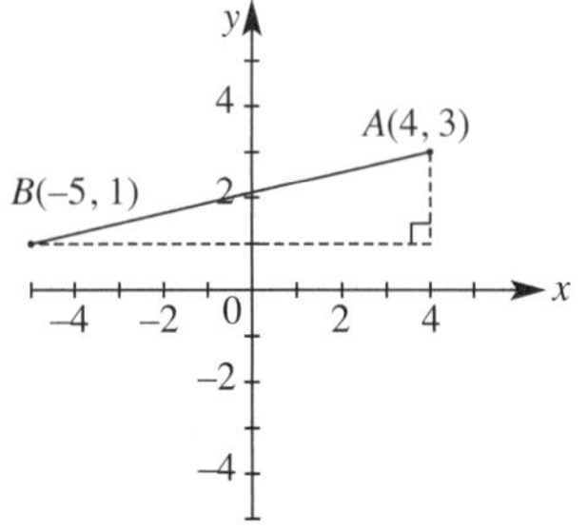

b

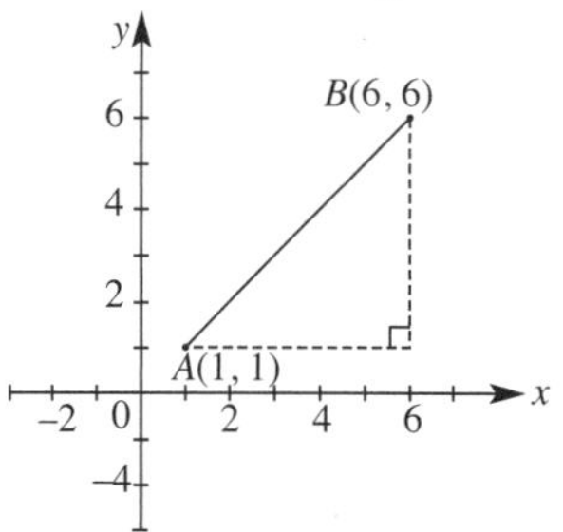

c

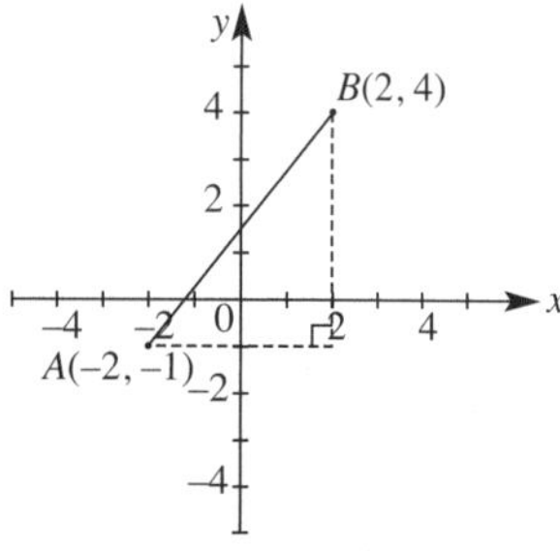

d

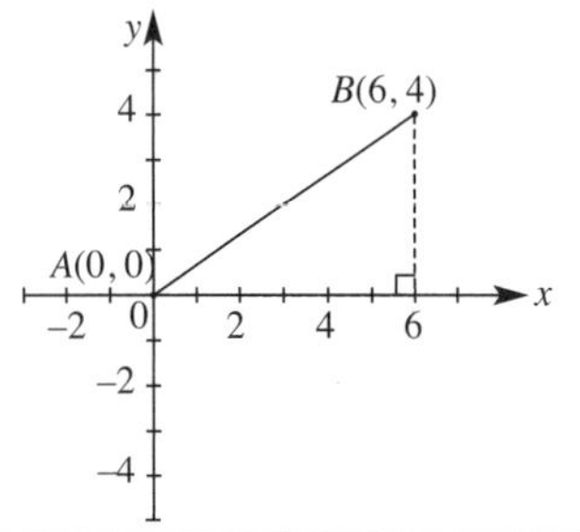

e

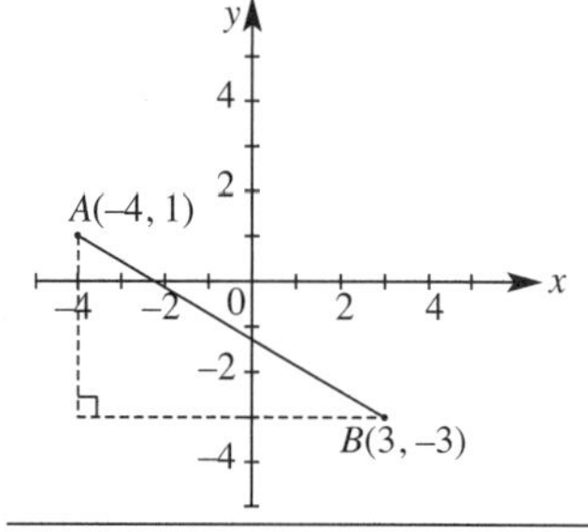

f

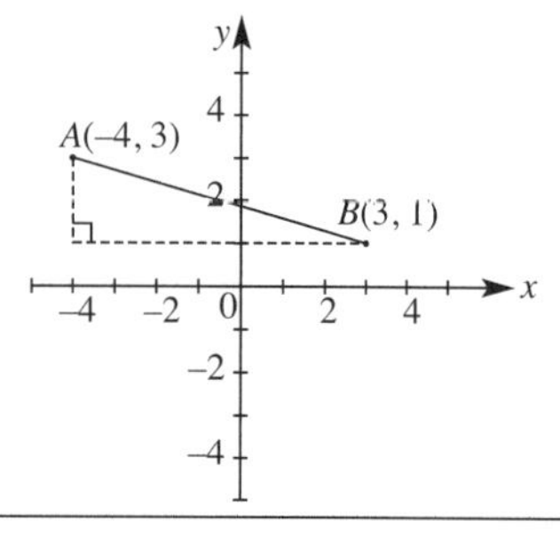

Linear and non-linear relationships

UNIT 3: The distance formula

QUESTION 1 Use the distance formula $d = \sqrt{(x_2 - x_1)^2 + (y_2 - y_1)^2}$ to find the distance between the following pairs of points. Leave your answer in surd form if necessary.

a $A(2, 5), B(7, 13)$

b $A(-1, -4), B(3, -8)$

c $A(0, 1), B(3, -4)$

d $A(3, 2), B(6, 6)$

e $A(4, 5), B(7, 9)$

f $A(5, -2), B(7, -5)$

g $P(4, 3), Q(3, 2)$

h $P(2, 5), Q(8, 12)$

i $P(-1, -3), Q(2, -5)$

j $P(-3, 2), Q(1, -6)$

k $P(1, 3), Q(3, 5)$

l $P(4, -5), Q(6, -9)$

QUESTION 2 Find the perimeter of a triangle whose vertices are $A(6, 2)$, $B(5, 2)$ and $C(-4, -5)$.

QUESTION 3 Find the square of the distance between the points $A(2, 5)$ and $B(5, 10)$.

Excel Mathematics Study Guide Years 9–10
Pages 52–69

UNIT 4: The midpoint of an interval

QUESTION 1

a What number is halfway between 6 and 10?

b What is the average of 6 and 10?

c Find $\frac{6+10}{2}$

d What number is halfway between –2 and 12?

e What is the average of –2 and 12?

f Find $\frac{-2+12}{2}$

QUESTION 2 What number is halfway between the point *A* and the point *B* on each number line?

a Number line from 0 to 7 with *A* at 1 and *B* at 5.

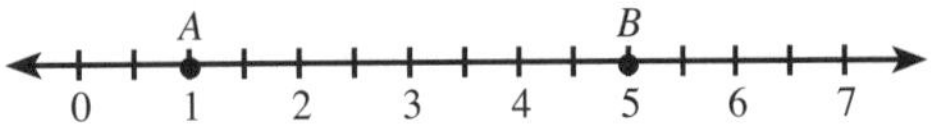

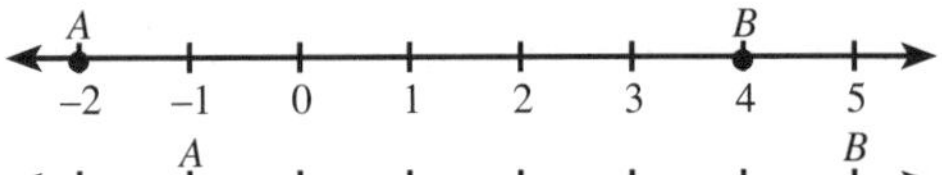

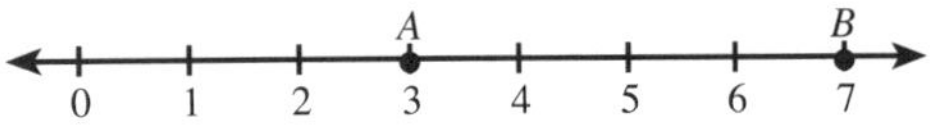

b Number line from –2 to 5 with *A* at –2 and *B* at 4.

c Number line from 0 to 7 with *A* at 3 and *B* at 7.

d Number line from –1 to 6 with *A* at 0 and *B* at 6.

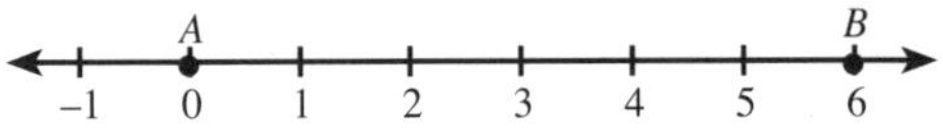

QUESTION 3 Find the number that is halfway between:

a 0 and 16 **b** 4 and 12 **c** 2 and 10

d 3 and 15 e 1 and 13 f –1 and 7

g –2 and 6 h –4 and 4 i 2 and 18

j –5 and 15 k 3 and 17 l 1 and 19

QUESTION 4 Consider the points *P*(4, 10) and *Q*(6, –2).

a Use the *x*-coordinates of the points *P* and *Q* to find the number halfway between 4 and 6

b Use the *y*-coordinates of the points *P* and *Q* to find the number halfway between 10 and –2

c What are the coordinates of the point *M*, which is halfway between *P* and *Q*?

QUESTION 5 If x_1 and x_2 are given, find the value of x when $x = \frac{x_1 + x_2}{2}$

a $x_1 = 3$ and $x_2 = 21$ **b** $x_1 = -2$ and $x_2 = 8$

c $x_1 = 5$ and $x_2 = 13$ **d** $x_1 = 4$ and $x_2 = 10$

e $x_1 = 1$ and $x_2 = 9$ **f** $x_1 = -6$ and $x_2 = 14$

g $x_1 = -4$ and $x_2 = 8$ **h** $x_1 = -6$ and $x_2 = 10$

i $x_1 = -5$ and $x_2 = 7$ **j** $x_1 = -2$ and $x_2 = 16$

k $x_1 = -7$ and $x_2 = -1$ **l** $x_1 = -8$ and $x_2 = -2$

UNIT 5: The midpoint formula

QUESTION 1 Use the midpoint formula $x = \frac{x_1 + x_2}{2}$, $y = \frac{y_1 + y_2}{2}$ to find the midpoint of the interval joining the following points.

a $A(0, 6), B(2, 4)$

b $A(4, 8), B(6, 10)$

c $A(-3, 2), B(-5, 0)$

d $A(3, 8), B(7, 2)$

e $A(7, 0), B(5, 0)$

f $A(2, 10), B(4, 4)$

g $P(-4, -11), Q(7, 4)$

h $P(10, 4), Q(8, 6)$

i $P(2, 10), Q(8, 8)$

j $P(4, 5), Q(6, 9)$

k $P(-3, -6), Q(1, 4)$

l $P(-8, 2), Q(4, -6)$

QUESTION 2 The vertices of ΔABC are $A(-2, 9)$, $B(10, 11)$ and $C(-7, 1)$. Find the midpoint of each side.

QUESTION 3 Show that the midpoint of $(7, -3)$ and $(-7, 3)$ is the origin.

UNIT 6: Finding an endpoint

QUESTION **1** For each diagram, find the coordinates of A, given that M is the midpoint of AB.

a

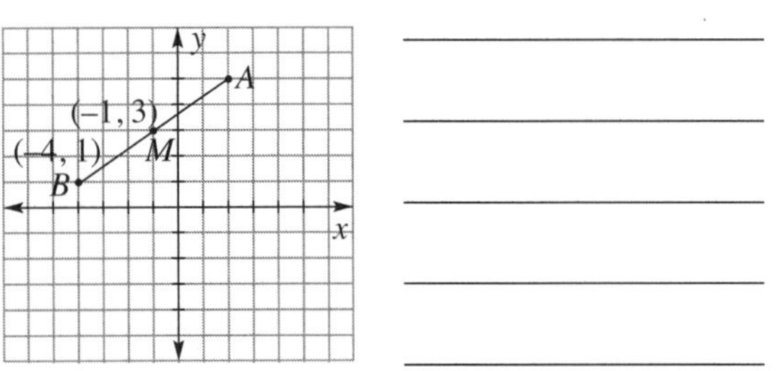

b

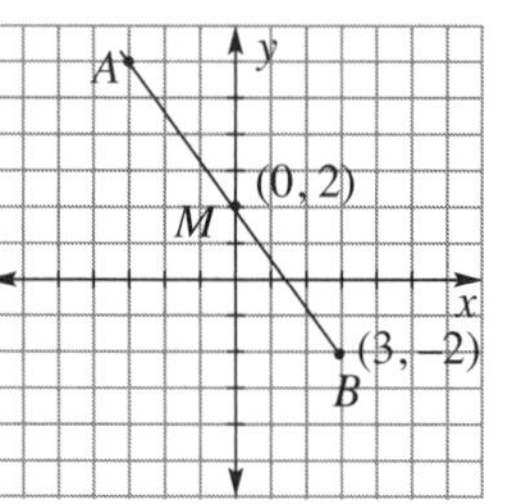

QUESTION **2** The coordinates of the midpoint M of an interval and one of its endpoints A, are given. Find the coordinates of the other endpoint B.

a $M(4, 7)$ and $A(1, 6)$ ____________ **b** $M(5, 9)$ and $A(1, 7)$ ____________

c $M(6, -3)$ and $A(4, 1)$ ____________ **d** $M(0, 8)$ and $A(4, 10)$ ____________

e $M(5, 9)$ and $A(1, 7)$ ____________ **f** $M(4, 3)$ and $A(0, 0)$ ____________

g $M(3, 9)$ and $A(-1, 5)$ ____________ **h** $M(7, 9)$ and $A(4, 5)$ ____________

i $M(2, 1)$ and $A(5, -5)$ ____________ **j** $M(4, 9)$ and $A(0, 2)$ ____________

k $M(8, 4)$ and $A(5, 2)$ ____________ **l** $M(8, 0)$ and $A(7, 3)$ ____________

QUESTION **3** Given the coordinates of the centre C, of a circle, and one endpoint B, of a diameter, find the coordinates of the other endpoint A, of the diameter.

a $C(2, 4)$ and $B(0, 1)$ ____________ **b** $C(3, 7)$ and $B(2, 6)$ ____________

c $C(-1, 2)$ and $B(-4, 4)$ ____________ **d** $C(-2, 5)$ and $B(-6, 4)$ ____________

e $C(0, 0)$ and $B(-4, -6)$ ____________ **f** $C(6, 9)$ and $B(4, 6)$ ____________

g $C(-1, 8)$ and $B(-4, 3)$ ____________ **h** $C(-3, 1)$ and $B(-7, 0)$ ____________

QUESTION **4**

a $(3, 6)$ is the midpoint of AB and A is the point $(0, 2)$. Find the coordinates of B.

____

b If the midpoint of (a, b) and $(9, 9)$ is $(6, 2)$. What are the values of a and b?

____

c If the midpoint of $(5, p)$ and $(7, -4)$ is $(6, 3)$. What is the value of p?

____

d If the midpoint of $(x, 5)$ and $(9, y)$ is $(1, 6)$. What are the values of x and y?

____

____

Linear and non-linear relationships

UNIT 7: The gradient of a line

QUESTION 1 State whether the gradient of the line is positive or negative.

a
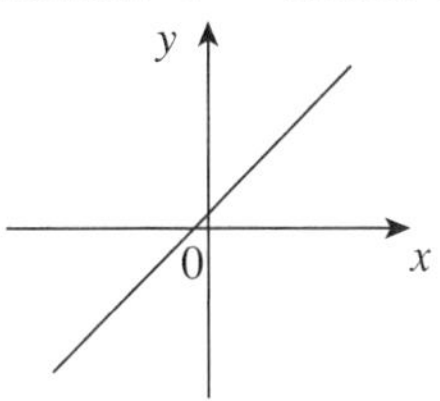

b
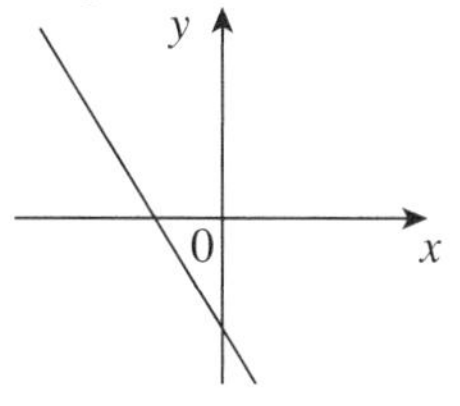

c
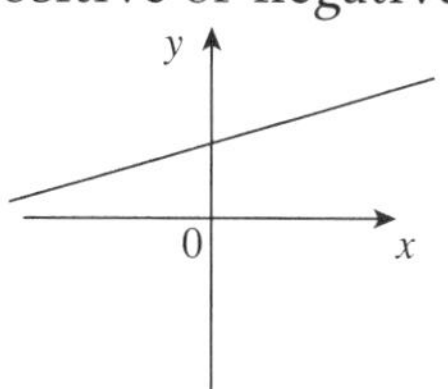

d
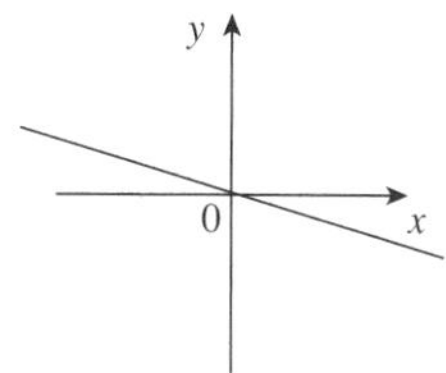

e
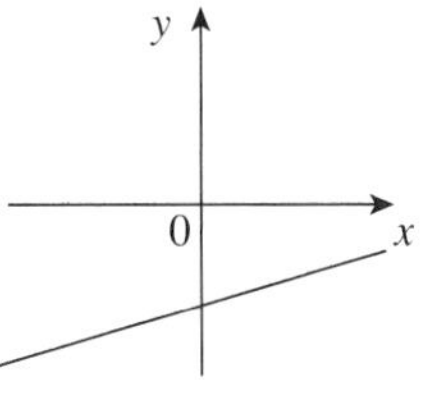

f
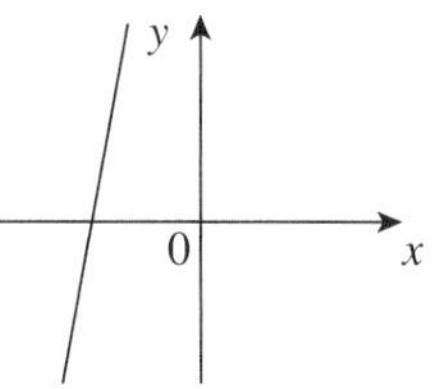

g
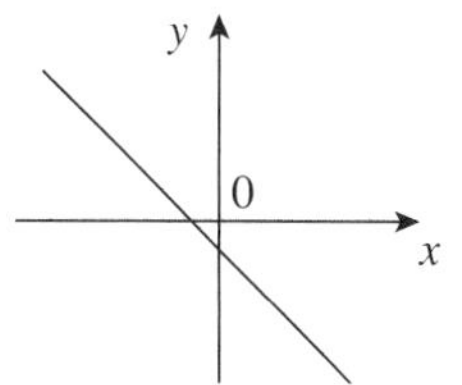

h
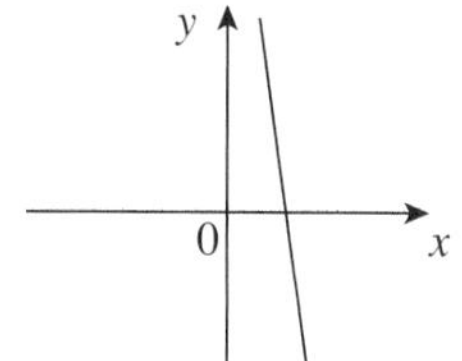

QUESTION 2 Use a right-angled triangle drawn from two points on the number plane and the relationship $m = \frac{\text{rise}}{\text{run}}$ to find the gradient of each interval below.

a *AD* **b** *BD*

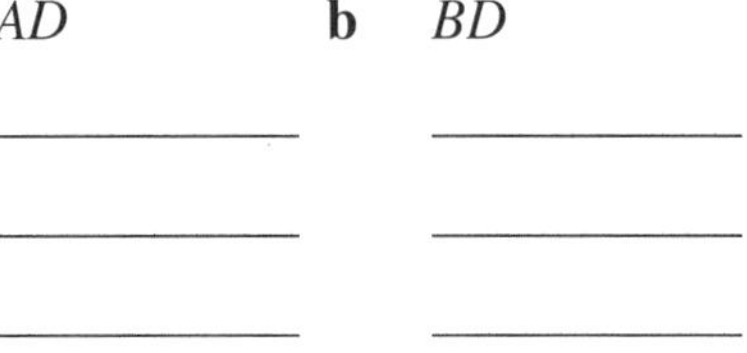

c *BE* d *ED*

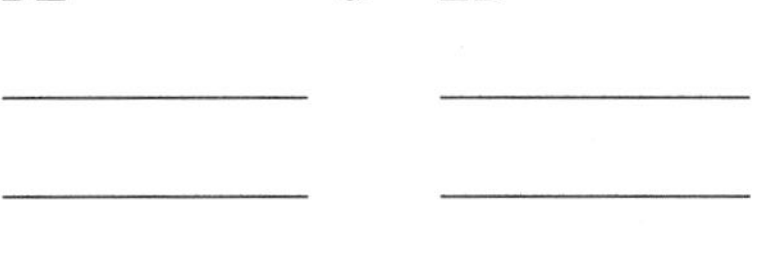

e *FC* f *AC*

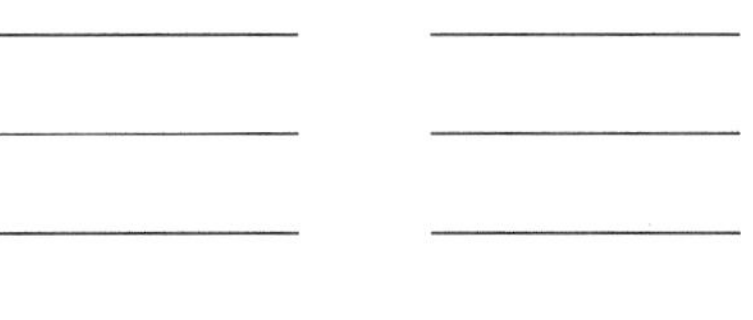

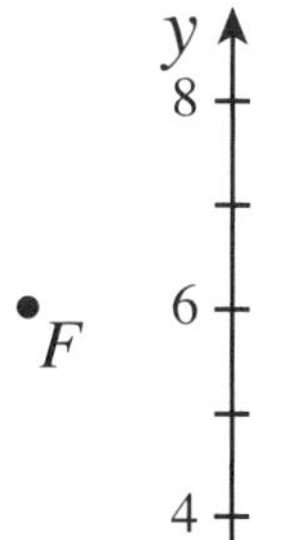

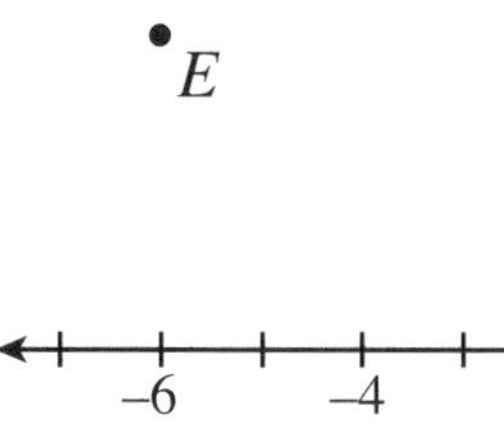

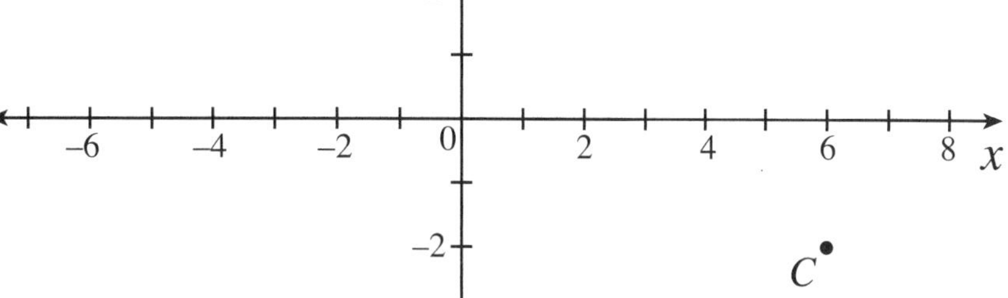

QUESTION 3 Referring to the number plane in Question 2, between which two points is the gradient?

a 5

b $\frac{1}{5}$

c $\frac{1}{2}$

d –1

e –3

f $-3\frac{1}{2}$

UNIT 8: The gradient formula

QUESTION 1 Use the gradient formula $m = \frac{y_2 - y_1}{x_2 - x_1}$ to find the gradient of the straight line passing through the following points.

a (1, 5) and (2, –7)

b (–1, –2) and (3, 4)

c (–2, –3) and (4, –7)

d (2, 4) and (–1, 3)

e (5, 4) and (–1, 5)

f (6, –2) and (8, –3)

g (–3, 6) and (–5, –1)

h (8, 10) and (5, 1)

i (3, 4) and (8, 6)

j (8, 1) and (4, 5)

k (–3, 6) and (2, 4)

l (0, 0) and (6, 9)

QUESTION 2 Show that (1, –1), (–1, 5) and (3, –7) are collinear.

QUESTION 3 Show that the four points $A(2, 6)$, $B(5, 2)$, $C(1, -1)$ and $D(-2, 3)$ are the vertices of a parallelogram.

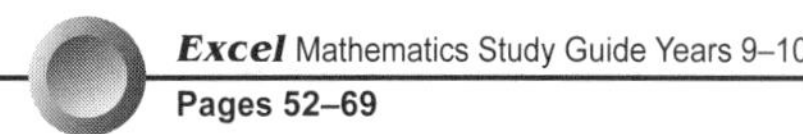

UNIT 9: Mixed questions on distance, midpoint and gradient

QUESTION 1 The vertices ΔPQR are $P(2, 3)$, $Q(10, 9)$ and $R(8, 0)$

a Find the midpoint, M, of PR.

b Find the midpoint, N, of QR.

c Find the gradient of PQ.

d Find the gradient of MN.

e What do you notice about the gradients of PQ and MN? ______

f Find the distance PQ.

g Find the distance MN.

h How many times larger is the distance PQ than the distance MN? ______

QUESTION 2 $P(2, 3)$, $Q(6, 0)$ and $R(-1, -1)$ are vertices of a triangle.

a Find the length PQ.

b Find the length QR.

c Find the length PR.

d Is ΔPQR right-angled?

e What other special type of triangle is ΔPQR?

QUESTION 3 The coordinates of the midpoint of AB are $(2, 3)$. If A is the point $(-3, -5)$, what are the coordinates of B?

QUESTION 4 Find the gradient of the straight line $4x - 3y + 9 = 0$

UNIT 10: Graphing lines

QUESTION 1 Complete the following tables of values and then graph the equation on the number plane.

a $y = x + 3$

x	-1	0	1	2
y				

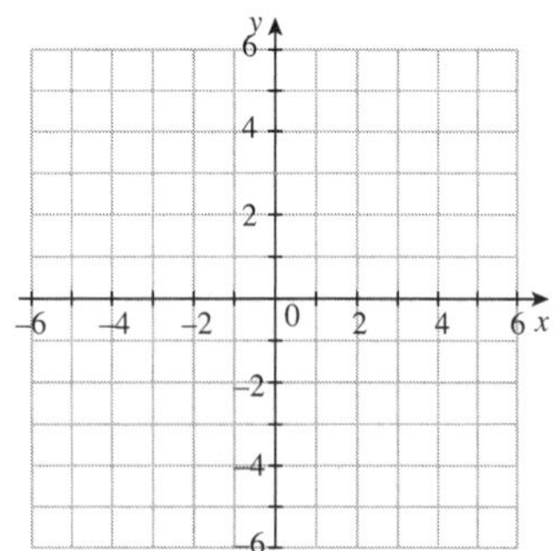

b $y = 2x - 1$

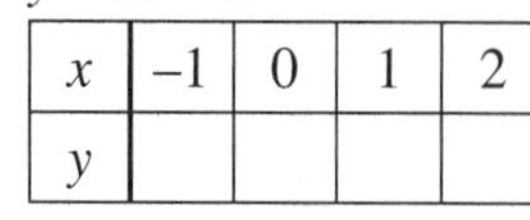

x	-1	0	1	2
y				

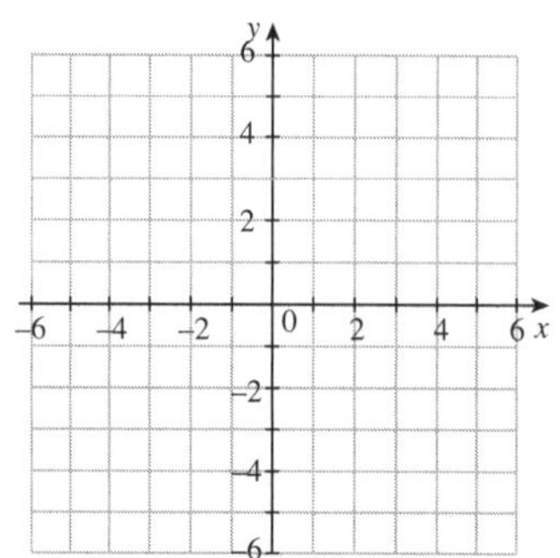

c $y = -x + 1$

x	-1	0	1	2
y				

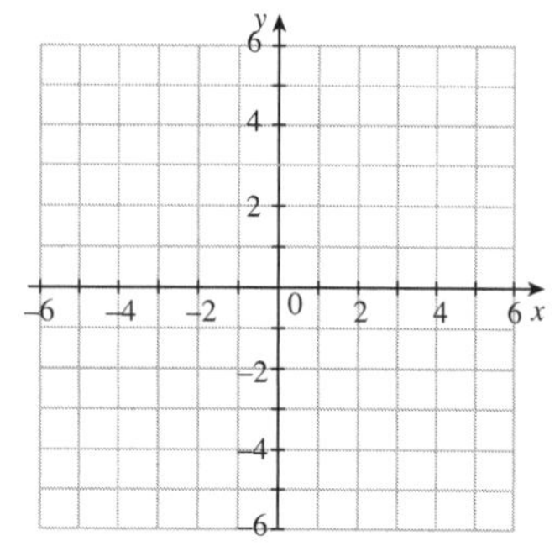

d $y = x - 1$

x	0	1	2	3
y				

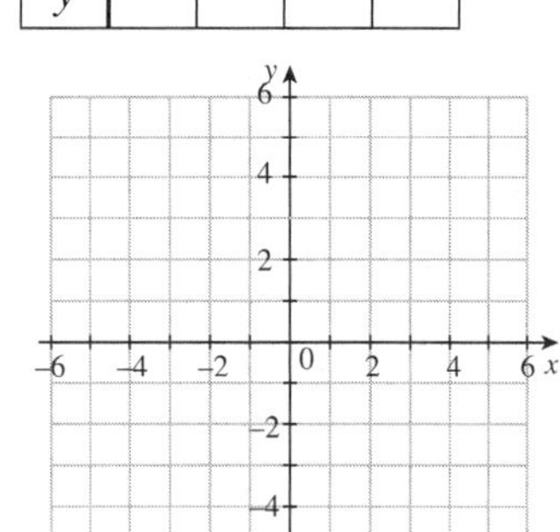

e $y = 3x$

x	0	1	2	3
y				

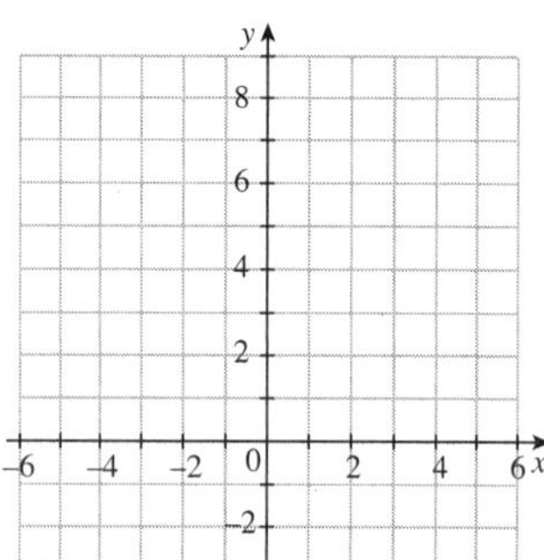

f $y = 2x + 2$

x	0	1	2	3
y				

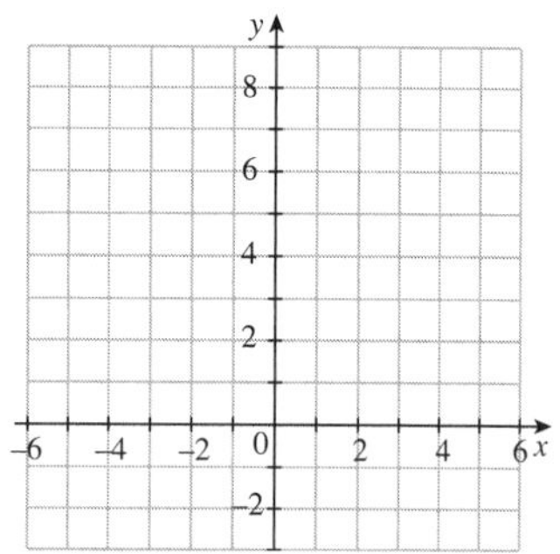

QUESTION 2 On the same number plane, draw the graphs of the following.

a $y = x$

b $y = -x$

c $y = 2x$

d $y = \frac{1}{2}x$

e $y = x + 2$

f $y = x - 2$

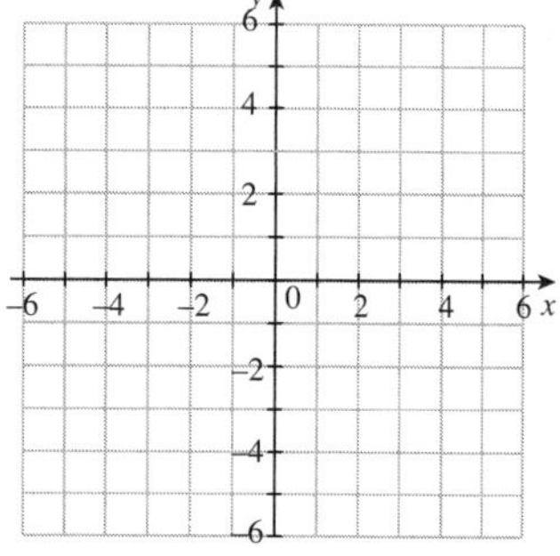

QUESTION 3 On the same number plane, sketch the graphs of the following.

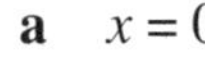

a $x = 0$

b $y = 0$

c $x = 3$

d $y = -2$

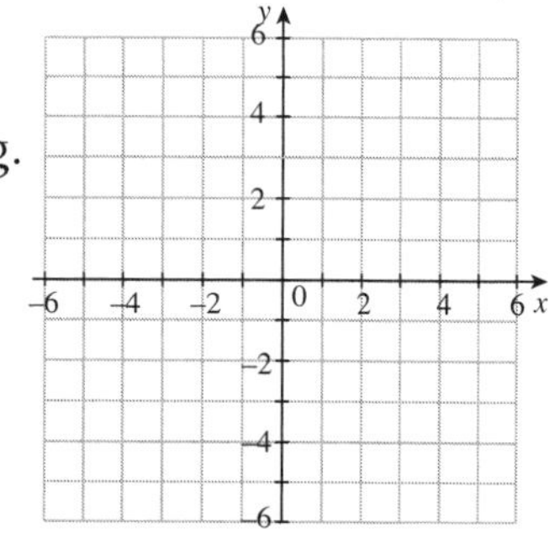

UNIT 11: Using the intercept method to graph lines

QUESTION 1 For each equation of a line find the x-intercept and the y-intercept.

a $x + y = 2$ **b** $x - y = 4$ **c** $2x + y = 6$ **d** $x - 3y = 6$

e $x - 2y = 4$ **f** $2x - y = 3$ **g** $3x - 4y = 12$ **h** $3x - y = 3$

QUESTION 2 Draw the graph of the line with the given x-intercept and y-intercept.

a x-intercept 3, y-intercept 2

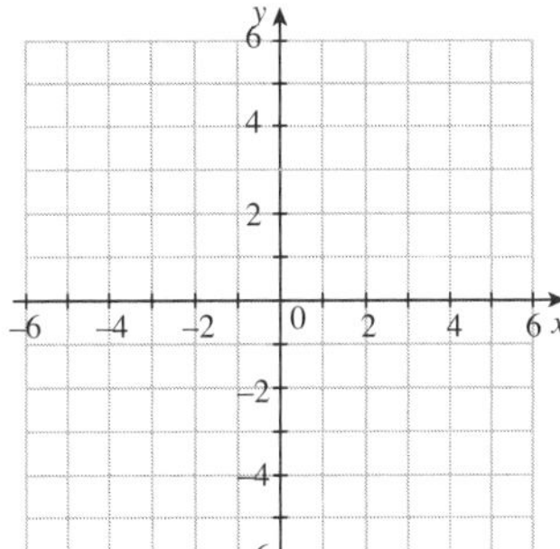

b x-intercept –1, y-intercept 3

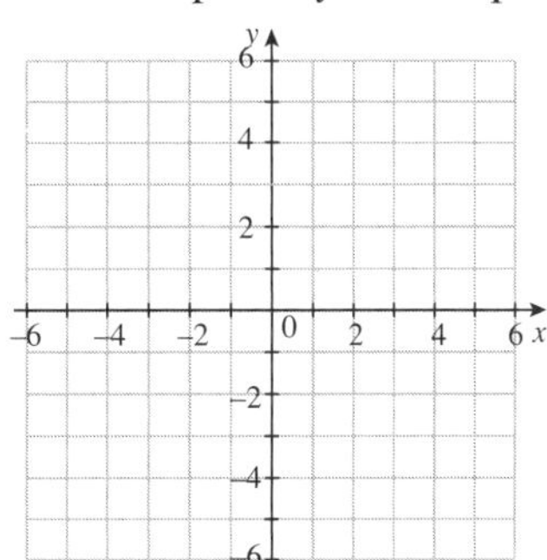

c x-intercept 1, y-intercept –4

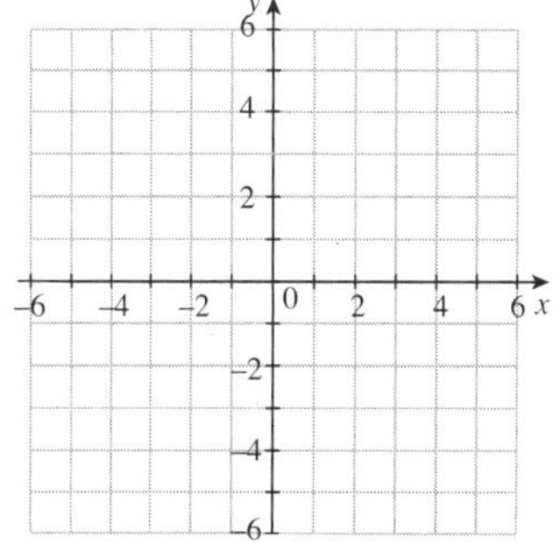

QUESTION 3 For each equation, find the x-intercept and the y-intercept and then draw its graph.

a $y = x - 1$

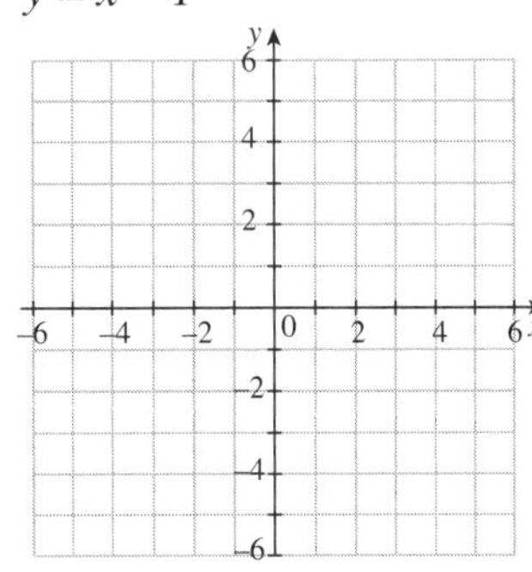

b $y = -2x + 3$

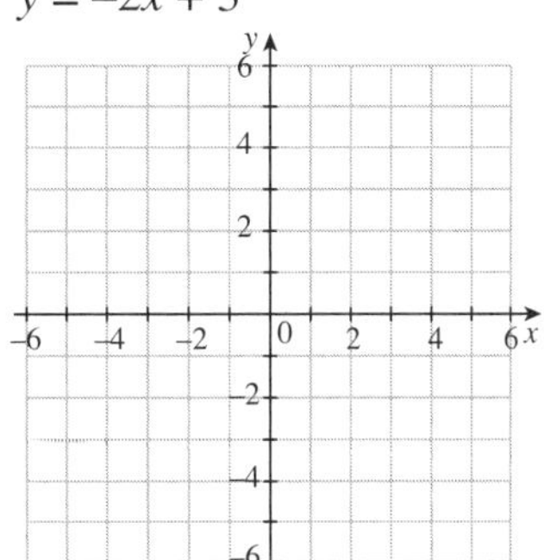

c $x + y - 5 = 0$

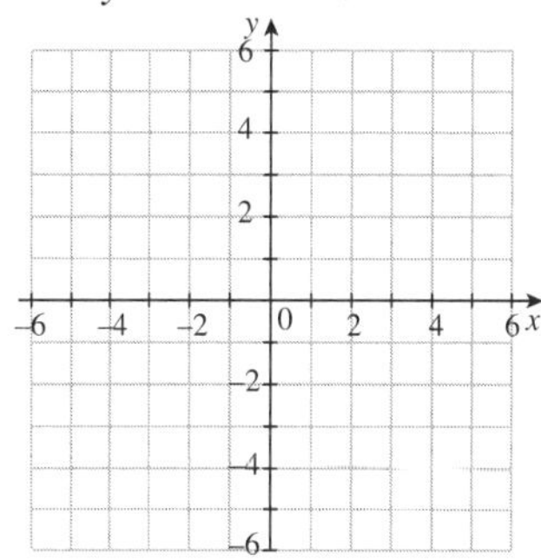

d $y = 2x - 3$

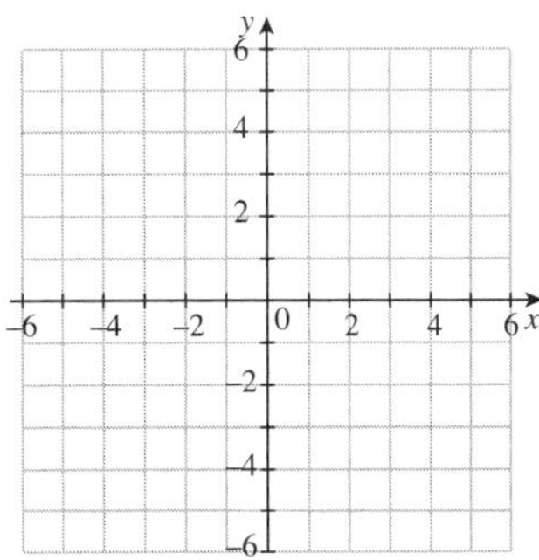

e $y = \frac{4}{3}x - 1$

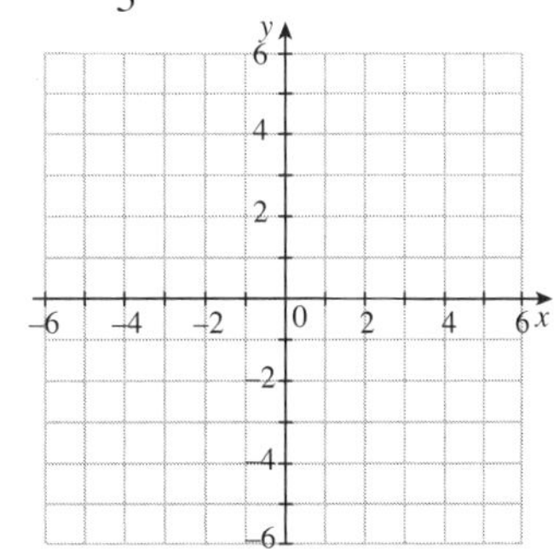

f $x - 2y = 4$

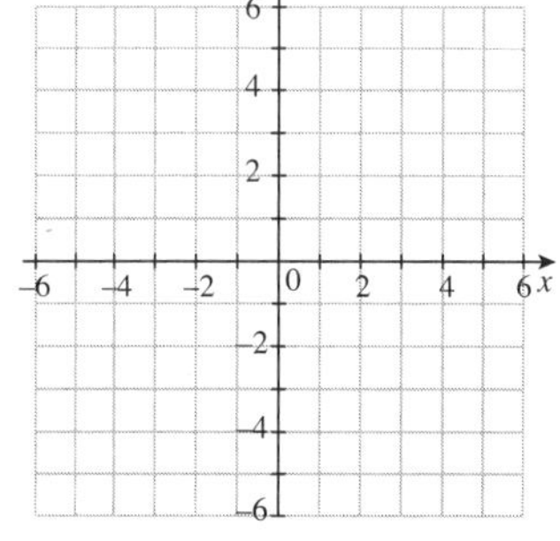

Linear and non-linear relationships

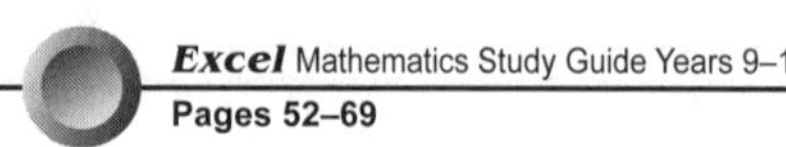

UNIT 12: The gradient and y-intercept of a line

QUESTION 1 Consider the equation $y = x + 2$

a Complete the table of values.

x	0	1	2
y			

b Draw the graph of the line.

c What is the gradient of the line? ________________

d Is the gradient positive or negative? ________________

e Does the line lean to the left or the right? ________________

f Write down the coefficient of x. ________________

g Is the coefficient of x the same as the gradient? ________________

h What is the y-intercept of the line? ________________

i Is the y-intercept the same as the constant term? ________________

QUESTION 2 Consider the equation $y = -2x + 1$.

a Complete the table of values.

x	0	1	2
y			

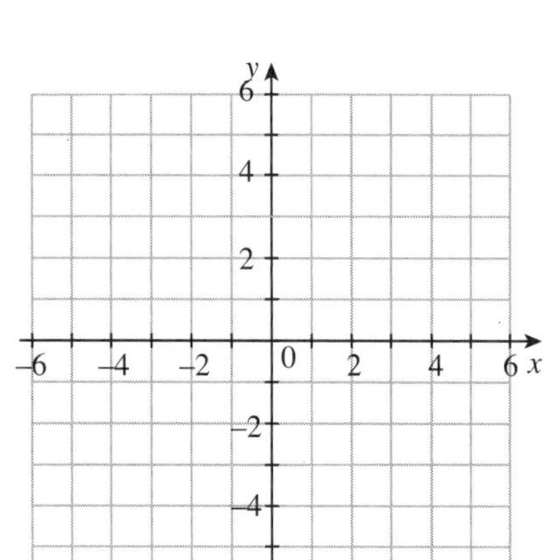

b Draw the graph of the line.

c What is the gradient of the line? ________________

d Is the gradient positive or negative? ________________

e Does the line lean to the left or the right? ________________

f Write down the coefficient of x. ________________

g Is the coefficient of x the same as the gradient? ________________

h What is the y-intercept of the line? ________________

i Is the y-intercept the same as the constant term? ________________

QUESTION 3 Complete.

In the equation $y = mx + b$, m is the ________________ of the line and b is the ________________.

QUESTION 4 For each given equation, write down the gradient and the y-intercept.

a $y = 3x - 8$

gradient ________________

y-intercept ________________

b $y = 4x + 7$

gradient ________________

y-intercept ________________

c $y = -2x + 5$

gradient ________________

y-intercept ________________

UNIT 13: The graph of $y = mx + b$

QUESTION 1 For each given equation, write down the gradient and y-intercept.

a $y = 2x + 7$

gradient __________

y-intercept __________

b $y = 3x + 1$

gradient __________

y-intercept __________

c $y = 7x$

gradient __________

y-intercept __________

d $y = 4x - 3$

gradient __________

y-intercept __________

e $y = \frac{1}{2}x + 6$

gradient __________

y-intercept __________

f $y = x + 4$

gradient __________

y-intercept __________

g $y = -3x + 8$

gradient __________

y-intercept __________

h $y = -x - 5$

gradient __________

y-intercept __________

i $y = 11 - 2x$

gradient __________

y-intercept __________

QUESTION 2 Find the y-intercept and the gradient and hence sketch the graph of each line.

a $y = 3x + 2$ __________ __________

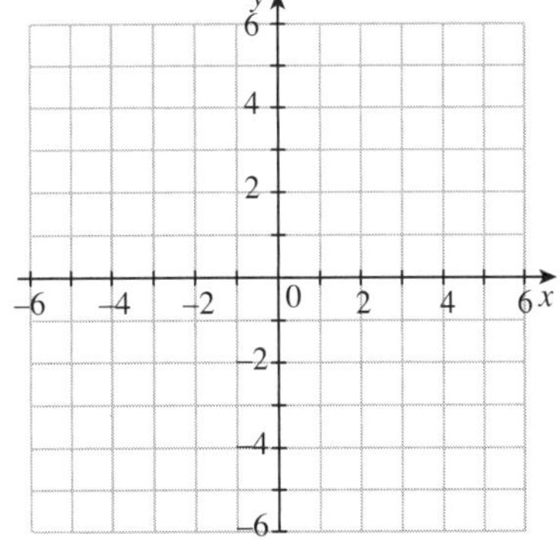

b $y = 2x - 1$ __________ __________

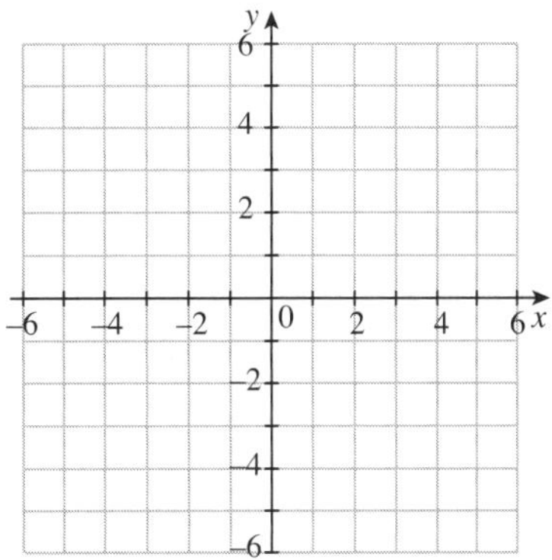

c $y = 3x - 5$ __________ __________

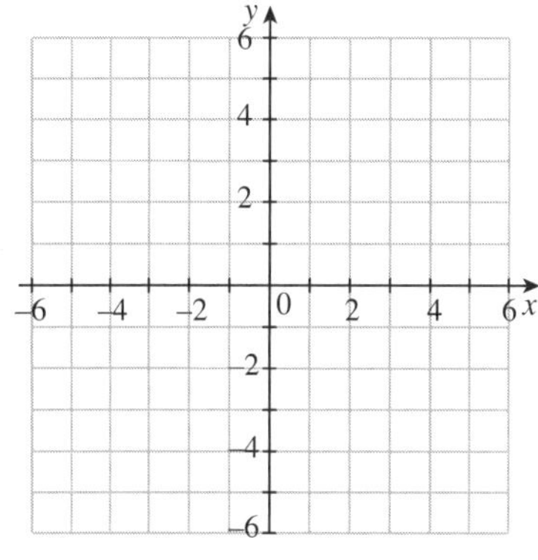

d $y = x$ __________ __________

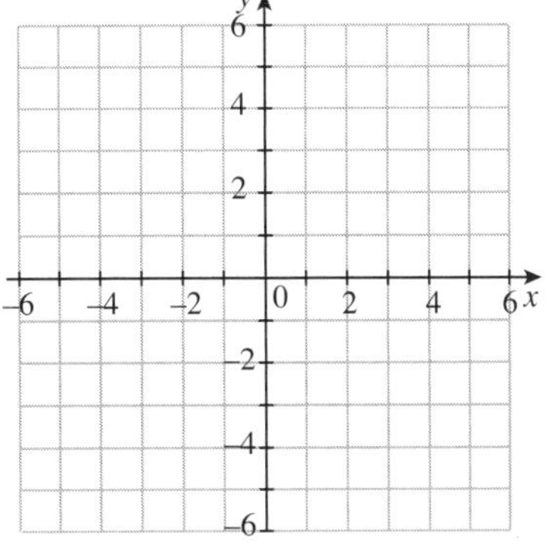

e $y = -2x + 1$ __________ __________

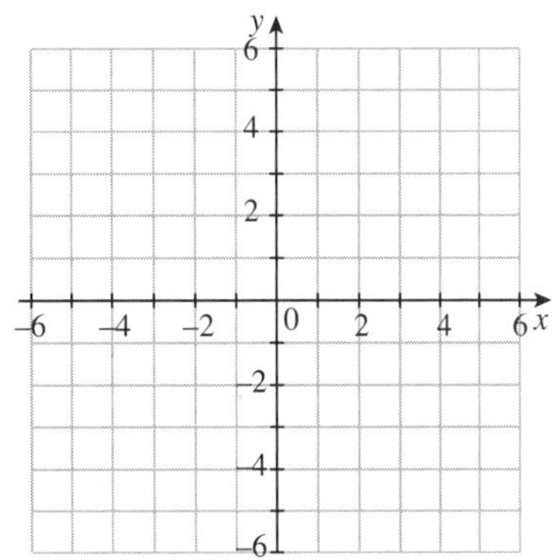

f $y = \frac{1}{2}x + 4$ __________ __________

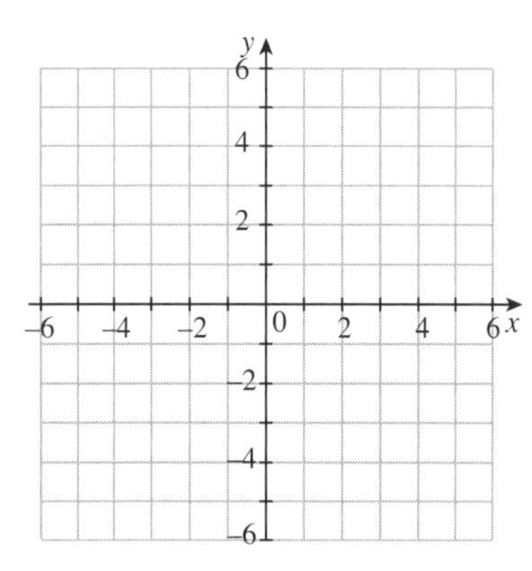

UNIT 14: General form of linear equations

QUESTION 1 Write each of the following linear equations in general form.

a $2x - 5y = 9$ **b** $3x + 4y = 8$ **c** $5x - 7 = 2y$

d $8y - 3 = 4x$ **e** $2x = 9 - y$ **f** $y = 8x + 7$

g $3y - 2x = 6$ **h** $9y = 8x + 12$ **i** $2y = \frac{x}{3} + 1$

QUESTION 2 Each of the following equations is in general form. Change it to gradient-intercept form.

a $2x + 3y - 8 = 0$ **b** $x + 5y - 7 = 0$ **c** $3x - 2y - 3 = 0$

d $x - y + 7 = 0$ **e** $2x + y - 9 = 0$ **f** $5x - 6y + 11 = 0$

g $3x - 2y - 6 = 0$ **h** $4x + 5y + 3 = 0$ **i** $2x - y + 6 = 0$

QUESTION 3 Write the equation of each line in gradient-intercept form and then change it to general form.

a $m = 4,\ b = 3$ **b** $m = 2,\ b = -5$ **c** $m = 3,\ b = 7$

d $m = \frac{1}{2},\ b = 4$ **e** $m = \frac{2}{3},\ b = 6$ **f** $m = -\frac{5}{6},\ b = 3$

UNIT 15: Determining whether or not a point lies on a line

QUESTION 1 Determine whether the following points lie on the line $3x + 4y = 12$

a (0, 3) **b** (0, 0) **c** (–4, 6)

d (4, 3) **e** (4, 0) **f** (8, –3)

QUESTION 2 Determine whether the following lines pass through the origin, (0, 0)

a $2x - y + 2 = 0$ **b** $2y = 3x$ **c** $x - 5y = 0$

d $2x + 3y = 6$ **e** $y = -2x$ **f** $y = 5x - 4$

QUESTION 3 Does the given point lie on the given line?

a $x + 2y = 3$ (3, 0) **b** $x + y = 2$ (0, 2) **c** $2x + 3y = 6$ (3, –2)

d $y = 5x - 3$ (1, 2) **e** $y = -x + 7$ (4, 3) **f** $2x + y = 5$ (2, –1)

QUESTION 4 A straight line $y = mx + 8$ passes through the point (–2, 2). Find the value of m.

QUESTION 5 If the point (–3, –6) is on the line $ax - 4y - 9 = 0$, what is the value of a?

QUESTION 6 Find the missing coordinates to make each of the following points satisfy the equation $y = 3x - 2$

a (0,) **b** (, 4) **c** (1,)

d (5,) **e** (, –5) **f** (, –8)

UNIT 16: Finding the equation of a line

QUESTION 1 Find the equation, in gradient-intercept form, of the line with gradient m and passing through the point P.

a $m = 2,\ P(1, 5)$ **b** $m = 3,\ P(-2, 7)$ **c** $m = -1,\ P(8, 0)$

d $m = \frac{1}{2},\ P(4, 6)$ **e** $m = -\frac{2}{3},\ P(3, -6)$ **f** $m = \frac{4}{3},\ P(-1, -4)$

QUESTION 2 Find the equation, in general form, of the line with gradient m and passing through the point P.

a $m = -2,\ P(1, -3)$ **b** $m = \frac{1}{4},\ P(2, 5)$ **c** $m = -\frac{3}{2},\ P(-3, -4)$

QUESTION 3 Find the equation, in gradient-intercept form, of the line that passes through the two given points.

a $(3, 5)$ and $(7, 7)$ **b** $(-2, 3)$ and $(5, 1)$ **c** $(4, 0)$ and $(-3, -2)$

QUESTION 4 Find the equation, in general form, of the line that passes through the two given points.

a $(6, -1)$ and $(2, -3)$ **b** $(-1, -2)$ and $(4, 7)$ **c** $(3, -2)$ and $(-2, 3)$

Linear and non-linear relationships

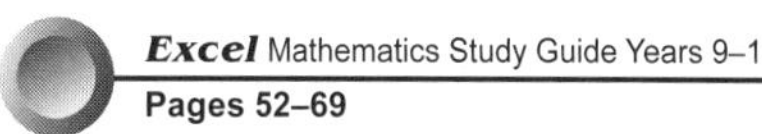

UNIT 17: Using graphs to solve linear equations

Question 1 The graph of $y = 2x - 4$ is shown at right.

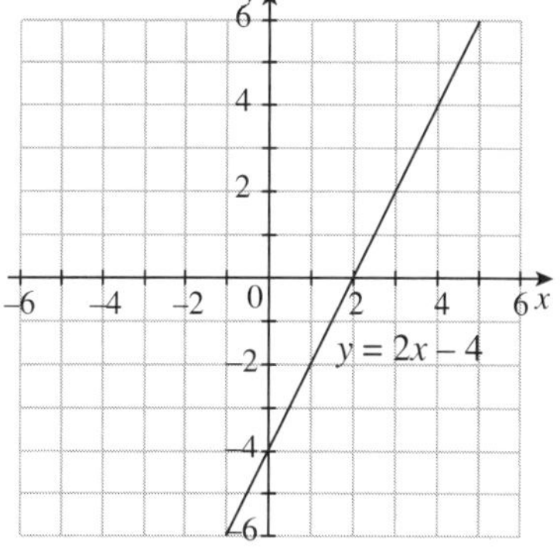

a Use the graph to find the y-value when:

i 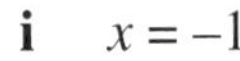 $x = -1$ **ii** $x = 2$ **iii** $x = 4$

__________ __________ __________

b Use the graph to find the x-value when:

i $y = -4$ **ii** $y = -2$ **iii** $y = 2$

__________ __________ __________

c Briefly explain why the answers to part b are the solutions to the equations $2x - 4 = -4$, $2x - 4 = -2$ and $2x - 4 = 2$.

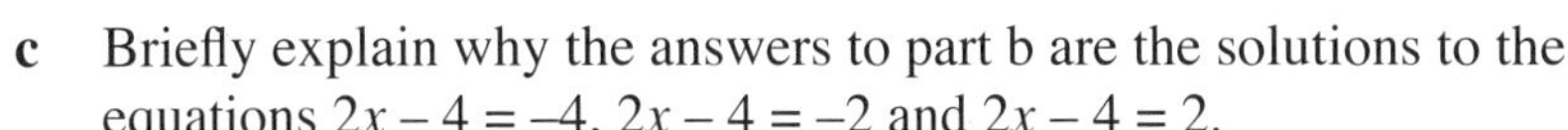

____

____

Question 2 The graph of $y = 3x + 1$ is shown at right. Use the graph to solve each of the following equations.

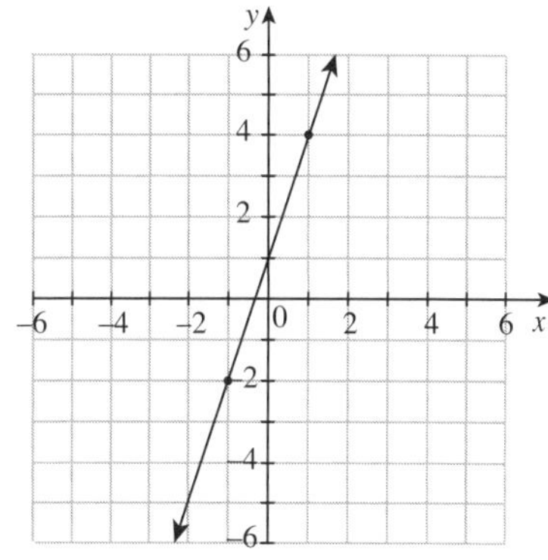

a $3x + 1 = 4$ **b** $3x + 1 = -2$ **c** $3x + 1 = 1$

__________ __________ __________

d $3x + 1 = -5$ **e** $3x + 1 = 7$ **f** $3x + 1 = 3$

__________ __________ __________

g $3x + 1 = 5$ **h** $3x + 1 = -3$ **i** $3x + 1 = -4$

__________ __________ __________

Question 3

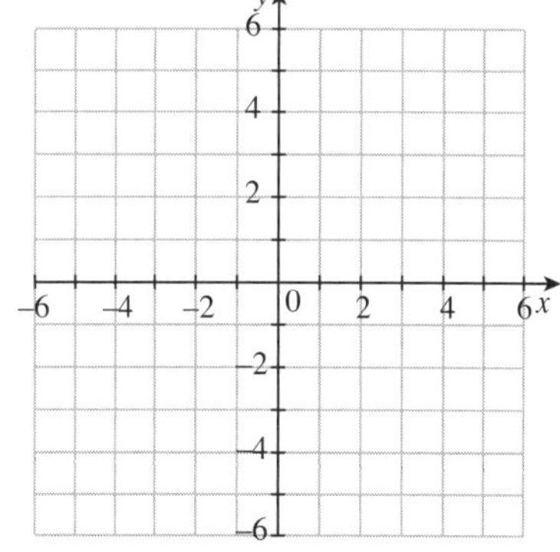

a Sketch the graph of $y = 2x - 1$ on the axes given.

b Use the graph to solve the following equations.

i $2x - 1 = 0$ **ii** $2x - 1 = -1$

__________ __________

iii $2x - 1 = 3$ **iv** $2x - 1 = -5$

__________ __________

Question 4

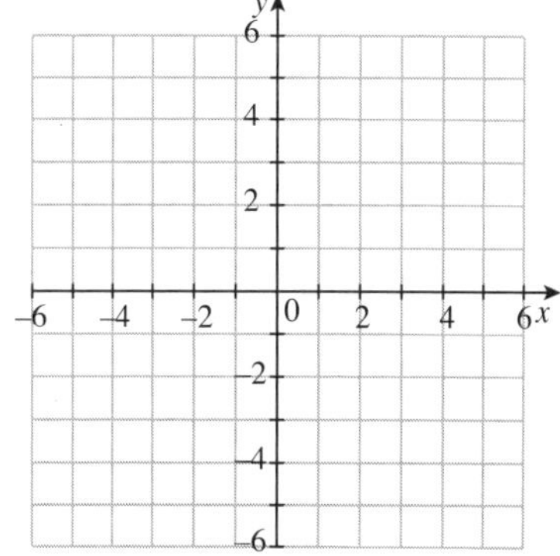

a Sketch the graph of $y = -3x + 5$ on the axes given.

b Use the graph to solve the following equations.

i $-3x + 5 = 2$ **ii** $-3x + 5 = -4$

__________ __________

iii $-3x + 5 = 5$ **iv** $-3x + 5 = -1$

__________ __________

UNIT 18: Graphs of parabolas

QUESTION 1 Complete the table of values and then, on the same number plane, sketch the graphs of $y = x^2$, $y = x^2 + 2$ and $y = x^2 - 2$

x	−3	−2	−1	0	1	2	3
$y = x^2$							
$y = x^2 + 2$							
$y = x^2 - 2$							

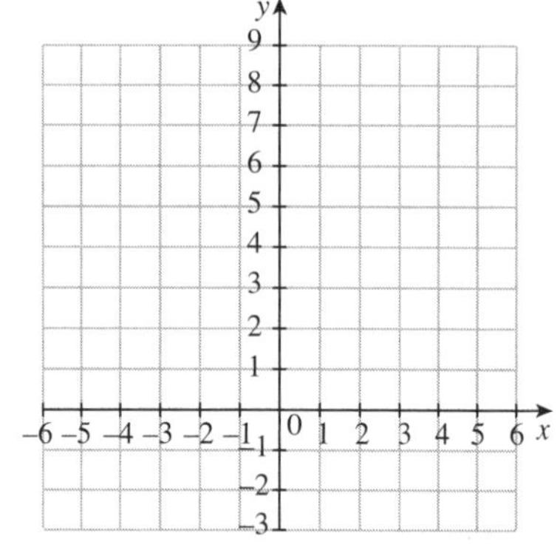

QUESTION 2

a Complete the table of values.

x	−3	−2	−1	0	1	2	3
$y = 2x^2$							
$y = \frac{1}{2}x^2$							

b On the same number plane sketch the graphs of $y = 2x^2$ and $y = \frac{1}{2}x^2$

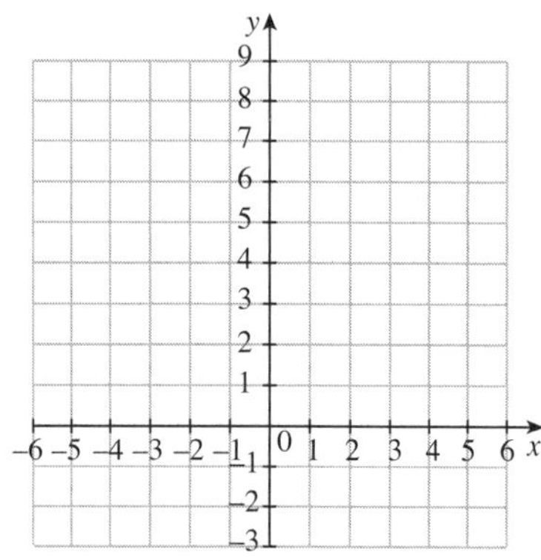

QUESTION 3

a Complete the table of values.

x	−3	−2	−1	0	1	2	3
$y = -x^2$							
$y = -x^2 + 4$							
$y = 9 - x^2$							

b On the same number plane sketch the graphs of $y = -x^2$, $y = -x^2 + 4$ and $y = 9 - x^2$.

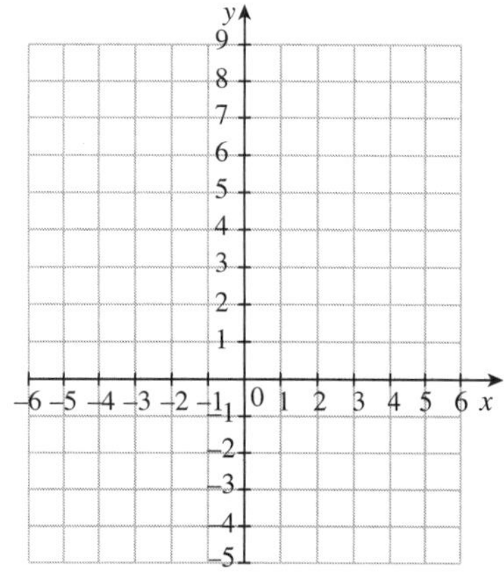

QUESTION 4 The diagram shows the graph of a parabola.

a Is the parabola concave up or concave down?

____

b What is the equation of the axis of symmetry of the parabola? __________

c What is the y-intercept? __________

d What are the x-intercepts? __________

e What is the equation of the parabola?

____

____

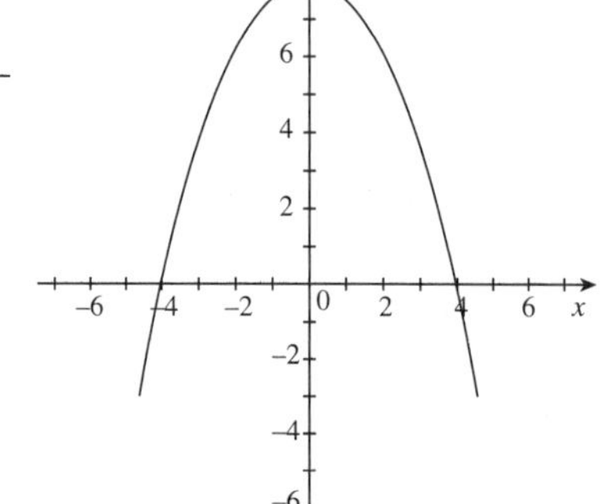

Linear and non-linear relationships

Excel Mathematics Study Guide Years 9–10
Pages 70–82

UNIT 19: Graphs of exponentials

Question 1

a Complete the table of values for the equation $y = 2^x$

x	-3	-2	-1	0	1	2	3
y							

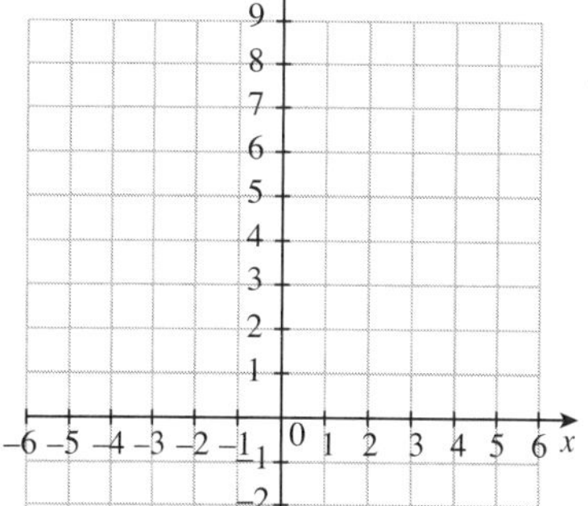

b Draw the graph of $y = 2^x$

c What is the value of 2^x when
i $x = -4$ **ii** $x = 4$?

______________________ ______________________

d Is there any value of x that would make 2^x negative? ______________

e What happens to the graph $y = 2^x$ as the value of x increases? ______________

f Where does this graph cut the y-axis? ______________

Question 2

a Complete the table of values for the equation $y = 3^x$

x	-3	-2	-1	0	1	2	3
y							

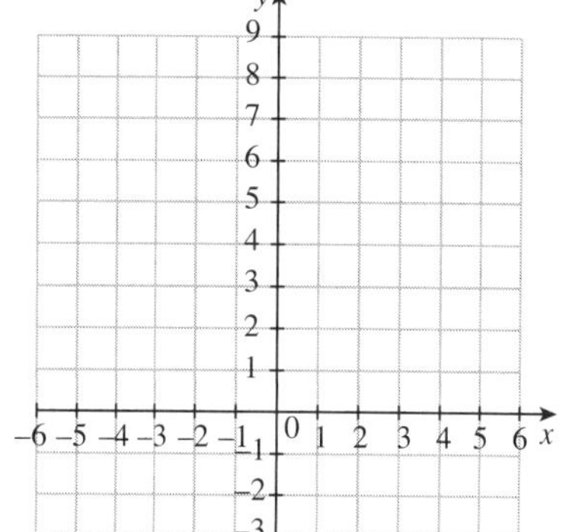

b Draw the graph of $y = 3^x$

c What is the value of 3^x when $x = 4$? ______________

d Is there any value of x that would make 3^x negative? ______________

e Where does this graph cut the y-axis? ______________

Question 3

a On the same diagram sketch the graphs of $y = 4^x$ and $y = 10^x$

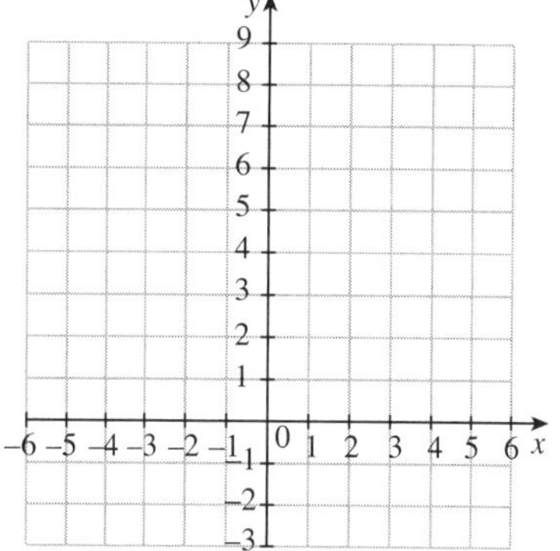

b What similarities and differences are there between the two curves?

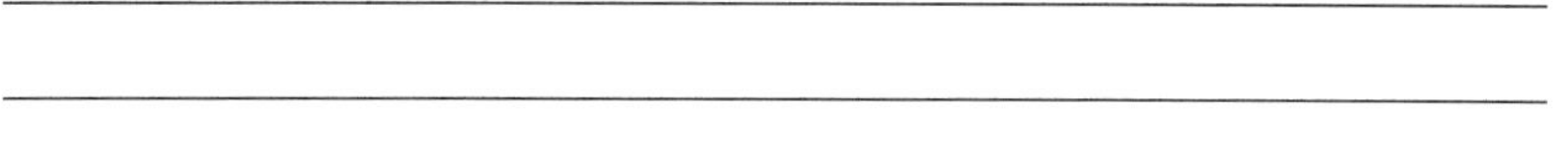

Question 4 Sketch graphs of:

a $y = 2^{-x}$

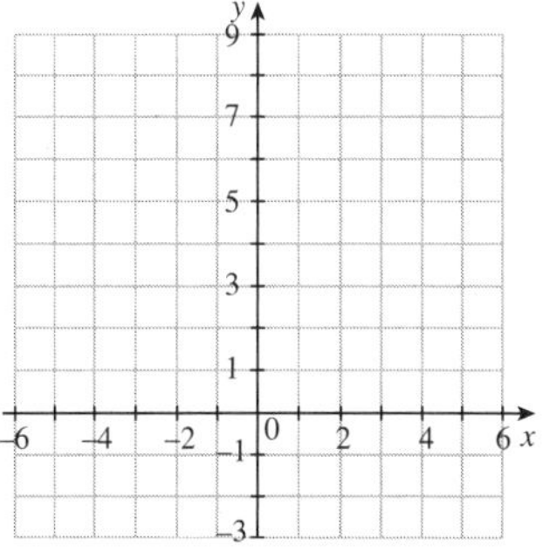

b $y = -2^x$

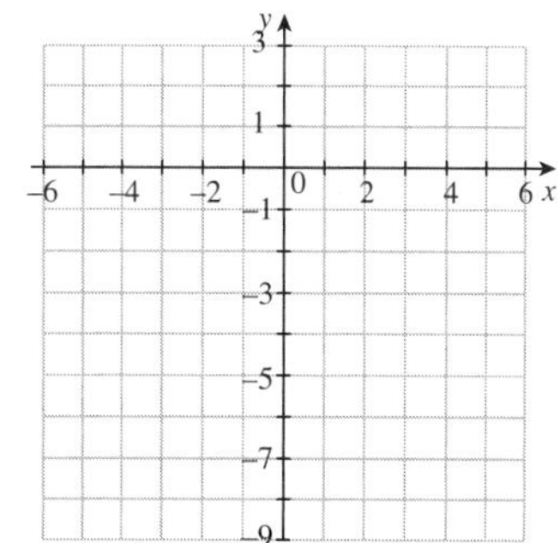

c $y = -2^{-x}$

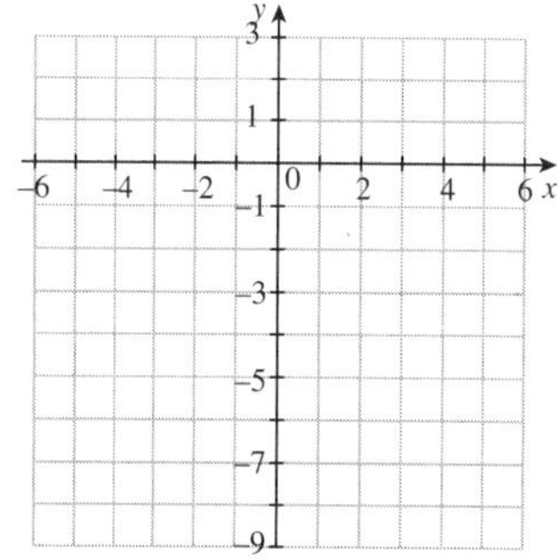

UNIT 20: Circles

QUESTION 1 What is the length of the radius of these circles?

a $x^2 + y^2 = 9$ **b** $x^2 + y^2 = 49$ **c** $x^2 + y^2 = 144$ **d** $x^2 + y^2 = 1$

______ ______ ______ ______

QUESTION 2 Write the equation of the circle with centre (0, 0) and radius of length:

a 10 units **b** 6 units **c** 13 units **d** 17 units

______ ______ ______ ______

QUESTION 3 Write the equation of these circles.

a

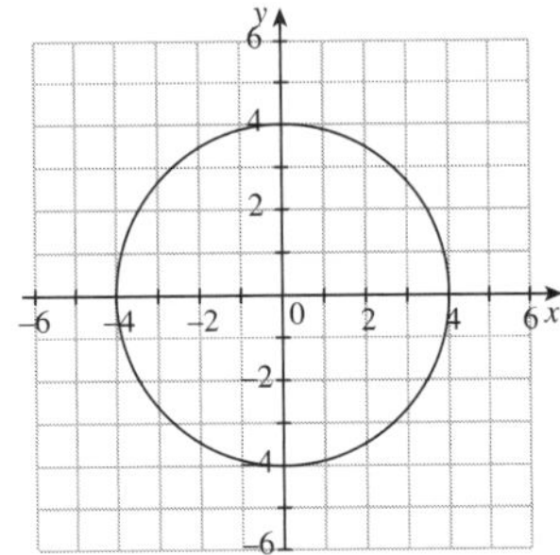

b

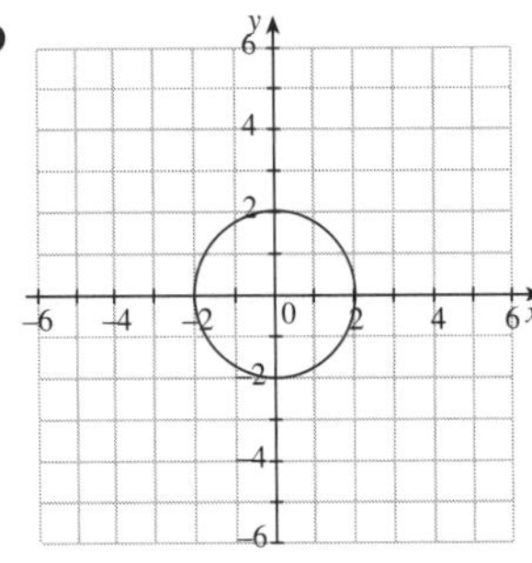

c

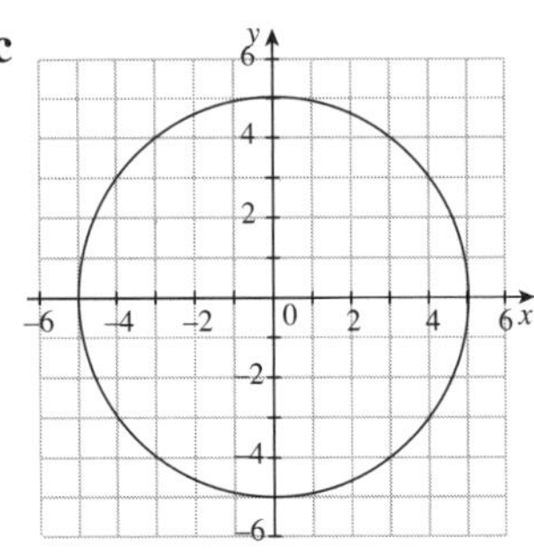

d

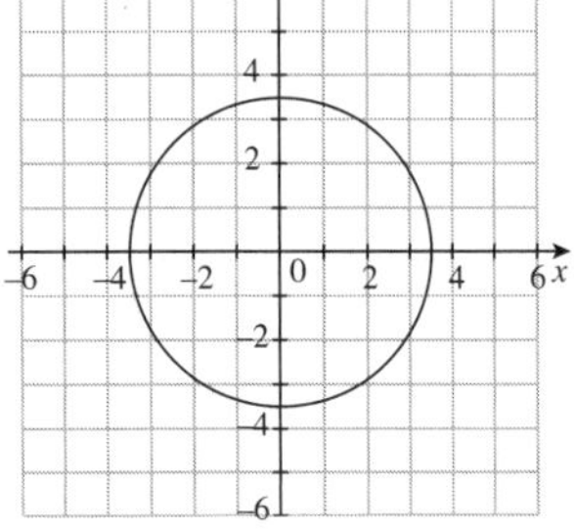

______ ______ ______ ______

QUESTION 4 Sketch these circles showing essential features.

a $x^2 + y^2 = 81$

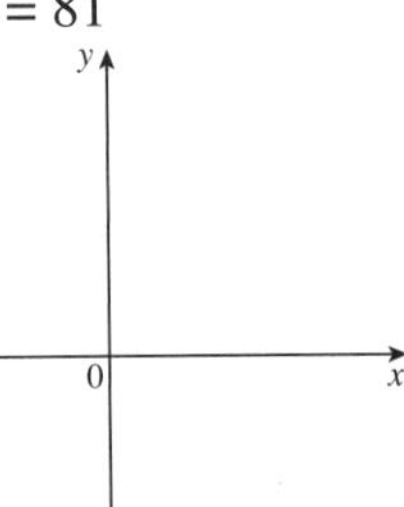

b $x^2 + y^2 = 121$

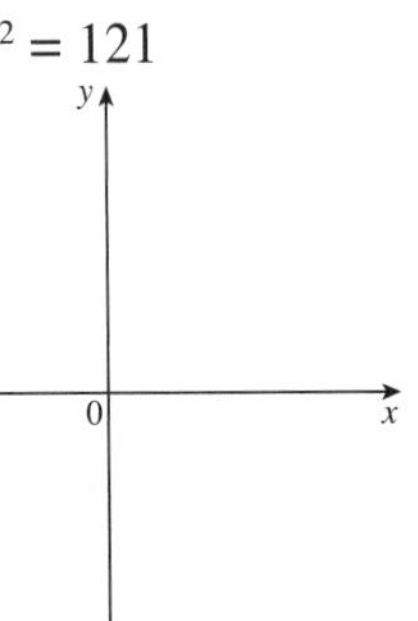

c $x^2 + y^2 = 196$

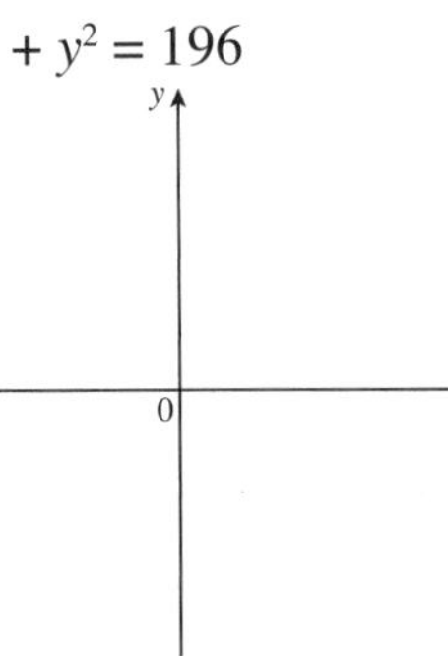

d $x^2 + y^2 = 6\frac{1}{4}$

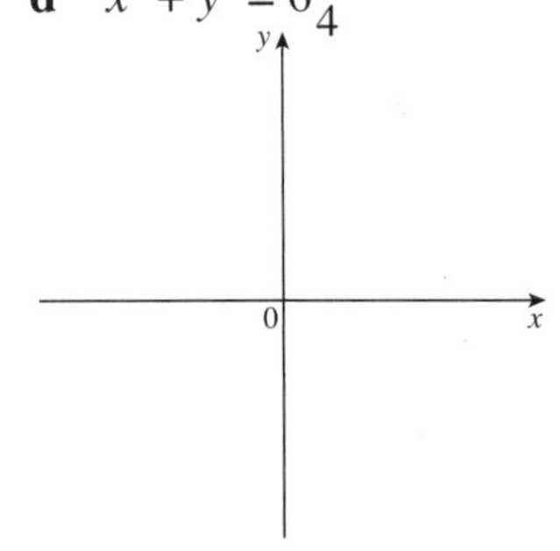

QUESTION 5 Consider the circle $x^2 + y^2 = 64$

a What are the coordinates of the centre of the circle? ______

b What is the length of the radius of the circle? ______

c Find the distance from the point (0, 0) to the point (5, 6).

______

______

______

______

d Does the point (5, 6) lie inside, on or outside the circle $x^2 + y^2 = 64$? ______

Linear and non-linear relationships

TOPIC TEST — PART A

Instructions
- This part consists of 10 multiple-choice questions.
- Fill in only ONE CIRCLE for each question.
- Each question is worth 1 mark.

Time allowed: 10 minutes — **Total marks: 10**

Marks

1 The point (6, 2) lies on the line

(A) $2x - 3y = 6$ (B) $2x + 3y = 6$ (C) $3x - 2y = 6$ (D) $3x + 2y = 6$ — 1

2 What is the gradient of the line that passes through the points (1, 3) and (2, –5)?

(A) –1 (B) 1 (C) –8 (D) 8 — 1

3 Find the equation of the line in the gradient–intercept form when the gradient (m) is $\frac{1}{2}$ and the y-intercept (b) is –5.

(A) $y = -\frac{1}{2}x + 5$ (B) $y = \frac{1}{2}x + 5$ (C) $y = \frac{1}{2}x - 5$ (D) $y = -\frac{1}{2}x - 5$ — 1

4 The midpoint of the interval joining the points (5, 9) and (–7, 1) is

(A) (1, –5) (B) (–1, –5) (C) (1, 5) (D) (–1, 5) — 1

5 Find the distance between the origin and the point (3, 4).

(A) 7 units (B) 5 units (C) $\sqrt{7}$ units (D) $\sqrt{5}$ units — 1

6 The straight line $y = 2x - 1$ passes through one of the following points. Which one?

(A) (0, 2) (B) (0, –2) (C) (0, 1) (D) (0, –1) — 1

7 The equation of the line l is

(A) $x = -3$ (B) $x = 3$ (C) $y = -3$ (D) $y = 3$ — 1

8 The equation of the line m on the diagram is

(A) $3x - 2y + 6 = 0$ (B) $3x + 2y - 6 = 0$ (C) $2x - 3y + 6 = 0$ (D) $2x + 3y - 6 = 0$ — 1

9 The graph shown could be a part of the graph with equation

(A) $y = -x^2$ (B) $y = x^2$ (C) $y = 2^x$ (D) $y = 2^{-x}$ — 1

10 Which of the following could be the equation of the graph?

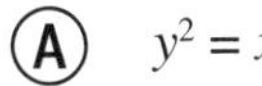

(A) $y^2 = x$ (B) $y = x^2$ (C) $y = 2^x$ (D) $x^2 + y^2 = 4$ — 1

Total marks achieved for PART A ___ / 10

Linear and non-linear relationships

TOPIC TEST — PART B

Time allowed: 20 minutes **Total marks: 15**

Marks

1 P is the point (–3, 6) and Q is (3, –2). For the interval PQ find the:

a gradient **b** midpoint **c** length

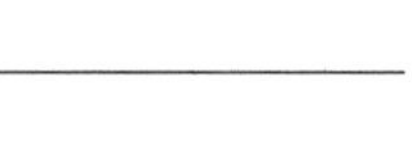

d Show the line joining P to Q on the axes at right.

e What is the y-intercept? ____________

f What is the equation, in gradient-intercept form of the line joining P to Q? ____________

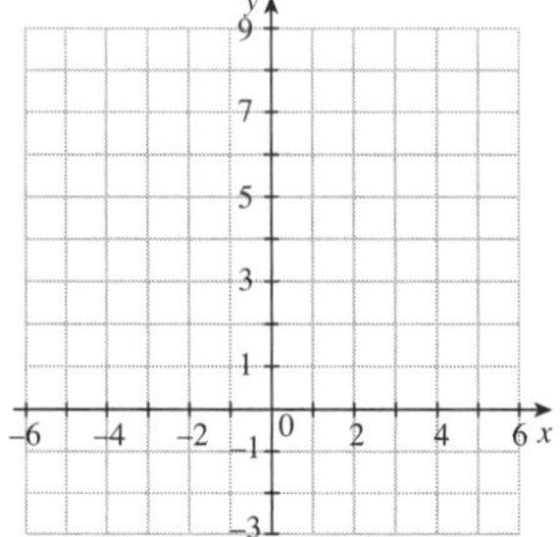

6

2 Sketch these graphs, showing essential features.

a $x^2 + y^2 = 9$ **b** $y = 1 - x^2$ **c** $y = 3^{-x}$

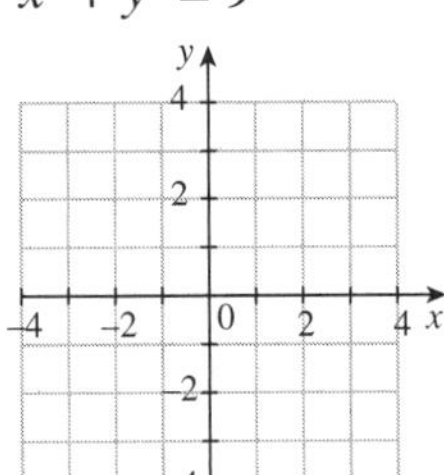

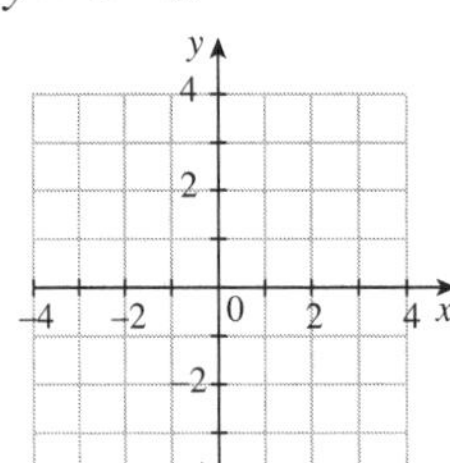

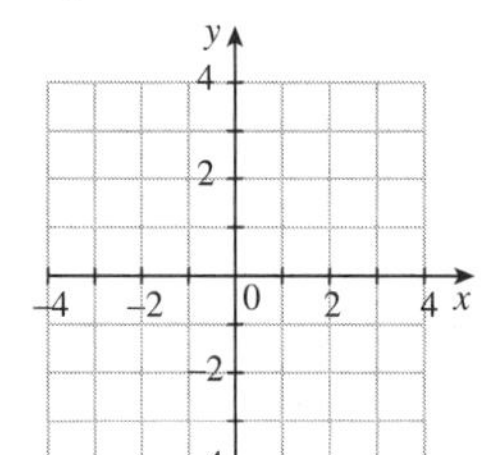

3

3 A line has equation $4x + y - 8 = 0$

a What is the x-intercept? **b** What is the y-intercept?

c Show the line on the axes at right.

d Does the point (5, –12) lie on the line?

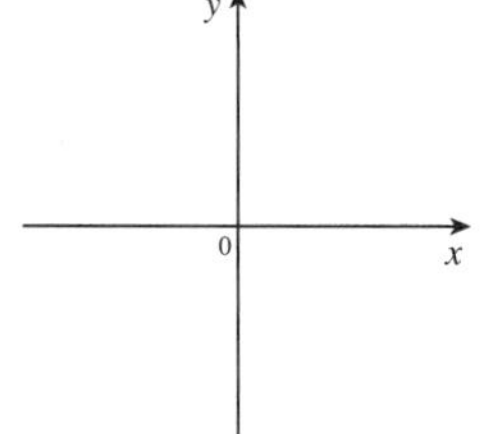

4

4 The equation of a straight line is $3x = y - 6$ Write the equation:

a in general form. **b** in gradient-intercept form.

2

Total marks achieved for PART B

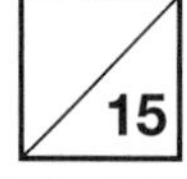

Chapter 6

Equations

Excel Mathematics Study Guide Years 9–10
Pages 38–51

UNIT 1: Simple equations

Question 1 Solve the following equations.

a $x + 3 = 7$ **b** $x - 2 = 8$ **c** $x - 9 = 5$ **d** $x + 5 = 9$ **e** $x + 1 = 10$

f $x + 16 = 7$ **g** $2 + x = 13$ **h** $3 = x + 1$ **i** $14 = 18 - x$ **j** $y - 17 = 37$

k $m - 13 = 27$ **l** $8 - a = 25$ **m** $16 - a = 56$ **n** $3 - p = 9$ **o** $8 + a = 64$

p $a + 4.2 = 9.8$ **q** $-6 - m = 16$ **r** $x + 3 = 11$ **s** $y - 1 = 19$ **t** $3 - y = 12$

Question 2 Solve the following equations.

a $6x = 24$ **b** $3x = 12$ **c** $2x = 14$ **d** $4x = 20$ **e** $\frac{x}{2} = 9$

f $\frac{x}{3} = 15$ **g** $\frac{x}{5} = 9$ **h** $\frac{x}{3} = -5$ **i** $\frac{x}{2} = 8$ **j** $2x = 20$

k $\frac{1}{3}x = 5$ **l** $8 = \frac{x}{3}$ **m** $2 = \frac{x}{7}$ **n** $\frac{m}{5} = 4$ **o** $\frac{x}{6} = 5$

Question 3 Solve the following equations.

a $\frac{2x}{5} = 4$ **b** $\frac{3x}{2} = 15$ **c** $\frac{4x}{3} = 20$ **d** $2x + 5 = 15$ **e** $3x + 7 = 13$

f $4x + 5 = 29$ **g** $8x - 7 = 49$ **h** $2x - 15 = 5$ **i** $6 - 3m = 18$ **j** $6y + 3 = 17$

k $9y + 8 = 21$ **l** $4p - 3 = 21$ **m** $\frac{2x}{3} + 5 = 15$ **n** $\frac{3x}{2} - 7 = 9$ **o** $\frac{4x}{5} + 15 = 45$

Equations

UNIT 2: Two-step equations

Excel Mathematics Study Guide Years 9–10
Pages 38–51

QUESTION 1 Solve the following equations.

a $3x + 1 = 7$

b $7y - 8 = 13$

c $4x + 7 = 19$

d $\frac{x}{5} - 1 = 3$

e $\frac{m}{2} + 5 = 7$

f $\frac{x}{3} - 5 = -2$

g $3k + 3 = 33$

h $4x - 7 = 33$

i $3x + 7 = 16$

j $\frac{7m}{2} = 14$

k $\frac{x-2}{3} = 6$

l $20 = 5x - 15$

m $2x + 3x = 15$

n $6a - a = 25$

o $10n - 3n = 28$

QUESTION 2 Solve.

a $2x - 4 = 8$

b $\frac{y}{3} + 6 = 15$

c $\frac{x-3}{8} = 2$

d $\frac{x-5}{6} = -1$

e $\frac{6m}{5} = 12$

f $18 - 3m = 0$

g $9y + 5 = -4$

h $3a - 2\frac{1}{2} = 6\frac{1}{2}$

i $5b + 0.3 = 4.8$

Equations

UNIT 3: Two- and three-step equations

QUESTION **1** Solve the following equations.

a $3x - 5 = 2x + 7$

b $2y - 1 = y + 9$

c $3m - 2 = 2m + 7$

d $4x + 9 = 3x - 12$

e $6x - 20 = 4x + 48$

f $6m + 7 = 7m + 10$

g $6t - 10 = 4t + 12$

h $7y - 14 = 5y + 20$

i $2x - 6 = 3 - x$

j $9m - 3 = 7m + 9$

k $12a - 3 = 7a + 32$

l $2x + 3 = x - 9$

m $3a + 5 = 21 - a$

n $6x - 4 = 2x + 16$

o $6x - 2 = 3x - 6$

QUESTION **2** Solve.

a $2x - 7 = x - 3$

b $4a - 3 = 3a + 9$

c $7y - 3 = 4y + 15$

d $11m - 6 = 7m + 14$

e $12p - 3 = 5p + 32$

f $2x - 14 = x - 12$

g $5x + 17 = 3 - 4x$

h $10y - 6 = 5y + 19$

i $4 + m = 16 - 3m$

Excel Mathematics Study Guide Years 9–10
Pages 38–51

UNIT 4: Equations with pronumerals on both sides

QUESTION **1** Solve the following equations.

a $7x - 3 = 6x + 7$ **b** $5x - 8 = 4x + 9$ **c** $9x - 7 = 8x + 9$ **d** $4x + 5 = 2x + 17$

e $7x - 11 = 5x + 19$ **f** $6x - 1 = 4x + 7$ **g** $5m - 6 = 3m$ **h** $10x + 3 = 7x + 24$

i $20x - 7 = 10x + 13$ **j** $33x - 64 = x$ **k** $11x - 90 = x$ **l** $5a - 8 = 3a + 84$

QUESTION **2** Solve the following equations.

a $8x + 20 = 6x + 6$ **b** $3x + 7 = x + 19$ **c** $5m + 12 = m + 3$

d $7t - 7 = 5t - 5$ **e** $4y - 11 = 3y + 16$ **f** $8y - 2 = 6y + 14$

g $5a = 18 + 2a$ **h** $5x = 32 - 3x$ **i** $6m = 9m + 27$

j $8n - 5 = n + 23$ **k** $7x = x + 18$ **l** $5x + 3 = 9 + 2x$

m $16 - 3m = m + 4$ **n** $7a - 10 = 5a + 12$ **o** $9x - 11 = 5x + 21$

UNIT 5: Equations involving grouping symbols (1)

Question 1 Solve the following equations.

a $6(m - 1) = 24$

b $4(a - 4) = 8$

c $8(3 - x) = 7(x - 6)$

d $5(a + 4) = 4(a - 3)$

e $2(a + 1) = a + 2$

f $5(a + 3) = 4(a + 9)$

g $2(m + 1) = 5$

h $3(x - 5) = 2(x + 4)$

i $6(a + 7) = 5(a - 3)$

j $7(x - 8) = 6(x + 2)$

k $3(x + 7) + x + 3 = 18$

l $3(x + 5) = 30$

Question 2 Solve the following equations.

a $5(2n - 1) = 25$

b $4(n - 3) = 36$

c $2(3x + 2) = 16$

d $2(3p - 1) = 22$

e $2(x + 5) = 18$

f $2(x - 7) = x - 12$

g $3(x + 4) = 18$

h $5(2x + 3) = 45$

i $7(y - 2) = 5(y + 4)$

Equations

Excel Mathematics Study Guide Years 9–10
Pages 38–51

UNIT 6: Equations involving grouping symbols (2)

QUESTION 1 Solve the following equations.

a $2(a+3)=9$

b $3(x+4)=36$

c $4(x+2)=48$

d $5(x-1)=25$

e $3(3-2x)=33$

f $5(8-2m)=100$

g $4(x-5)=3x+9$

h $2(4x-3)=7x-6$

i $3(2x-5)=5x+23$

j $6(x-7)=4x-8$

k $9+6x=2(2x+1)$

l $3(2x-7)=8x-5$

QUESTION 2 Solve the following equations.

a $4(a+3)=3(a+2)$

b $5(x-2)=4(x-1)$

c $5(m-3)=4(m+2)$

d $3(y+2)=2(y-1)$

e $8(2t+5)=4(3t+8)$

f $4(3x-1)=2(3x+1)$

g $6(3a+2)=5(2a+9)$

h $3(4m+6)=4(2m-1)$

i $5(3\text{a}+1)=2(2a-1)$

j $5+2(a+1)=3(a+2)$

k $3x+2(x+1)+3(x+2)=8$

l $7m-(6m-9)=5$

UNIT 7: Equations with one fraction (1)

Question 1 Solve the following equations.

a $\frac{x}{2} = \frac{2}{3}$

b $\frac{a}{7} = \frac{1}{5}$

c $\frac{2y}{3} = 2\frac{1}{2}$

d $\frac{y}{3} + 1 = 4$

e $\frac{p}{2} + 5 = -7$

f $\frac{m}{3} - 4 = 3$

g $\frac{a+2}{4} = 8$

h $\frac{2m+3}{6} = 4$

i $\frac{2x+9}{3} = 10$

j $\frac{x}{5} + 3 = x$

k $\frac{m+4}{3} = m$

l $\frac{m}{7} + m = 8$

Question 2 Solve.

a $\frac{7x-3}{8} = -4$

b $\frac{3x-2}{4} = -5$

c $\frac{m+9}{4} = m$

d $\frac{5x}{3} - 7 = 8$

e $\frac{5x}{3} - 6 = 10$

f $\frac{8}{x} = 4$

g $\frac{6}{5x} = 12$

h $\frac{6x}{5} - 7 = 4x$

i $\frac{3x}{2} + 4 = 7$

j $\frac{5a-2}{3} = 18$

k $\frac{k+5}{7} = 8$

l $\frac{m}{5} - 4 = 2$

Equations

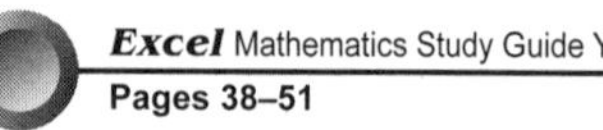

Excel Mathematics Study Guide Years 9–10
Pages 38–51

UNIT 8: Equations with one fraction (2)

QUESTION 1 Solve the following equations.

a $\frac{x}{2} + 1 = 5$

b $\frac{p}{3} - 2 = 7$

c $\frac{y}{5} - 6 = 3$

d $\frac{a + 2}{3} = 7$

e $\frac{4m + 3}{7} = 5$

f $\frac{2x + 9}{3} = 5$

g $\frac{7x - 3}{8} = -5$

h $\frac{5x - 2}{3} = -4$

i $\frac{9 + 3x}{6} = 6$

j $\frac{x}{7} + 5 = 6$

k $\frac{x}{3} - 8 = 12$

l $\frac{x}{5} + 3 = 7$

QUESTION 2 Solve the following equations.

a $\frac{2x}{3} + 5 = 20$

b $\frac{3x}{7} - 6 = 5$

c $\frac{4x}{5} - 3 = 10$

d $\frac{5}{x} = 10$

e $\frac{8}{m} = 4$

f $\frac{3}{4x} = 6$

g $\frac{5x}{6} - 5 = 3x$

h $\frac{2x}{3} + 4 = 3x$

i $\frac{3y}{4} - 5 = 2y$

j $9 - \frac{3}{y} = 6$

k $\frac{1}{2x + 3} = 3$

l $\frac{3}{x + 1} = 2$

Equations

UNIT 9: Harder equations

QUESTION **1** Solve the following equations.

a $\frac{a}{3}+\frac{a}{4}=6$

b $\frac{x}{3}-\frac{x}{2}=5$

c $\frac{2x}{3}-\frac{x}{6}=20$

d $\frac{a}{4}-\frac{a}{8}=16$

e $\frac{y}{2}-\frac{y}{6}=9$

f $\frac{t}{3}+\frac{t}{4}=3$

g $\frac{p}{5}-\frac{3}{5}=\frac{9}{10}$

h $\frac{2x}{3}+\frac{3x}{5}=\frac{2}{3}$

QUESTION **2** Solve these equations.

a $\frac{3x}{5}+\frac{5x}{2}=\frac{3}{10}$

b $\frac{x+2}{3}=\frac{x-5}{5}$

c $\frac{3m+5}{8}=\frac{2m+9}{4}$

d $\frac{t-1}{3}=\frac{2t+1}{2}$

e $\frac{x}{2}+\frac{x}{3}=\frac{3x}{4}+\frac{1}{6}$

f $\frac{m+1}{2}+\frac{m+3}{3}=6$

g $\frac{6m}{7}-\frac{m-2}{5}=3$

h $\frac{7a}{10}-\frac{a+1}{5}=2$

i $\frac{x+7}{3}=\frac{x-7}{2}$

j $\frac{x}{7}-\frac{x}{8}=10$

k $\frac{1}{x}+\frac{1}{5x}=6$

l $\frac{1}{x}+\frac{1}{2x}+\frac{1}{3x}=2$

Excel Mathematics Study Guide Years 9–10
Pages 38–51

UNIT 10: Solving problems using equations

Question 1

a If 12 is added to the product of 7 and a number, the result is 47. What is the number?

b If 18 is subtracted from 4 times a number, the result is 62. What is the number?

c The sum of 3 consecutive odd numbers is 57. Find the numbers.

d 5 more than three times the number equals the number plus 25. What is the number?

e The length of a rectangle is 3 times the width of the rectangle and the perimeter is 48 cm. Find the width and length of the rectangle.

Question 2

a The angles of a triangle are in the ratio 1:2:3. Find the size of each angle.

b 12 more than 3 times a number equals the number plus 48. What is the number?

c If 8 years are added to a man's present age and this value is doubled, the result if 100. Find the man's present age.

d Melissa's age is two times Steven's age. If Melissa is 12 years older than Steven, what are their ages?

e Kristina spent $\frac{1}{4}$ of her money, then she spent $\frac{1}{4}$ of the remainder. Altogether she spent $126. How much money did she have to start with?

Equations

TOPIC TEST

PART A

Instructions
- This part consists of 10 multiple-choice questions.
- Fill in only ONE CIRCLE for each question.
- Each question is worth 1 mark.

Time allowed: 10 minutes **Total marks: 10**

	Marks
1 If $7x - 3 = 81$, what is the value of x? (A) $\frac{78}{7}$ (B) 27 (C) 12 (D) 9	1
2 If $10x - 2 = 6x$, then x is equal to (A) $-\frac{1}{2}$ (B) $\frac{1}{8}$ (C) $\frac{1}{2}$ (D) 2	1
3 If $\frac{m}{3} - 2 = 4$, then $m =$ (A) 2 (B) 6 (C) 14 (D) 18	1
4 The value of x that satisfies the equation $4(x - 4) = 20$ is (A) 1 (B) 6 (C) 9 (D) 24	1
5 When $m + 3 = \frac{5m}{2}$, the value of m is (A) 0 (B) 2 (C) 6 (D) 30	1
6 When $3(a + 7) = 42$, the value of a is (A) 5 (B) 6 (C) 7 (D) 8	1
7 If $\frac{x+1}{5} - 2 = 3$ then x is equal to (A) 5 (B) 24 (C) 6 (D) 26	1
8 If $4(3m - 5) = 6m - 14$ then m equals (A) 2 (B) 1 (C) −2 (D) −1	1
9 If $12x - 4 = 8$, then x is equal to (A) $\frac{1}{3}$ (B) $\frac{2}{3}$ (C) 1 (D) −1	1
10 Three more than twice a number equals the number plus 7. What is the number? (A) 2 (B) 4 (C) 5 (D) 10	1

Total marks achieved for PART A ___ / 10

Equations

TOPIC TEST — PART B

Time allowed: 20 minutes **Total marks: 15**

Marks

1 Solve these equations.

a $x - 11 = 24$

b $\frac{y}{8} = -9$

c $\frac{m}{2} - 8 = 16$

3

2 15 more than 4 times a number equals the number plus 45. What is the number?

1

3 Solve these equations.

a $2x - 3 = 8$

b $3x + 8 = 5x + 2$

c $3x - 7 = 1 - x$

3

d $\frac{5a - 1}{3} = a + 2$

e $2(5 - 3x) = 3$

f $\frac{2y + 3}{3} = 4y$

3

g $m - 6 = 2(m - 7)$

h $\frac{6a + 2}{4} = 5$

i $\frac{3x - 1}{4} + x = 2$

3

j $2m + 3(m - 1) = 7$

1

k $3x - 1 + 4x = 8$

1

Total marks achieved for PART B

Chapter 7
Area and volume

Excel Mathematics Study Guide Years 9–10
Pages 124–138

UNIT 1: Areas of triangles and quadrilaterals

Question 1 Find the area of each triangle.

a

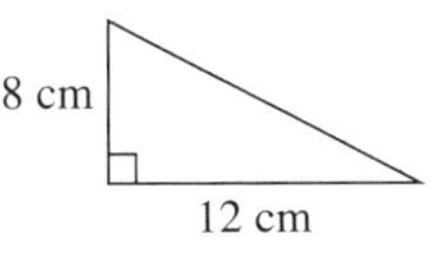

b

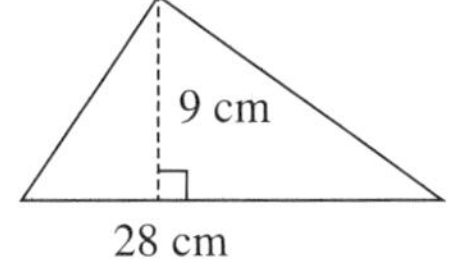

c

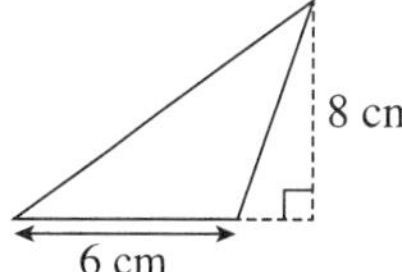

Question 2 Find the area of each quadrilateral.

a

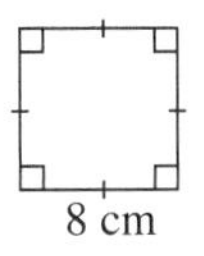

b

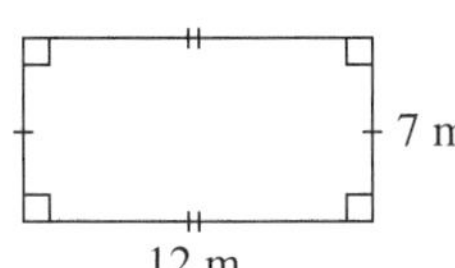

c

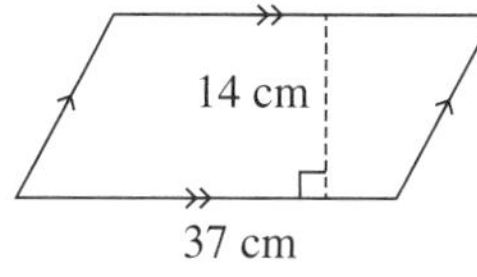

d

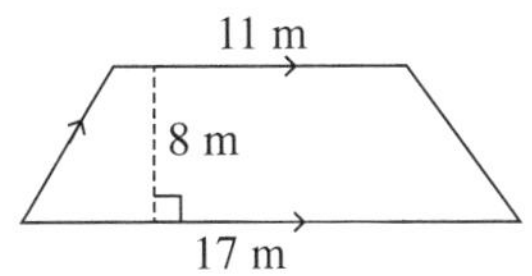

e

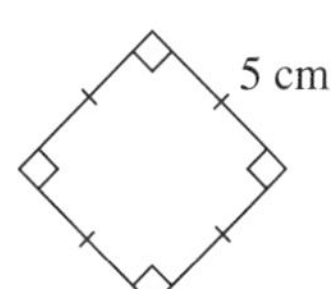

f

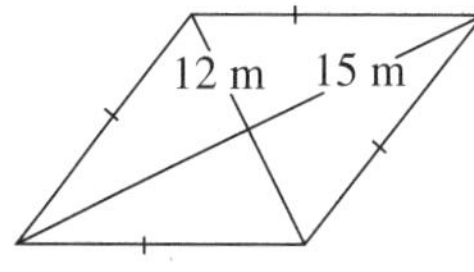

Question 3 Find these areas.

a

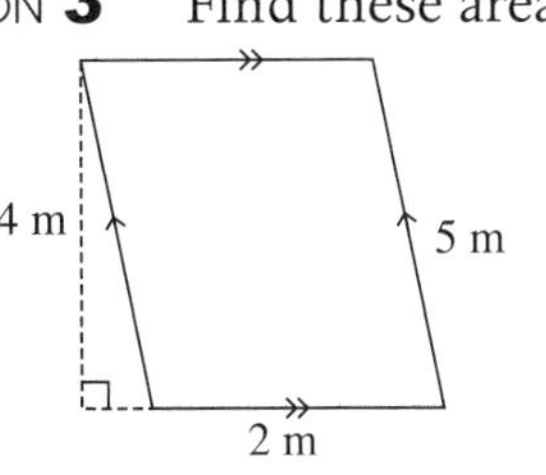

b

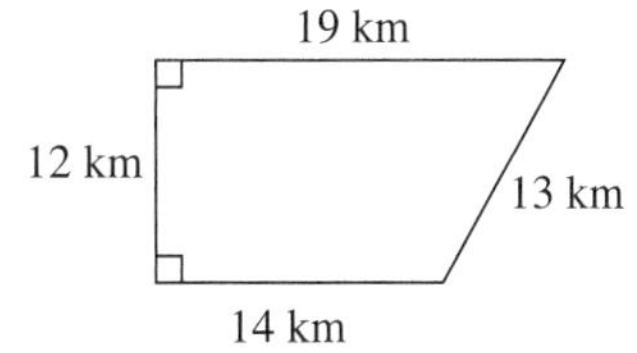

c

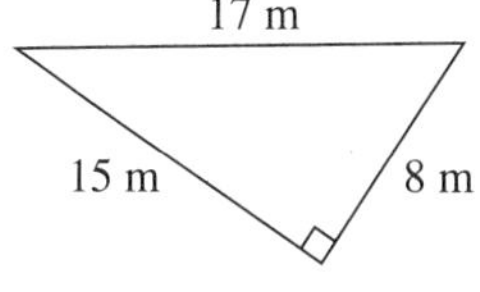

Area and volume

Excel Mathematics Study Guide Years 9–10
Pages 124–138

UNIT 2: Areas of plane shapes

QUESTION 1 Write the area formula for each shape below.

a

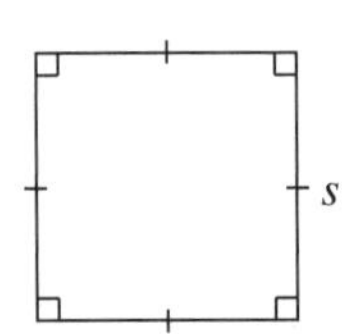

b

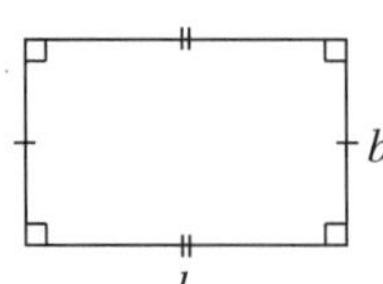

c

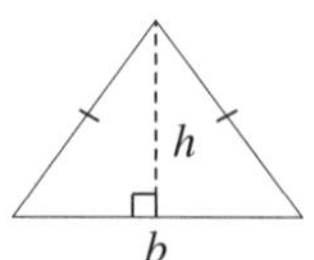

d

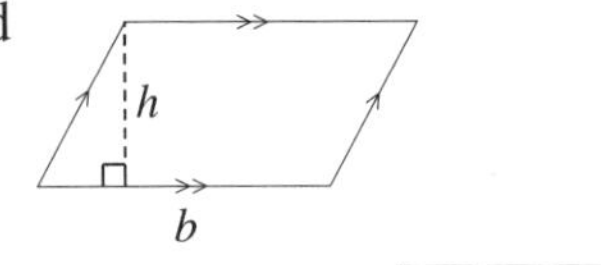

e

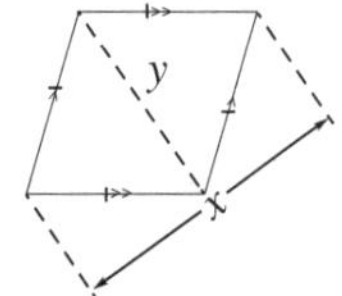

f

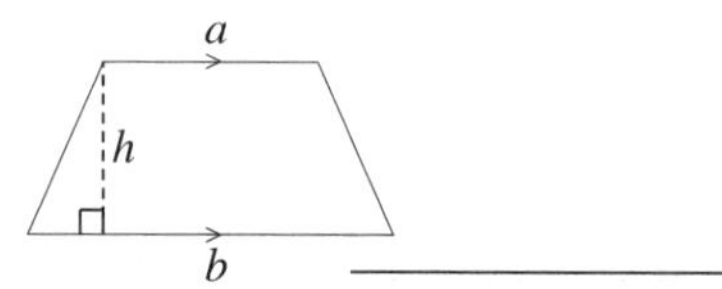

g

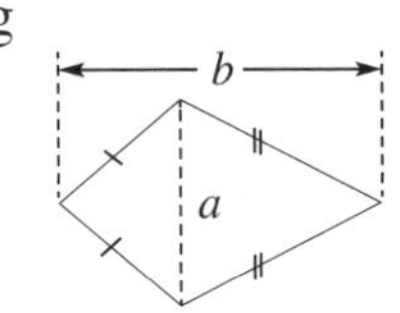

h

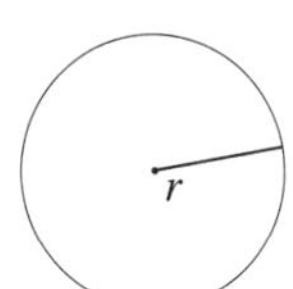

i

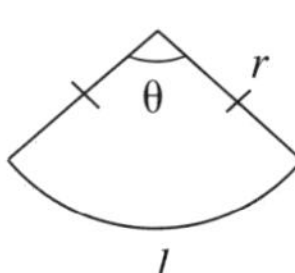

QUESTION 2 Find the area of the following shapes. All measurements are in centimetres.

a

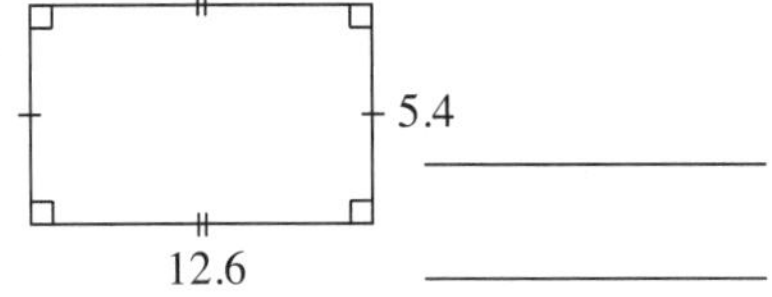

b

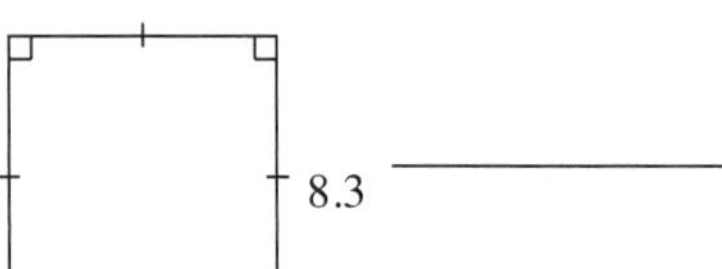

c

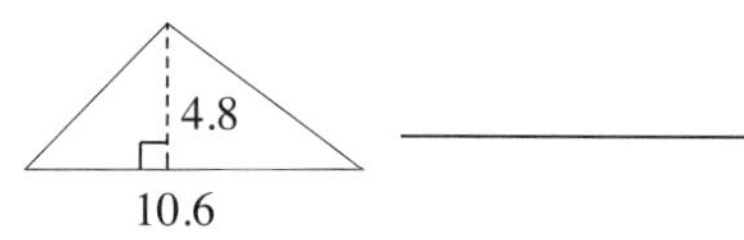

d

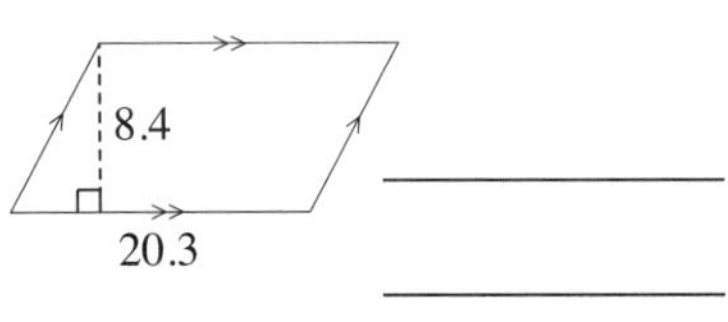

e

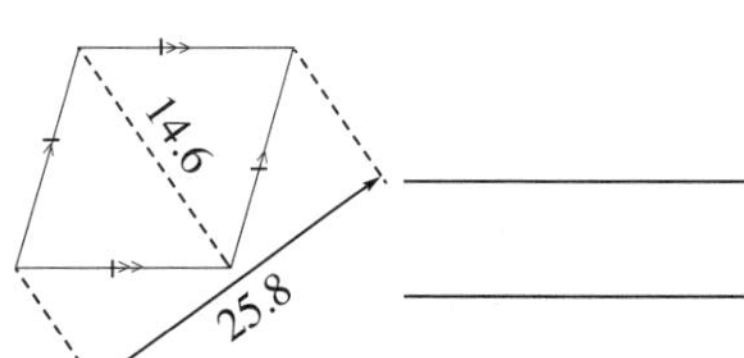

f

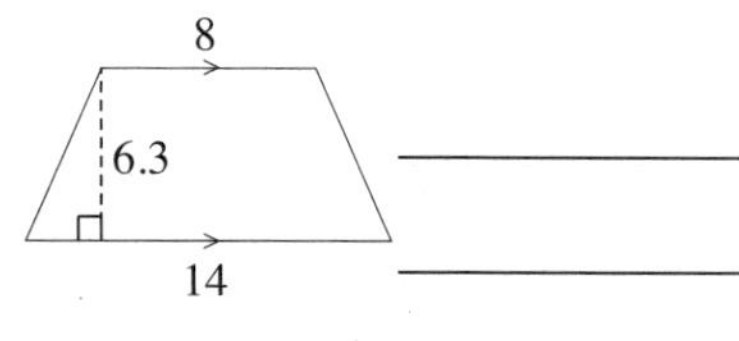

QUESTION 3 Find the area of each shape (Give answers to one decimal place.).

a

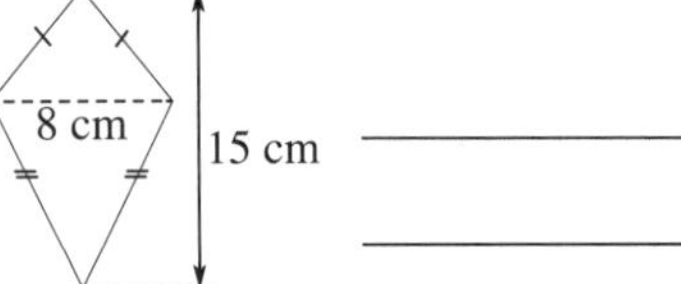

b

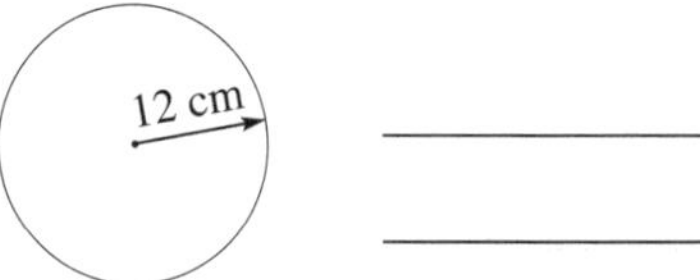

c

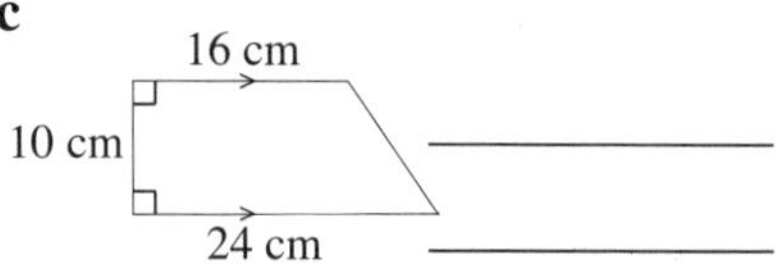

d

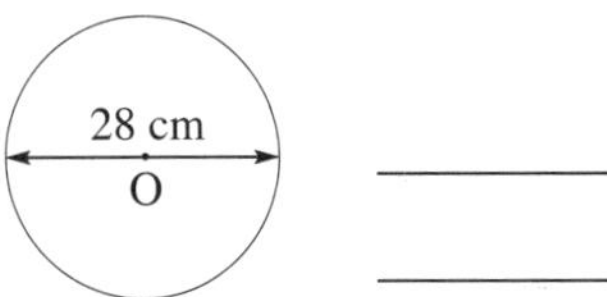

e

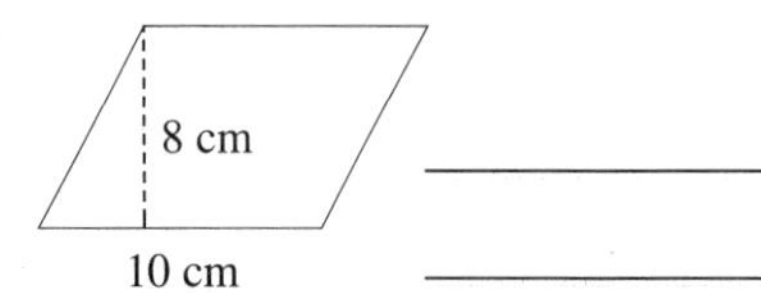

f

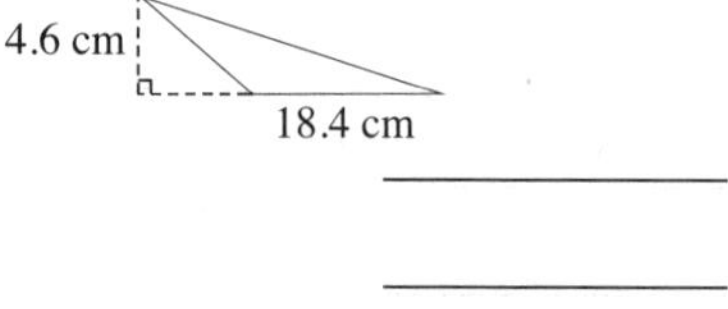

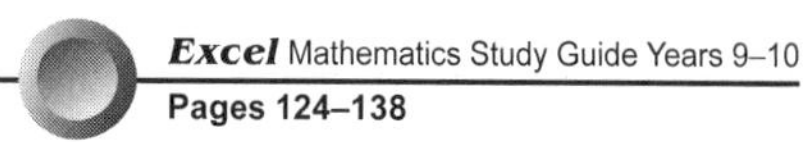

UNIT 3: The area of parts of a circle

QUESTION 1 Calculate the area of the following circles, correct to 1 decimal place.

a

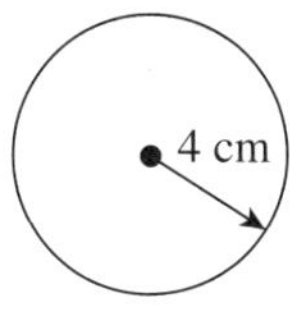

b

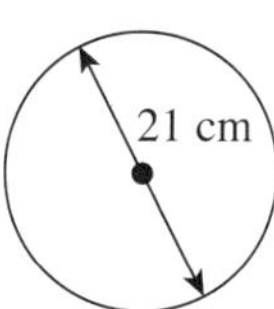

QUESTION 2 Find the fraction of the circle given in the following diagrams.

a

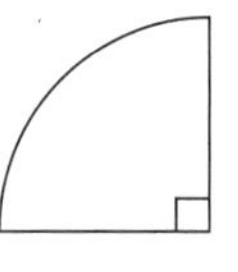

b

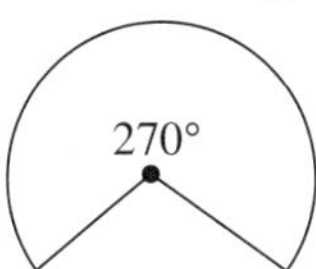

c

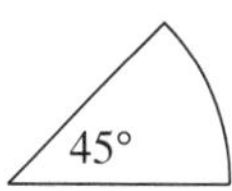

d

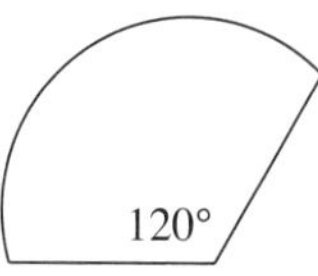

QUESTION 3 Find the area of the sectors below, correct to 1 decimal place.

a

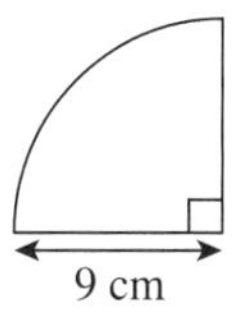

b

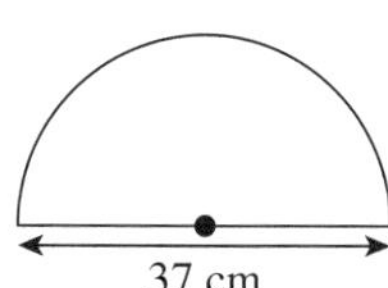

c

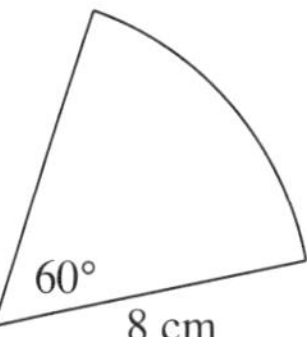

d

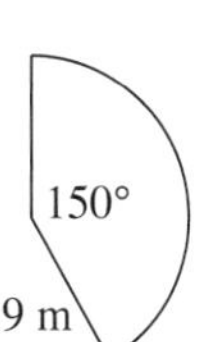

e

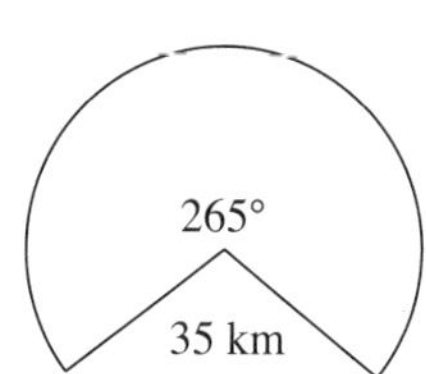

f

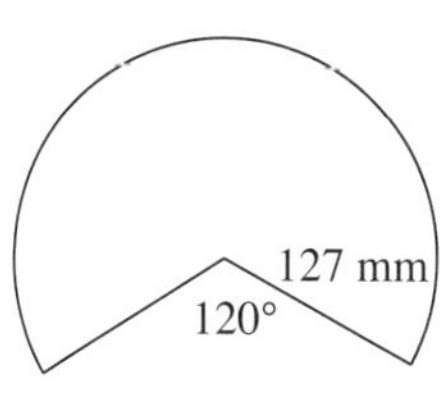

Area and volume

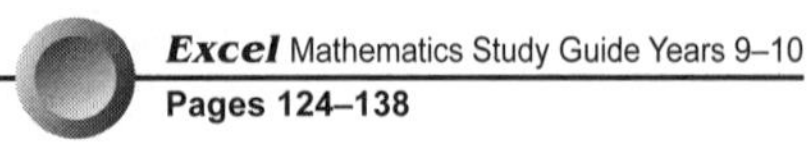

UNIT 4: Area of an annulus

QUESTION 1 Find the area of the shaded region, correct to 1 decimal place.

a

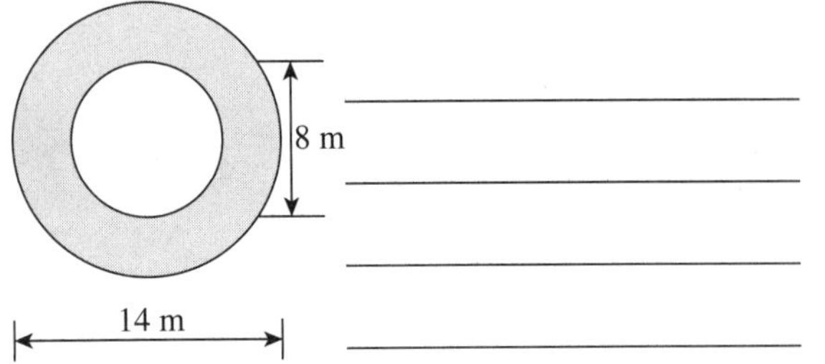

b

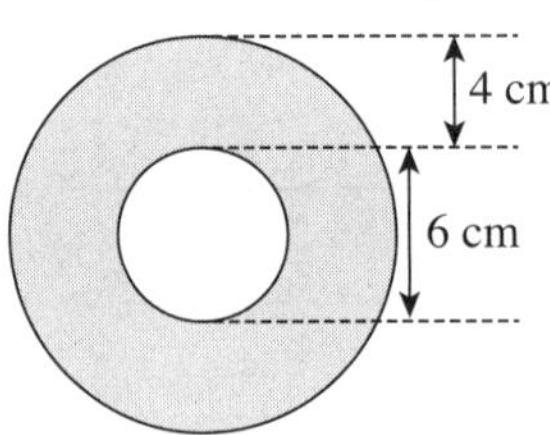

QUESTION 2 Find the area of the shaded region correct to 2 significant figures.

a

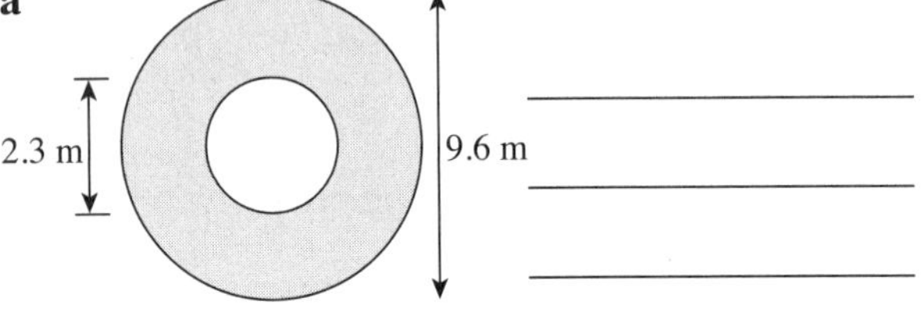

b

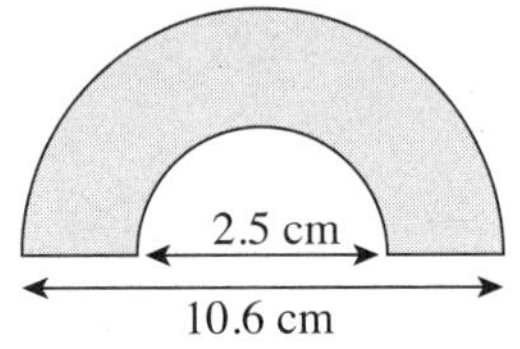

QUESTION 3 A 4 metre wide path is built around a circular pool that has a diameter of 9 metres. Find the area of the path correct to 1 decimal place.

QUESTION 4 A clock has a minute hand that is 8 cm long and an hour hand that is 3.5 cm long. In one complete revolution of each hand, find the difference in the area they cover (Give the answer to the nearest square centimetre.).

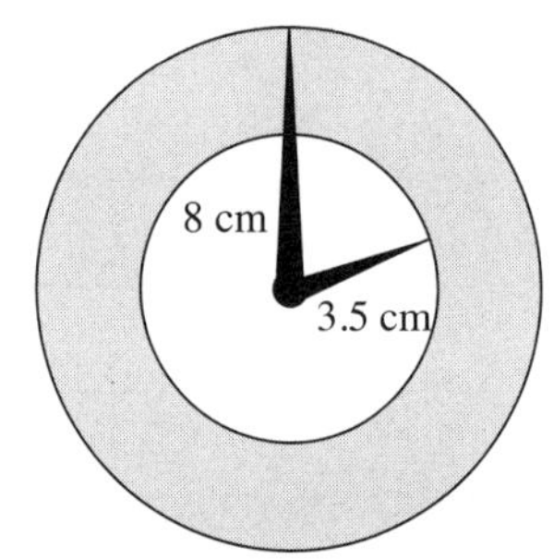

Area and volume

UNIT 5: Composite areas (1)

QUESTION 1 Find the area of each composite figure by dividing it into different shapes (all measurements are in cm).

a

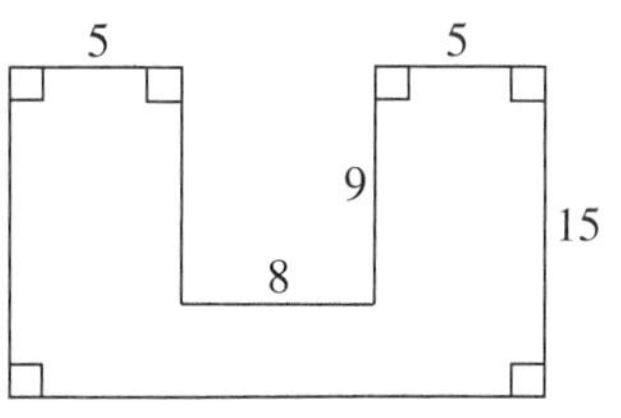

b

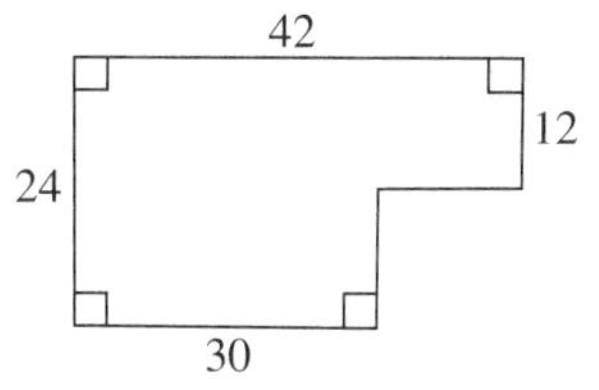

c

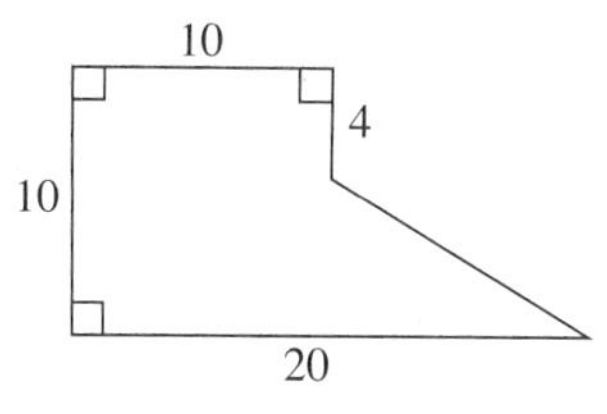

d

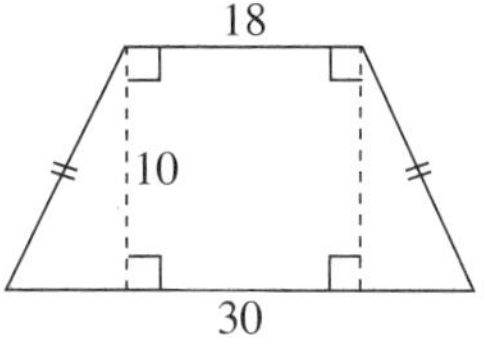

e

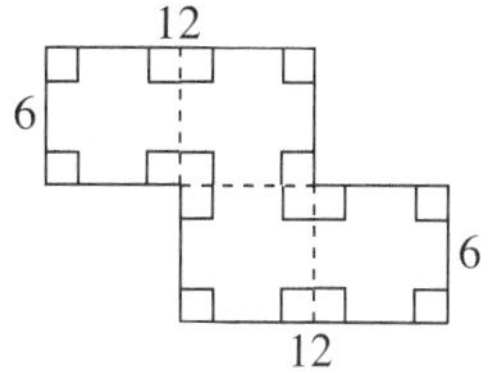

f

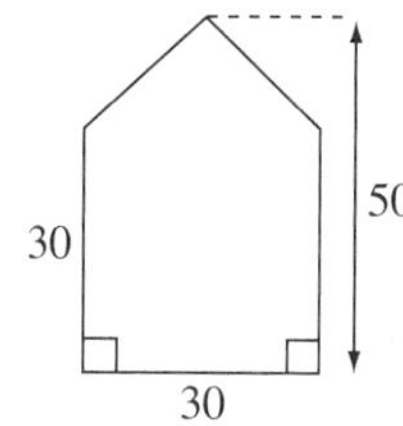

g

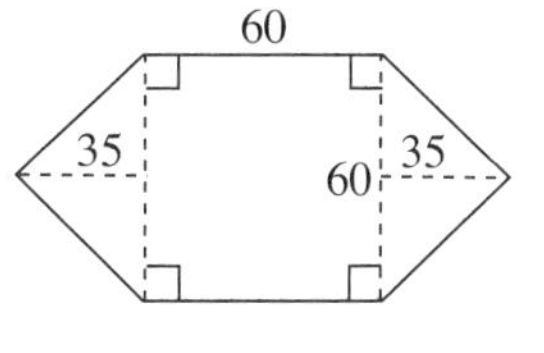

h

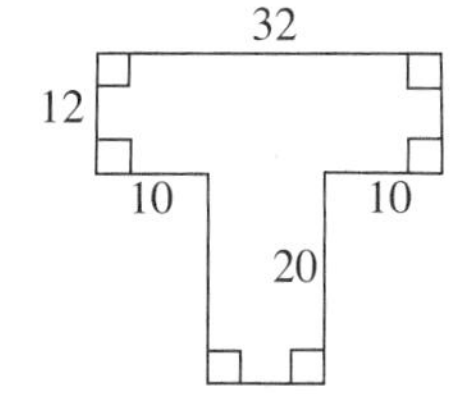

i

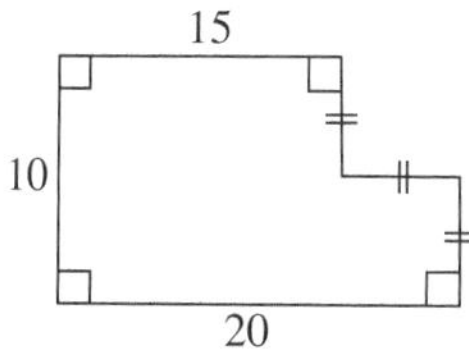

j

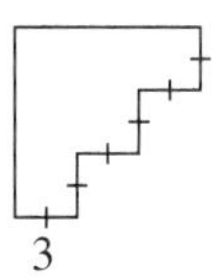

k

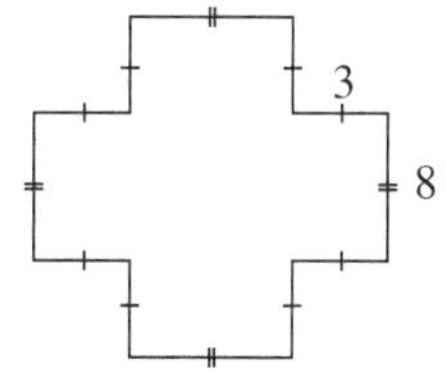

l

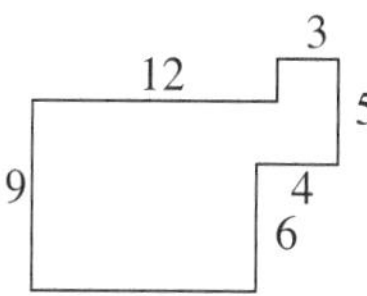

m

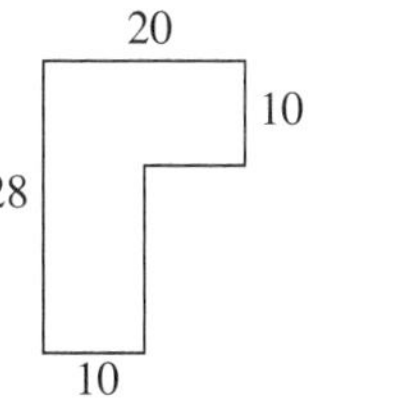

n

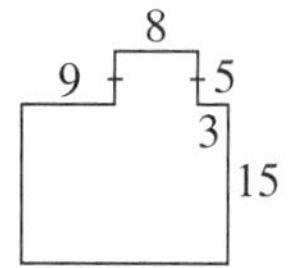

o

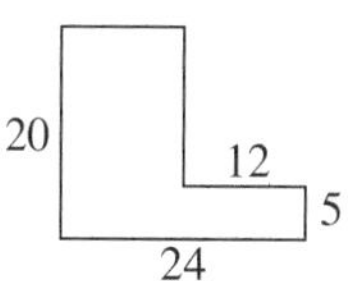

Area and volume

Excel Mathematics Study Guide Years 9–10
Pages 124–138

UNIT 6: Composite areas (2)

QUESTION **1** Find the area of the following, correct to the nearest cm².

a

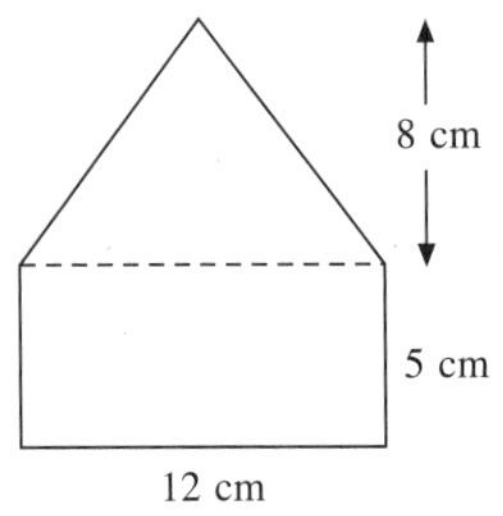

b

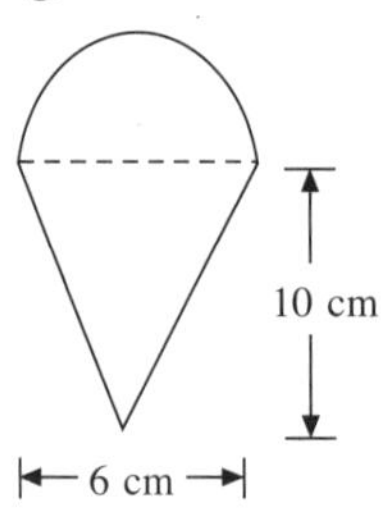

c

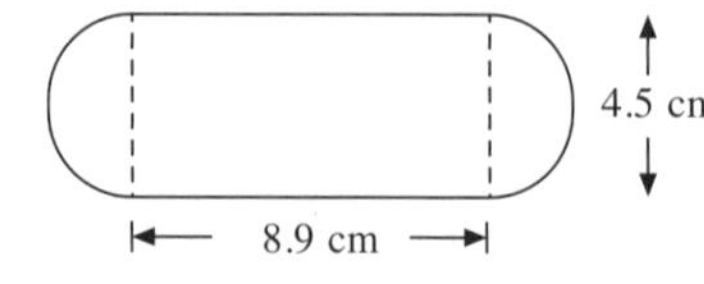

d

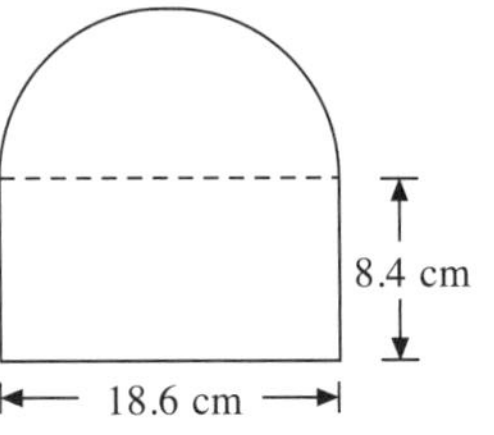

e

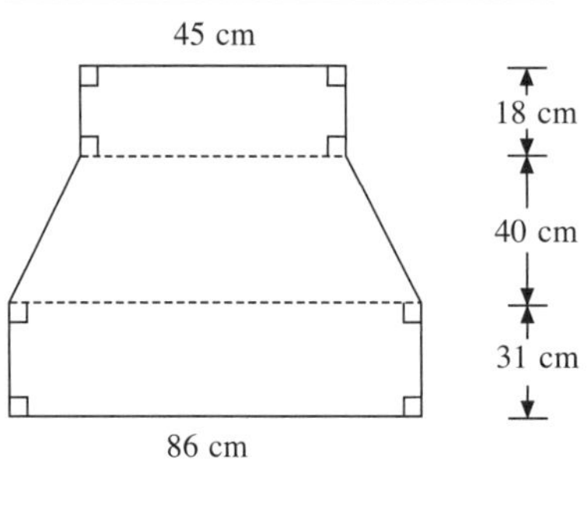

f

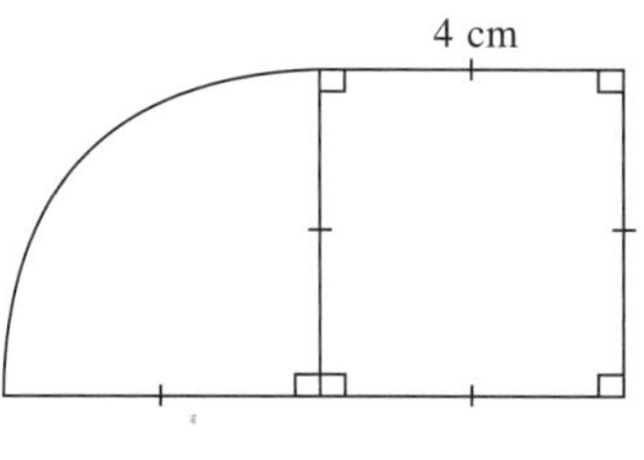

QUESTION **2** The rectangle in the diagram is inscribed inside the circle.

a Find the diameter of the circle.

b Find the area shaded to 1 decimal place.

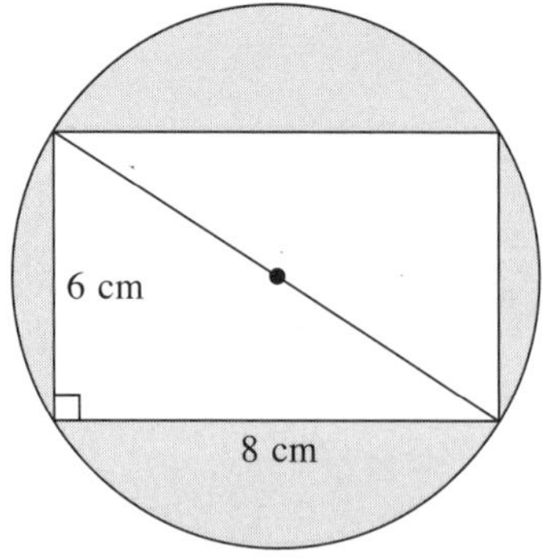

Area and volume

UNIT 7: Shaded areas (1)

QUESTION **1** Find the shaded area of each shape (all measurements are in cm).

a

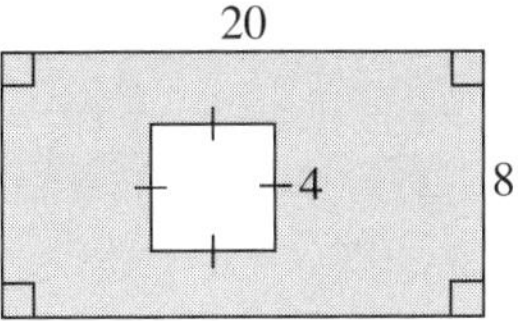

b

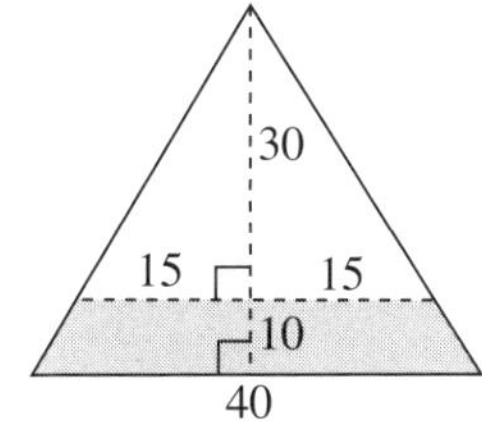

c

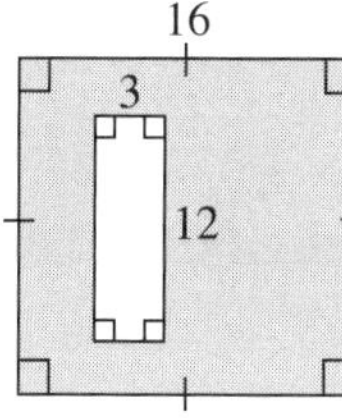

d

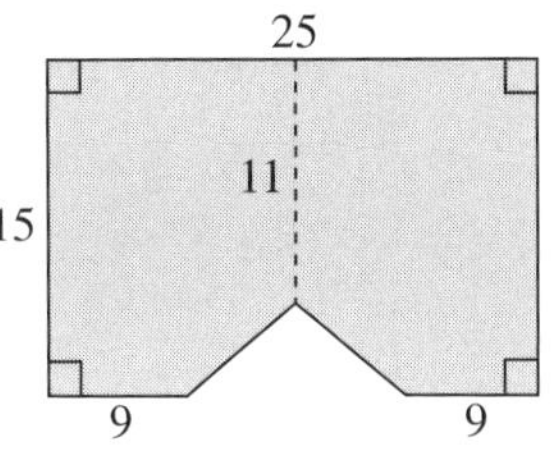

e

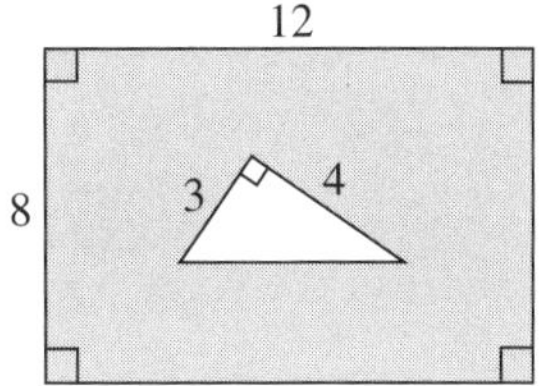

f

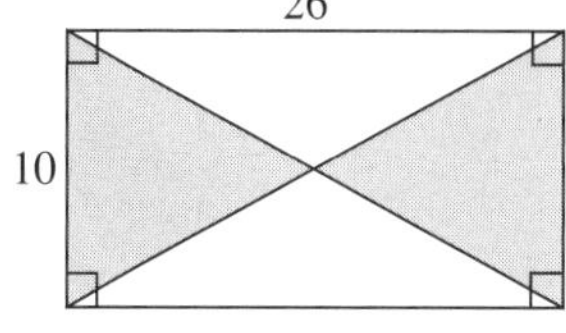

QUESTION **2** Find the shaded area correct to 1 decimal place.

a

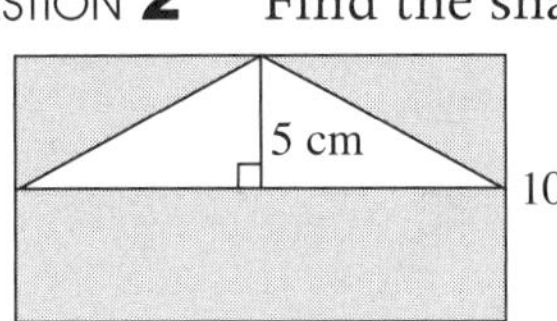

b

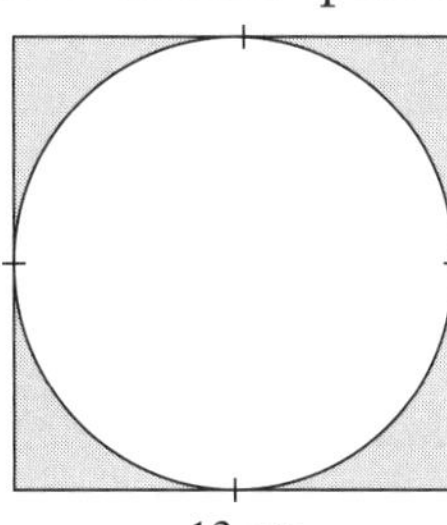

c

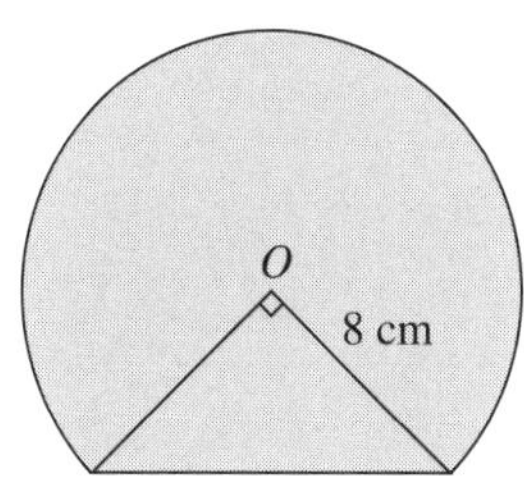

Area and volume

UNIT 8: Shaded areas (2)

Excel Mathematics Study Guide Years 9–10
Pages 124–138

QUESTION 1 Find the area of the following shaded shapes. All measurements are in centimetres.

a

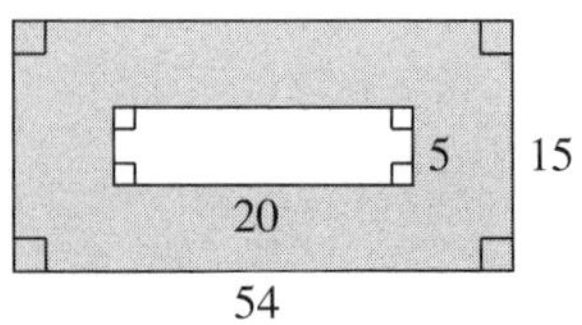

b

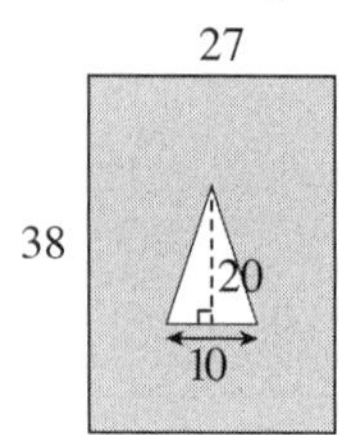

c

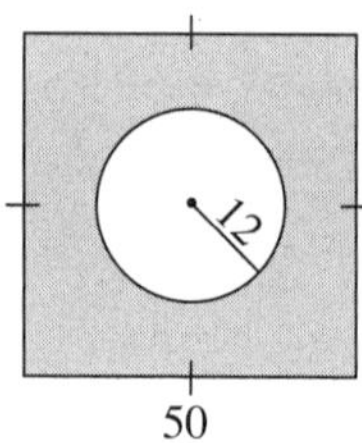

d

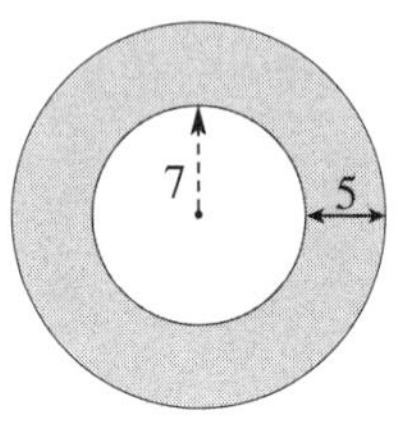

e

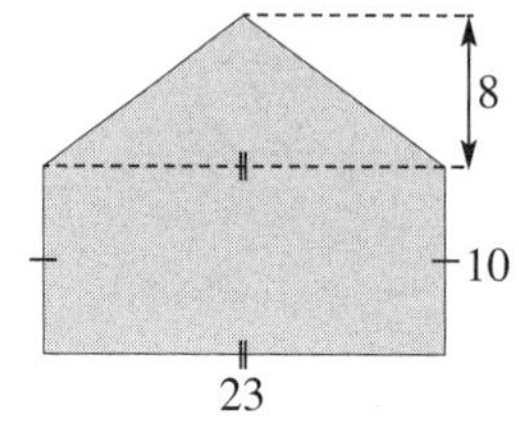

f

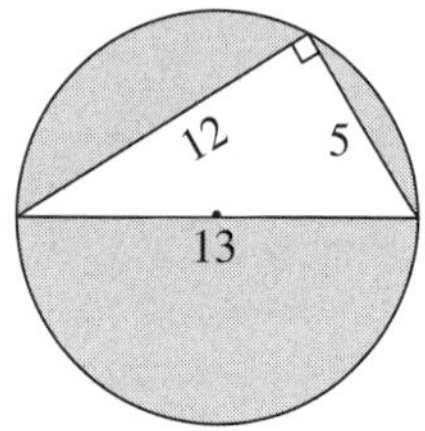

QUESTION 2 Find the area of each shaded shape.

a

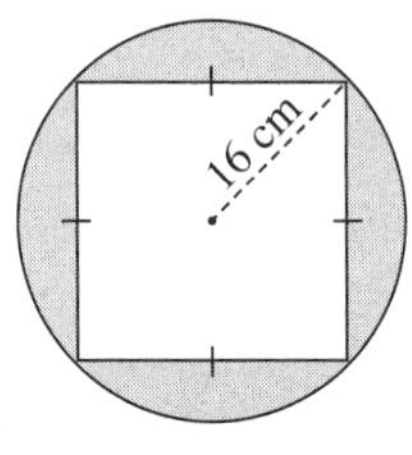

b

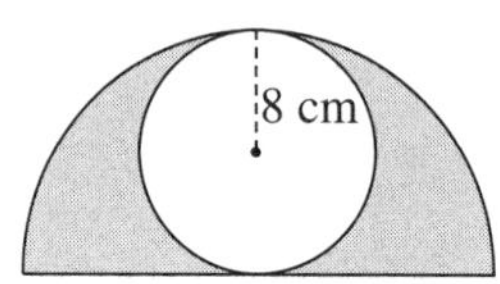

c

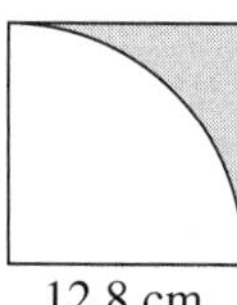

d

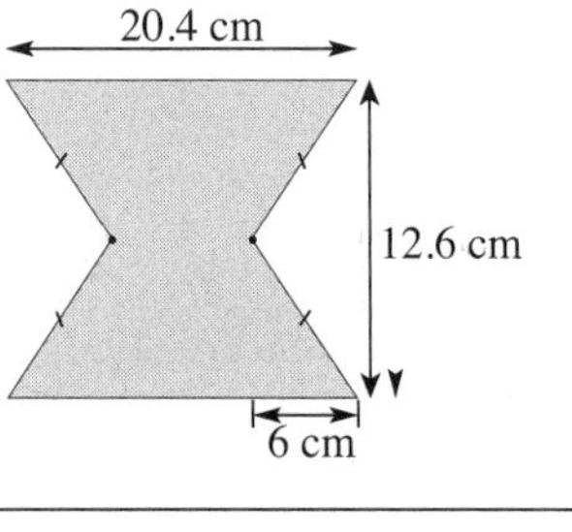

e

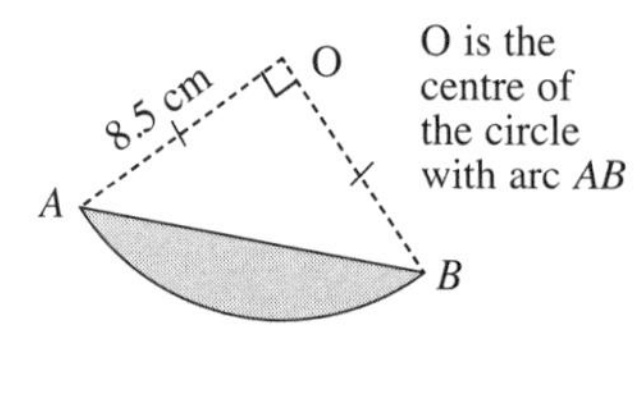

f

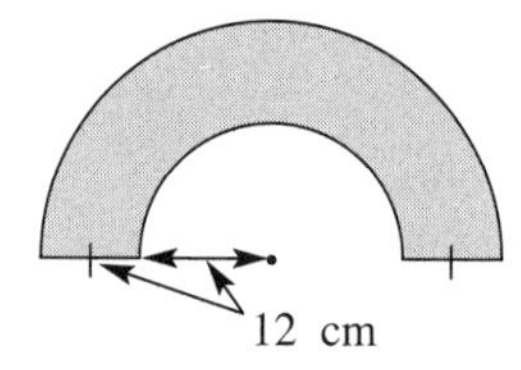

Area and volume

UNIT 9: Surface area of rectangular prisms

QUESTION 1 Find the surface area of each cube to 1 decimal place.

a

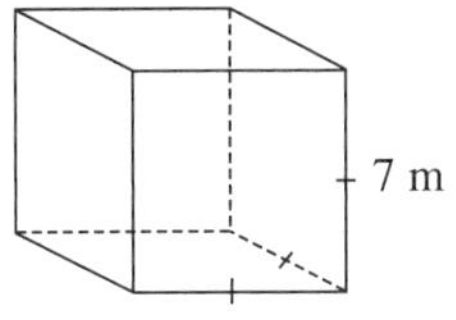

b

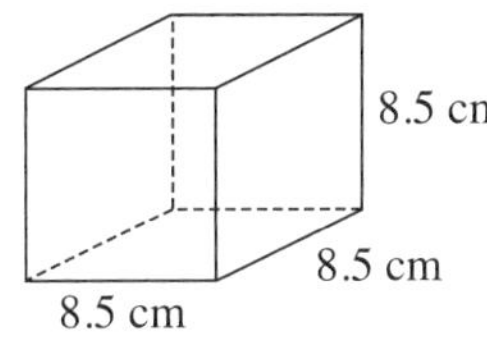

c

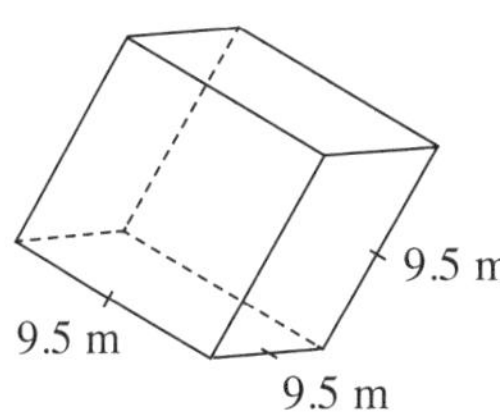

QUESTION 2 Find the surface area of each rectangular prism to 1 decimal place.

a

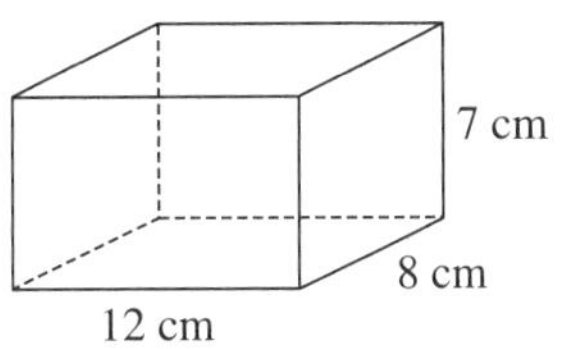

b

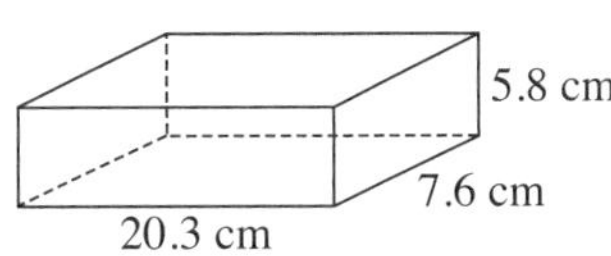

c

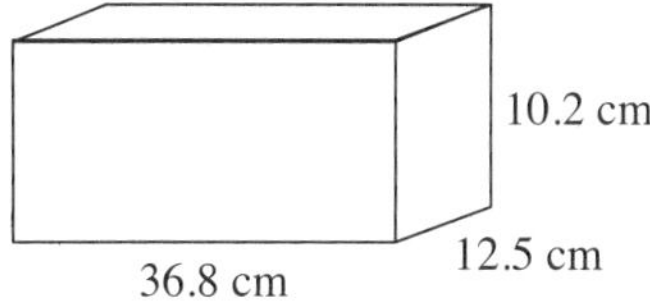

d

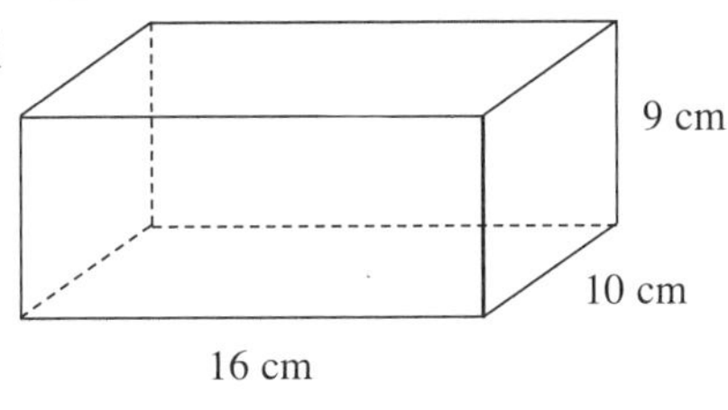

e

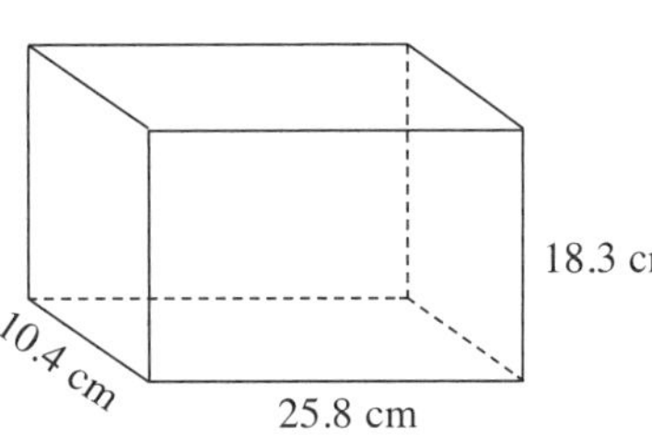

f

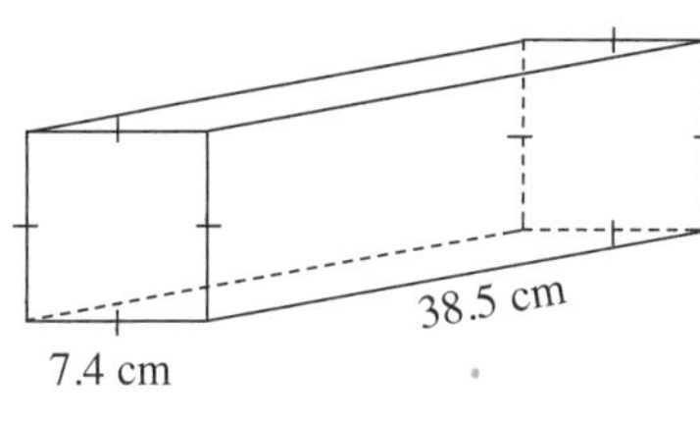

QUESTION 3 Find the surface area of these rectangular prisms with length (l cm), breadth (b cm) and height (h cm) to the nearest whole number.

a $l = 13,\ b = 15,\ h = 9$

b $l = 87,\ b = 65,\ h = 41$

c $l = 56,\ b = 28,\ h = 11$

Area and volume

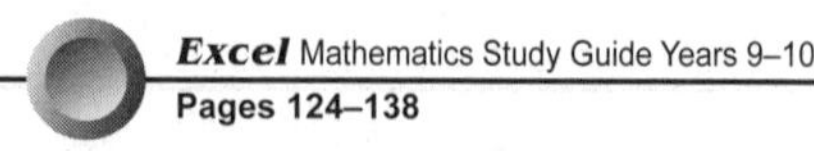

UNIT 10: Surface area of triangular prisms

QUESTION **1** Find the surface area of each triangular prism.

a

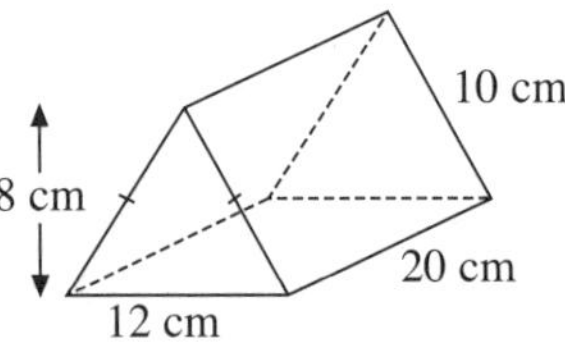

b

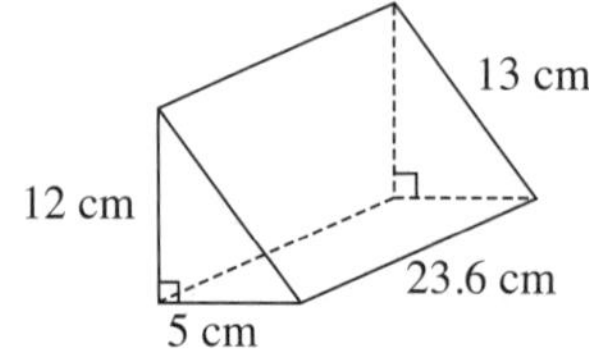

c

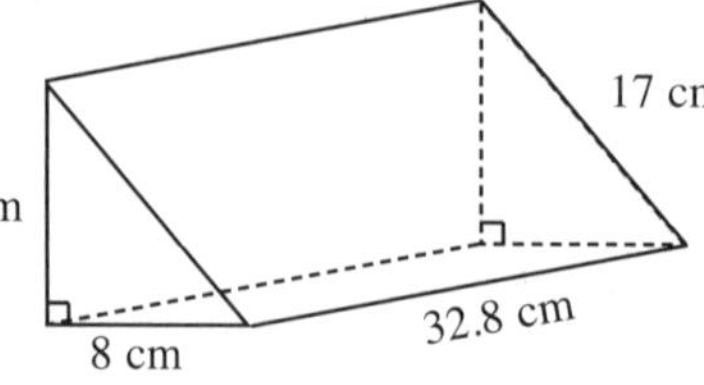

d

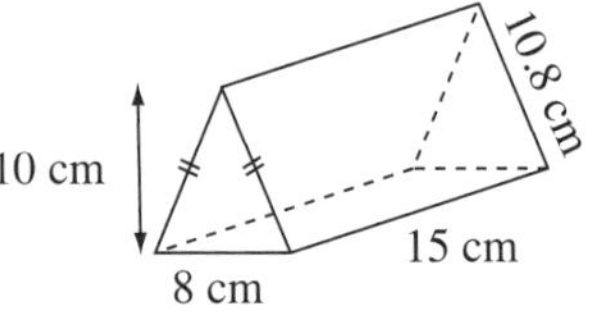

e

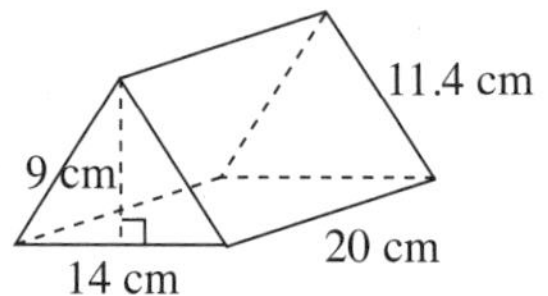

f

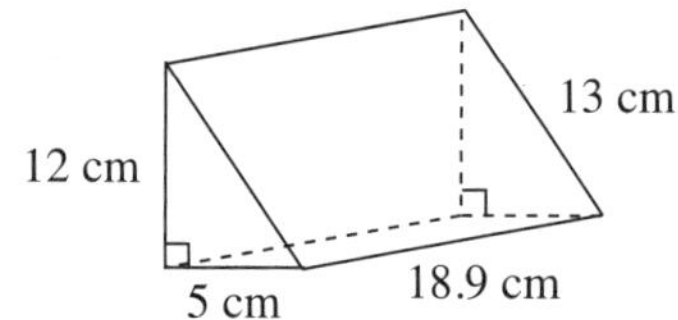

QUESTION **2** Find the surface area of these triangular prisms. Use Pythagoras' theorem to find the missing sides.

a

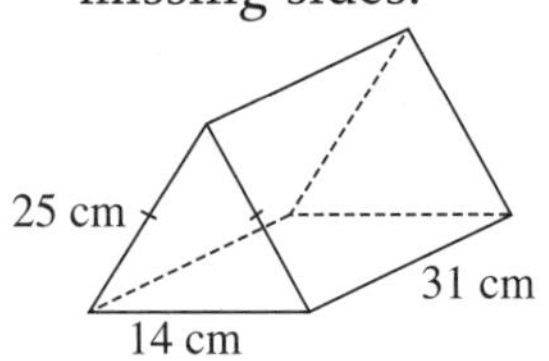

b

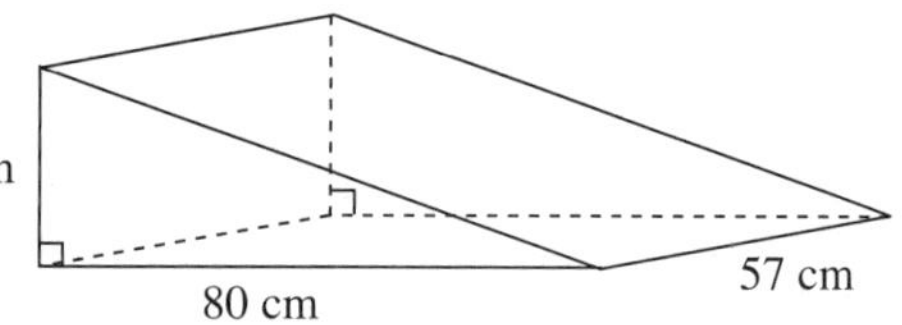

Area and volume

UNIT 11: Surface area of cylinders

QUESTION 1 For this cylinder find, to two decimal places, the area of the:

a circular base

b curved surface

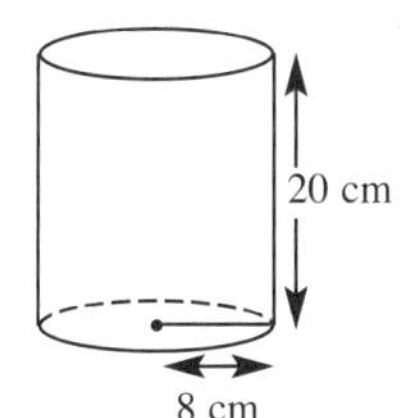

QUESTION 2 Calculate the surface area of the following closed cylinders correct to 2 significant figures. All measurements are in centimetres.

a

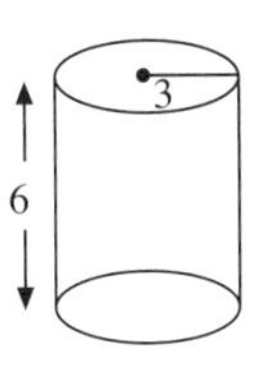

b

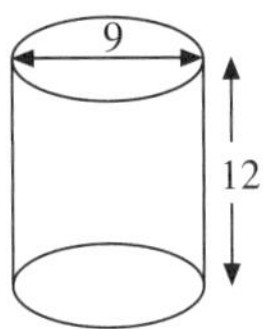

c

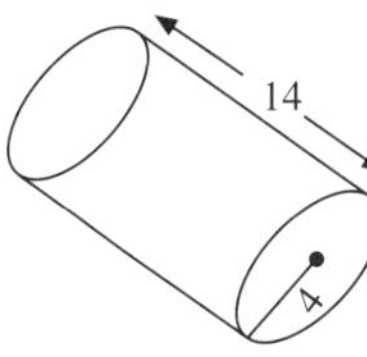

d

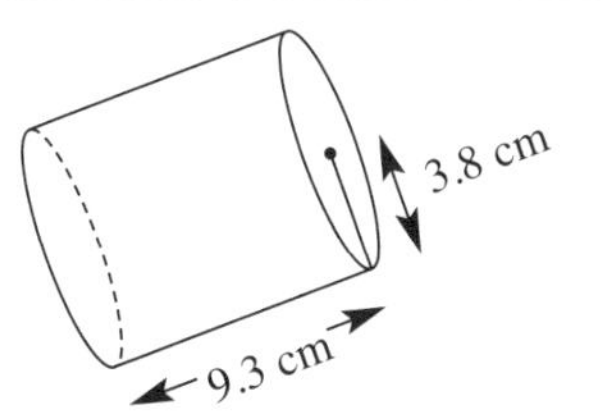

e

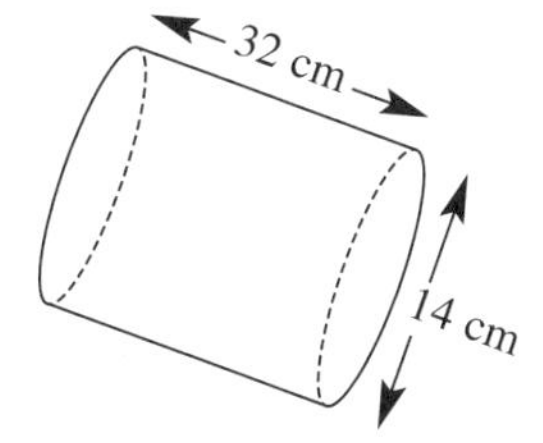

f

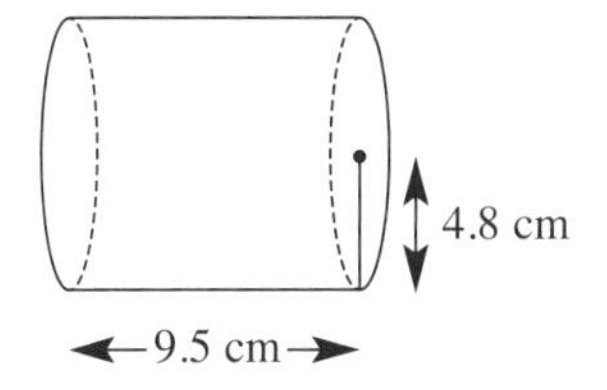

QUESTION 3 Calculate the exterior surface area of the following cylinders correct to 1 decimal place. All measurements are in centimetres.

a

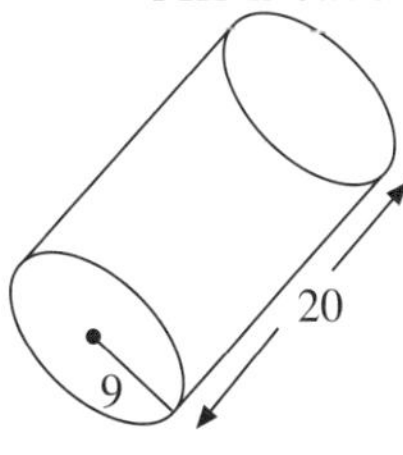

open at top

b

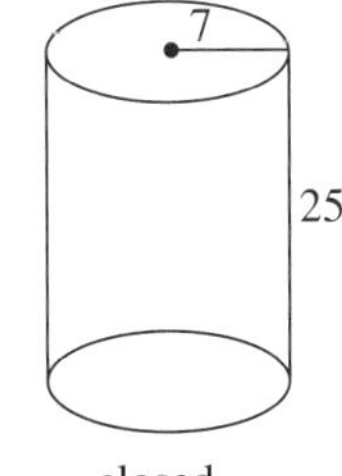

closed

c

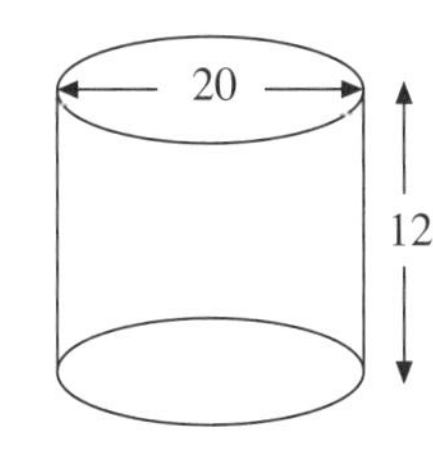

open at top and bottom

UNIT 12: Further surface area

QUESTION 1 Find the surface area of these prisms, given the area of the shaded face.

a

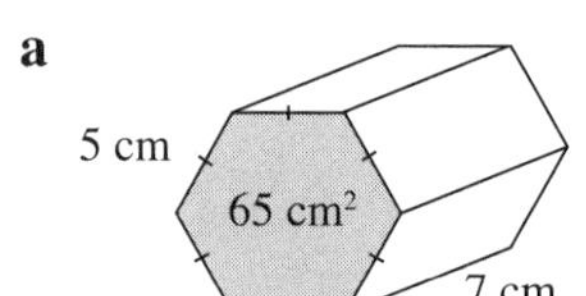

b

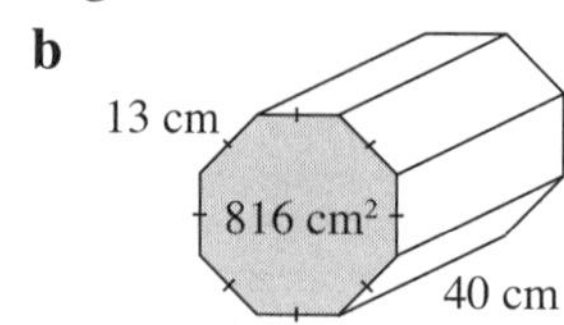

QUESTION 2 Find the surface area of these trapezoidal prisms.

a

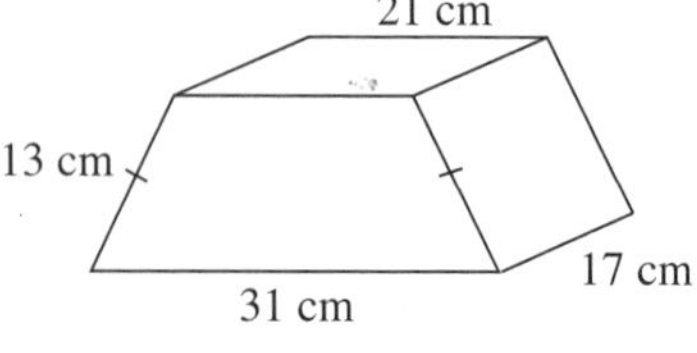

b

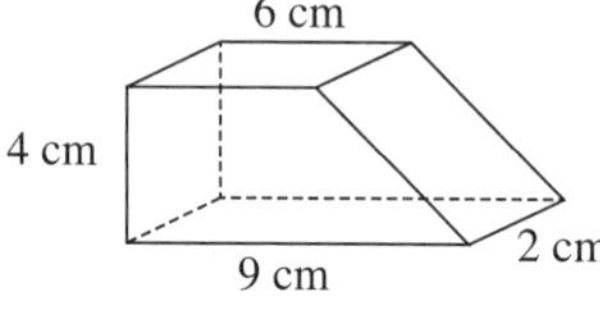

c

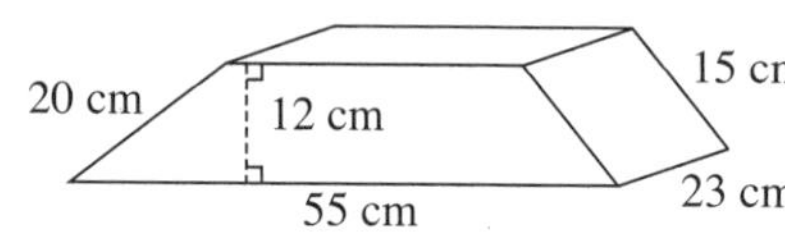

QUESTION 3 Find the surface area.

a

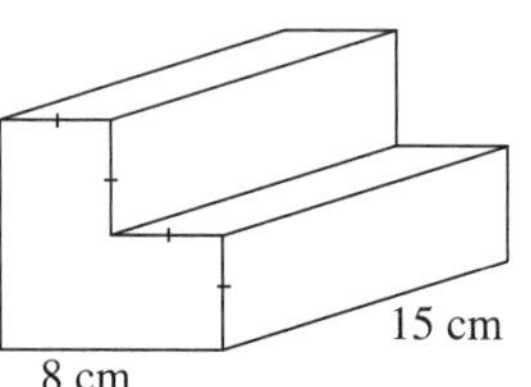

b

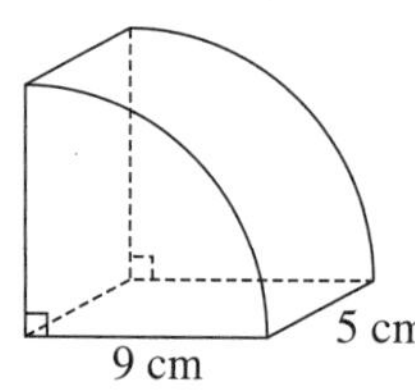

c

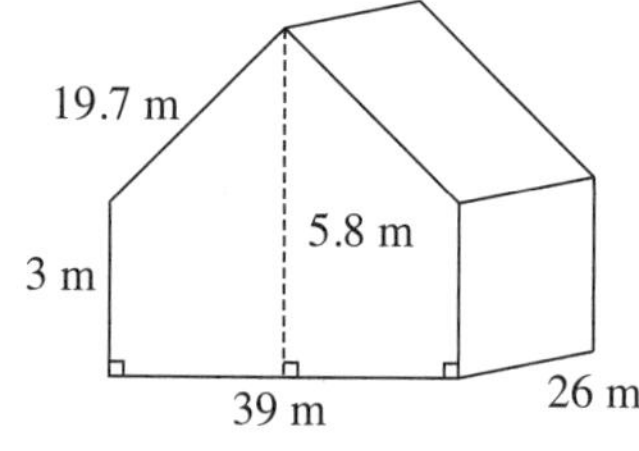

Area and volume

UNIT 13: Volume of right prisms (1)

QUESTION 1 Find the volume of each cube.

a

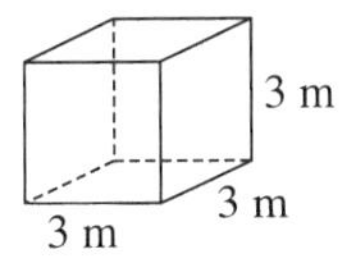

b

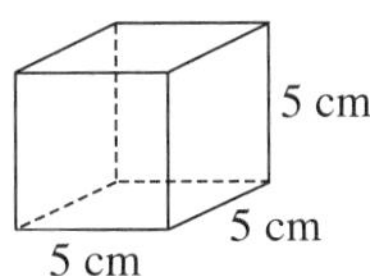

c

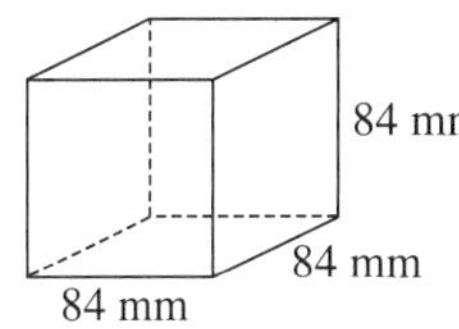

QUESTION 2 Find the volume of each rectangular prism.

a

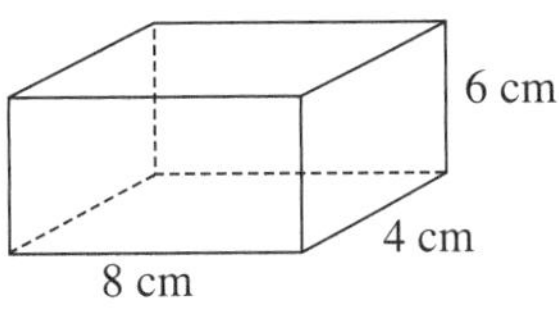

b

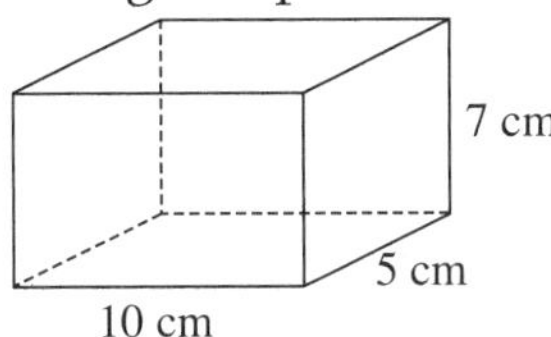

c

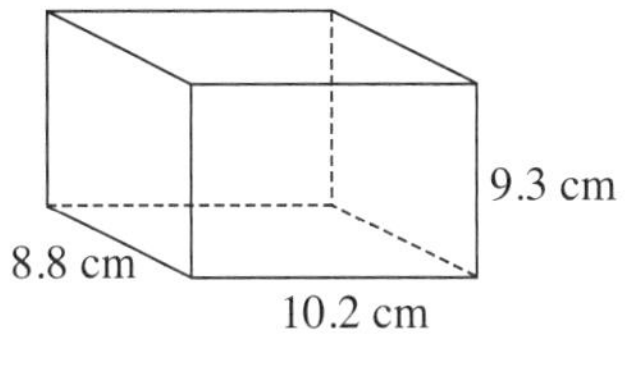

QUESTION 3 Find the volume of each prism, given the area of the shaded face.

a

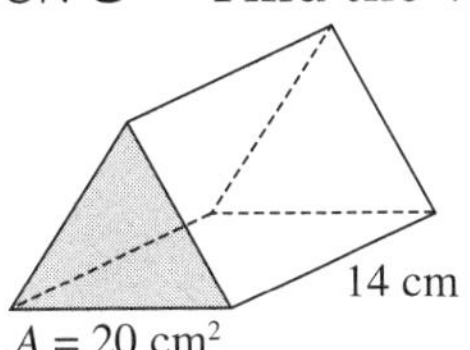

b

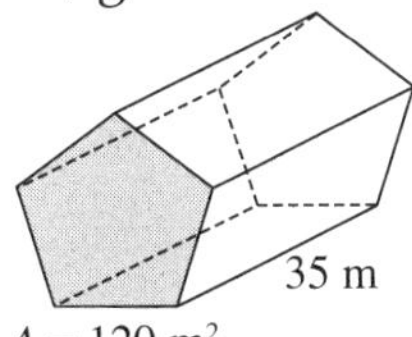

c

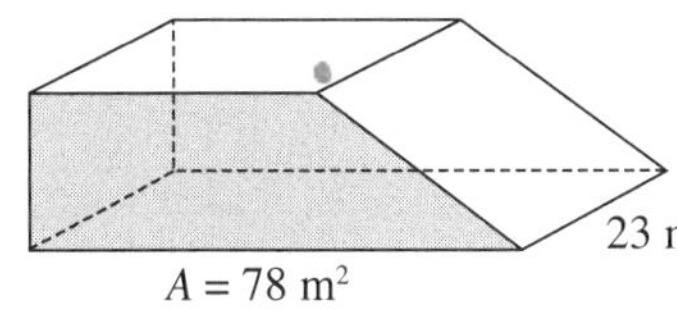

QUESTION 4 For the triangular prism, find:

a the area of the shaded face

b the volume of the prism

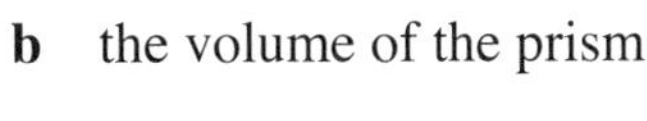

5 m

4 m

7 m

Excel Mathematics Study Guide Years 9–10
Pages 124–138

UNIT 14: Volume of right prisms (2)

QUESTION 1 Find the volume of these prisms (give answers to four significant figures).

a

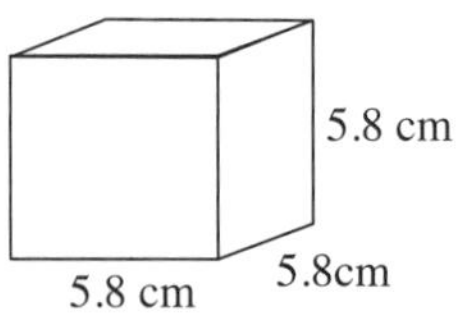

b

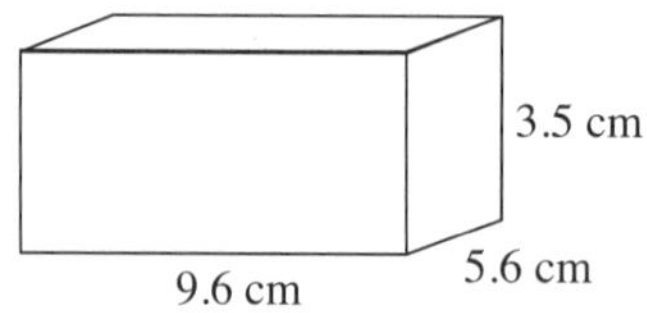

c

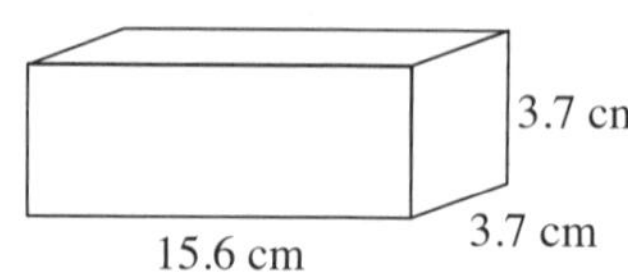

d

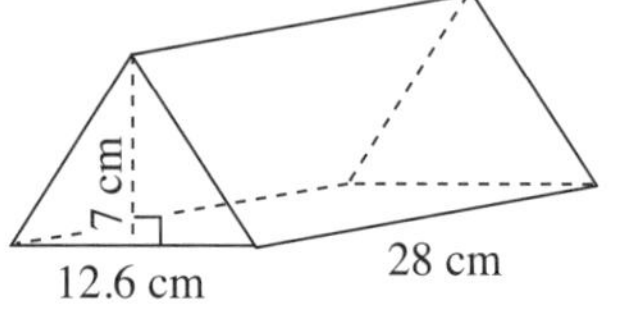

e

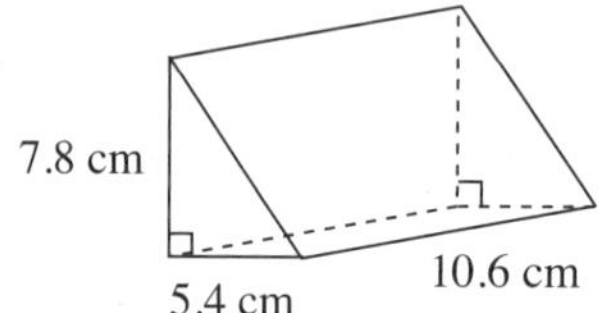

f

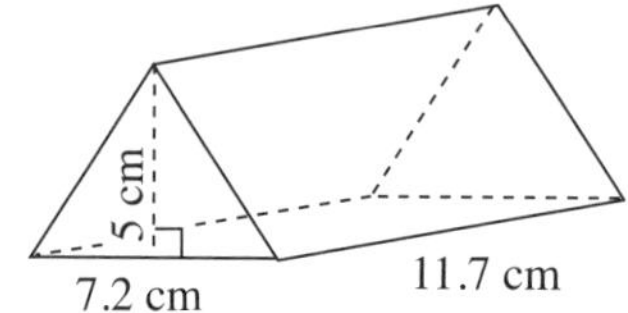

g

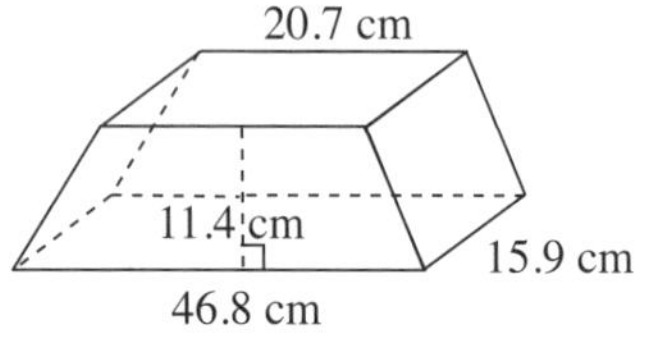

h

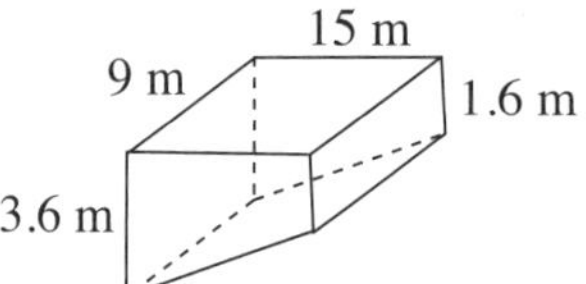

i

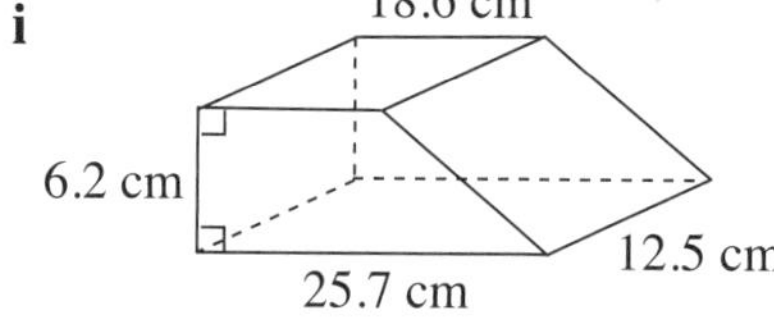

QUESTION 2 Find the volume of the following solids.

a

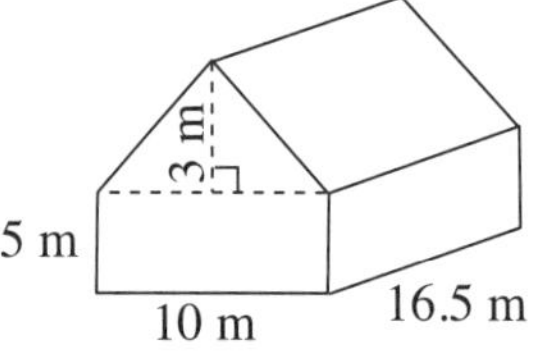

b

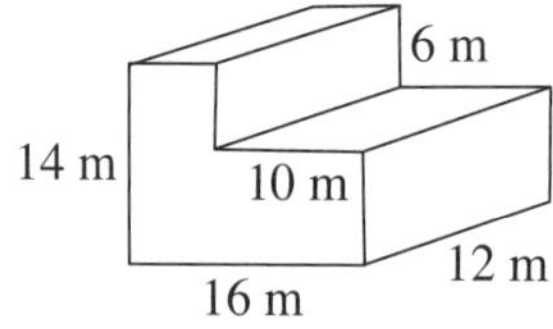

c

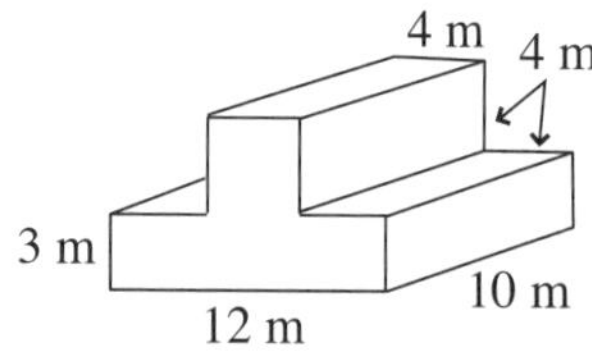

Area and volume

UNIT 15: Volume of cylinders (1)

QUESTION 1 Find the volume of each cylinder correct to two significant figures.

a radius 6 cm, height 20 cm

b radius 9.6 cm, height 18 cm

c radius 20.8 cm, height 30.4 cm

QUESTION 2 Find the volume of each cylinder correct to two decimal places.

a

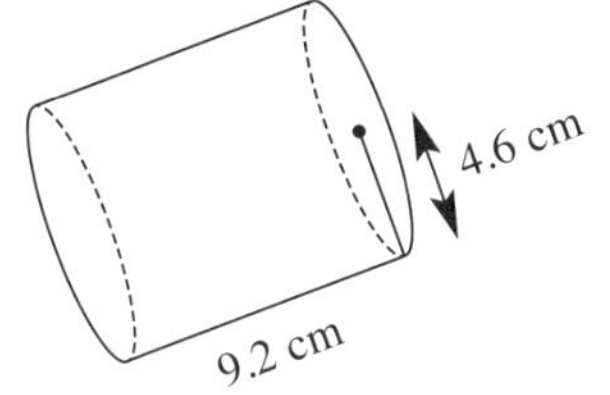

b

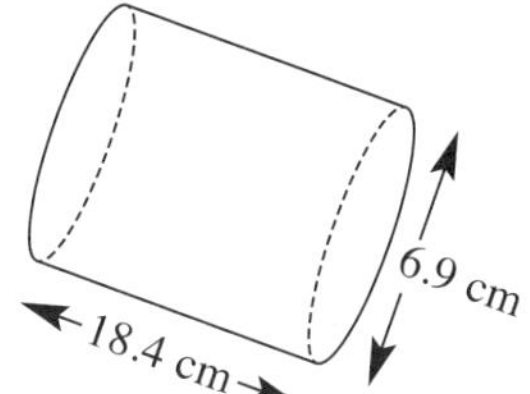

c

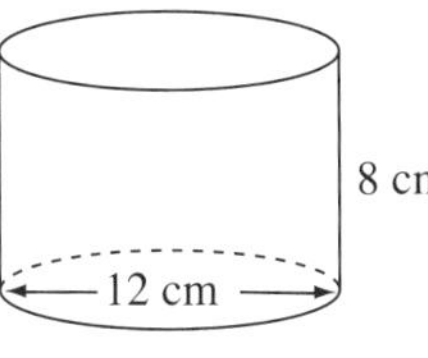

d

12.5 cm

7 cm

e

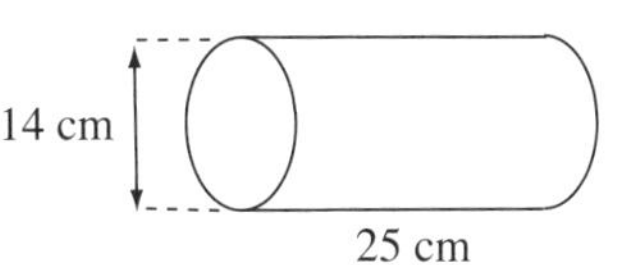

f

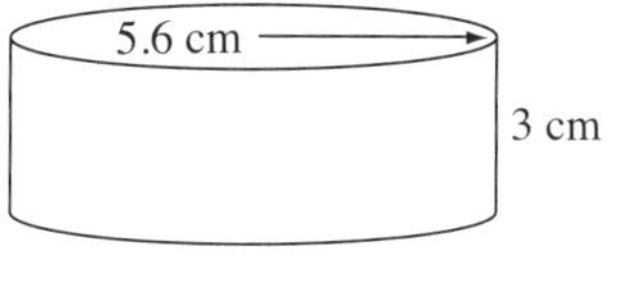

QUESTION 3 Find the volume of this cylinder in cubic metres correct to three significant figures.

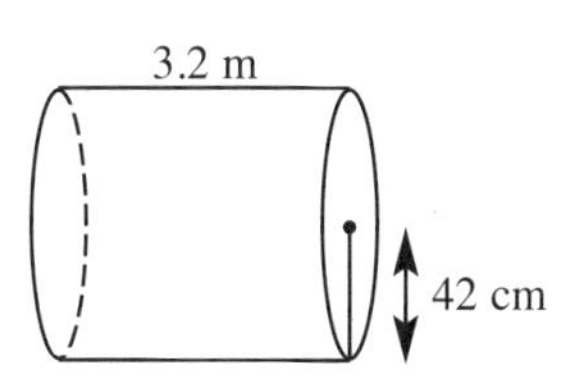

Area and volume

Excel Mathematics Study Guide Years 9–10
Pages 124–138

UNIT 16: Volume of cylinders (2)

QUESTION 1 Find the volume in cubic centimetres, correct to one decimal place, of a soft drink can with height 115 mm and radius 30 mm.

QUESTION 2 Without actually calculating the volume determine which of these cylinders has the larger volume. Justify your answer.

A

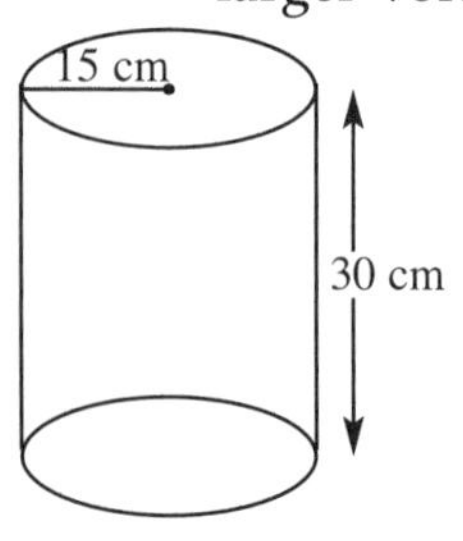

B

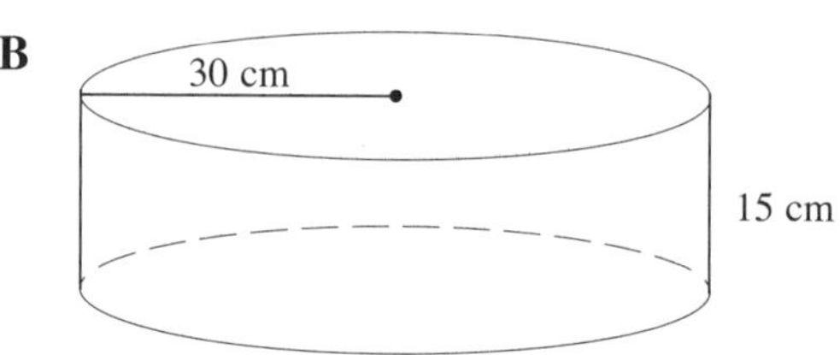

QUESTION 3 Find the volume of the following solids (correct to one decimal place).

a

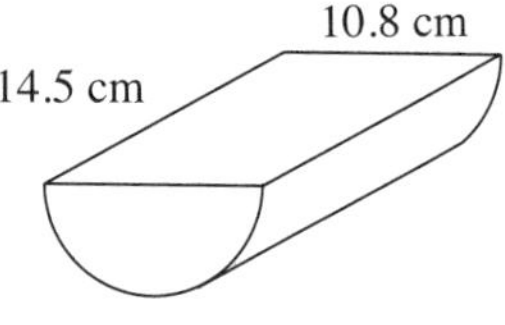

b

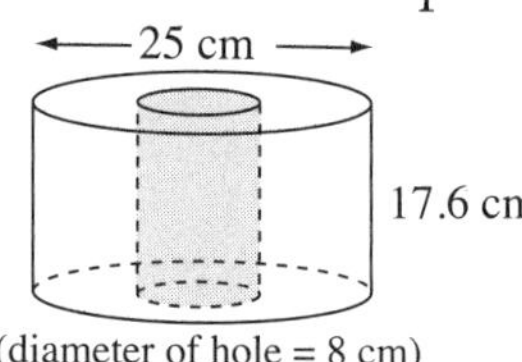

c

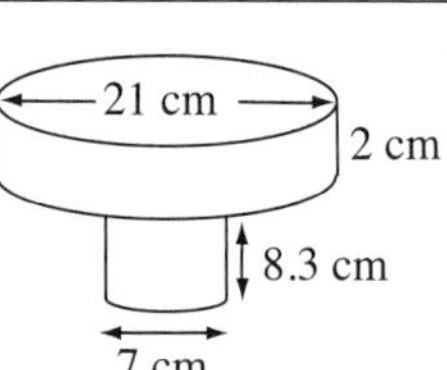

d

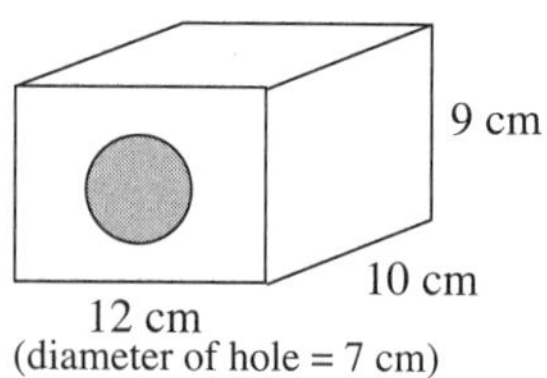

Area and volume

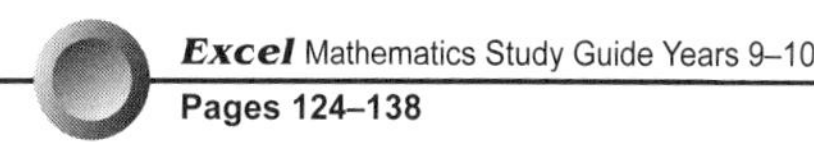

UNIT 17: Problem solving (1)

QUESTION 1 Complete the following.

a 1 cm^3 = ______________ mL **b** 1000 cm^3 = ______________ L **c** 1 m^3 = ______________ L

QUESTION 2 A pot has a volume of 15 000 cm^3. How many litres of water can it hold?

QUESTION 3 A milk carton measures 7.1 cm by 7.1 cm by 19.9 cm.

a What is the volume of the carton to the nearest cubic centimetre?

b What volume of milk does this container hold? Round this value to a commonly used capacity.

QUESTION 4 A rectangular roof is 28 m long and 12 m wide.

a What volume of water, in m^3 will fall on the roof if we receive 20 mm of rain?

b A tank catches all the rain that falls on the roof. How many litres of water will flow into the tank from 20 mm of rain?

c The tank holds 35 000 litres. How much rain, to the nearest mm, would need to fall to fill the tank if it is empty and only catches rain from the above roof?

QUESTION 5 A rectangular swimming pool with uniform depth is 30 metres long, 8 metres wide and 2.8 metres deep. It is to be tiled. Calculate the following.

a the cost of tiling it at $53 per square metre

b its capacity in litres

Area and volume

UNIT 18: Problem solving (2)

QUESTION **1** Round your answers to a reasonable number of decimal places where necessary.

a What is the volume of this cylindrical water storage tank?

b A farm household used 180 L of water, on average, each day. How long would a tankful of water last from this tank?

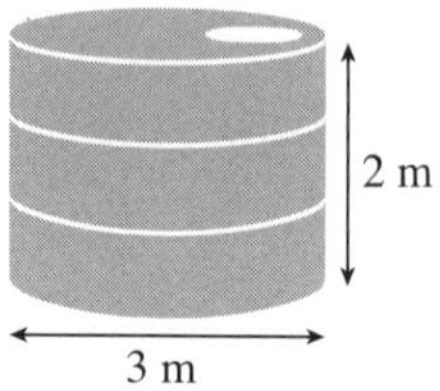

2 m

3 m

QUESTION **2** A water pipe's diameter is 5 cm. Water flows through it at the rate of 8 cm each second. How much water, in litres, will pass through:

a in each minute?

b in each hour?

QUESTION **3** A rectangular fish tank is 25 cm wide, 85 cm long and 45 cm high. It is filled with water to within 3 cm of the top.

a How much water does it hold?

b It is recommended that a certain species of fish have at least 8 L of water each in which to swim. What is the maximum number of these fish that should be placed in the tank?

QUESTION **4** A gold bar has the shape of a rectangular prism as shown.

a What is the volume of gold in the bar?

b The mass of each cubic centimetre of gold is 19.3 g. What is the mass of this gold bar?

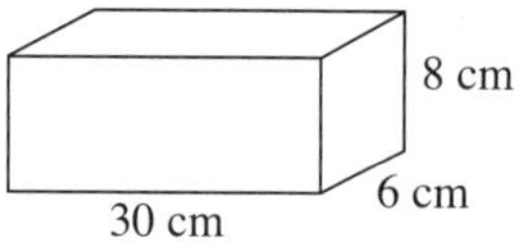

8 cm

6 cm

30 cm

QUESTION **5** A pipe is to be made from copper with the dimensions shown.

a What volume of copper is in a 1 m length of pipe?

b What is the mass of this length of copper, given that the mass of 1 cm^3 of copper is 8.96 g?

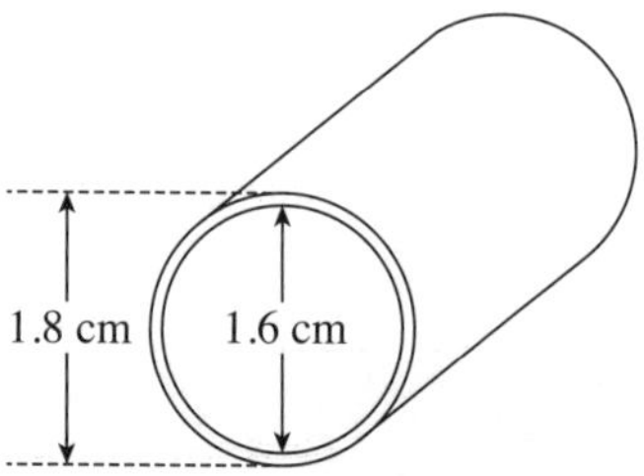

1.8 cm

1.6 cm

Area and volume

TOPIC TEST — PART A

Instructions
- This part consists of 10 multiple-choice questions.
- Fill in only ONE CIRCLE for each question.
- Each question is worth 1 mark.

Time allowed: 10 minutes — **Total marks: 10**

Marks

1 A rectangular prism is 10 cm long, 8 cm wide and 4 cm high. Its surface area is — 1

(A) 152 cm^2 (B) 304 cm^2 (C) 320 cm^2 (D) 640 cm^2

2 Give the total surface area in cm^2 correct to one decimal place of a closed cylinder with dimensions of radius 6 cm and height 15 cm. — 1

(A) 226.2 cm^2 (B) 565.5 cm^2 (C) 791.7 cm^2 (D) 678.6 cm^2

3 A cube has a volume of 729 cm^3. Find the length of each side of the cube. — 1

(A) 6 cm (B) 9 cm (C) 18 cm (D) 27 cm

4 A cylinder has height 9 m and radius 6 m. Its volume is closest to — 1

(A) 113 m^3 (B) 452 m^3 (C) 2036 m^3 (D) 1018 m^3

5 The volume of a rectangular prism with base area of 75 cm^2 and vertical height of 8 cm is — 1

(A) 200 cm^3 (B) 400 cm^3 (C) 600 cm^3 (D) 800 cm^3

6 A triangular prism has base area 24 m^2 and perpendicular height 4 m. What is its volume? — 1

(A) 48 m^3 (B) 96 m^3 (C) 24 m^3 (D) 32 m^3

7 What is the surface area of a cube of side length 7 cm? — 1

(A) 196 cm^2 (B) 245 cm^2 (C) 294 cm^2 (D) 343 cm^2

8 The shaded area shown is closest to — 1

(A) 298 m^2 (B) 20 m^2
(C) 101 m^2 (D) 75 m^2

7 m
12 m

9 Which is closest to the curved surface area of a cylinder with diameter 23 cm and height 20 cm? — 1

(A) 723 cm^2 (B) 2890 cm^2 (C) 8310 cm^2 (D) 1445 cm^2

10 The shaded area shown is closest to — 1

(A) 9.8 cm^2 (B) 4.9 cm^2
(C) 45.4 cm^2 (D) 40.5 cm^2

70°
4 cm

Total marks achieved for PART A — /10

Area and volume

TOPIC TEST — PART B

Time allowed: 20 minutes — **Total marks: 15**

Marks

1 For a closed cylinder of height 16.4 cm and diameter 12.8 cm, find to two decimal places:

a area of circular base **b** the total surface area **c** its volume

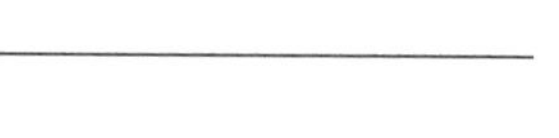

3

2 Find the volume of each of these solids.

a 2 m, 4 m, 80 cm

b 3.5 m, 4.2 m, 3.1 m

c

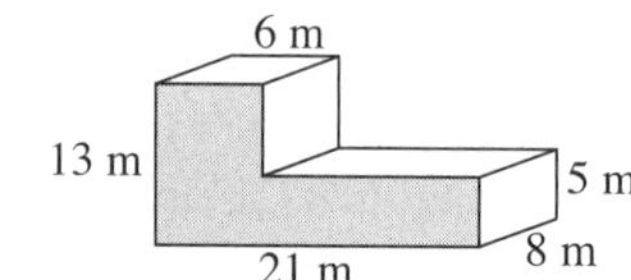

3

3 Find the surface area of each of the solids in question 2.

a **b** **c**

3

4 Find the shaded area to one decimal place.

a

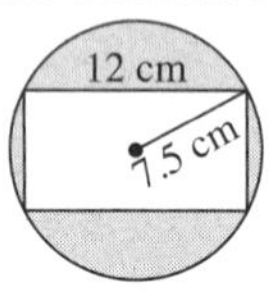

b

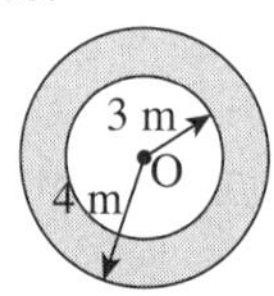

c

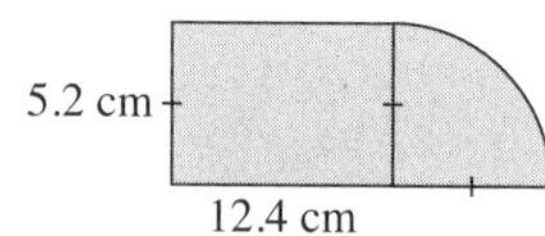

3

5 A swimming pool has the shape of a trapezoidal prism as shown.

28.6 m, 2 m, 1.2 m, 9.8 m

a Find the volume of the pool in m^3

b The mass of 1 kL of water is 1 t. How many tonnes of water can the pool hold?

c Over a period of hot weather, the level of the pool dropped by 50 cm, The pool was originally full. What volume of water, in litres, evaporated?

3

Total marks achieved for PART B

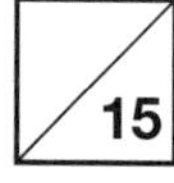

CHAPTER 8

Similarity

Excel Mathematics Study Guide Years 9–10
Pages 151–153

UNIT 1: The enlargement factor

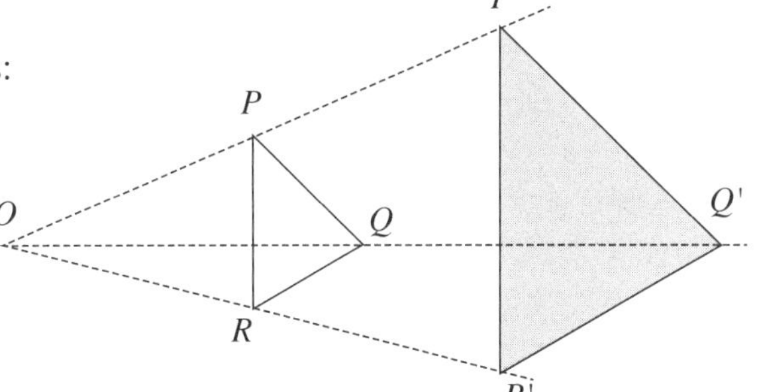

QUESTION 1 $\Delta P'Q'R'$ is an enlargement of ΔPQR.

a Use a ruler to measure the length, in millimetres, of sides:

i PQ ________ **ii** $P'Q'$ ________

iii QR ________ **iv** $Q'R'$ ________

v PR ________ **vi** $P'R'$ ________

b Find in simplest form.

i $\frac{P'Q'}{PQ}$ = ____________ **ii** $\frac{Q'R'}{QR}$ = ____________ **iii** $\frac{P'R'}{PR}$ = ____________

iv $\frac{O'P'}{OP}$ = ____________ **v** $\frac{O'Q'}{OQ}$ = ____________ **vi** $\frac{O'R'}{OR}$ = ____________

c What is the enlargement factor? ________ **d** Where is the centre of enlargement? ________

QUESTION 2 $ABCD$ has been reduced.

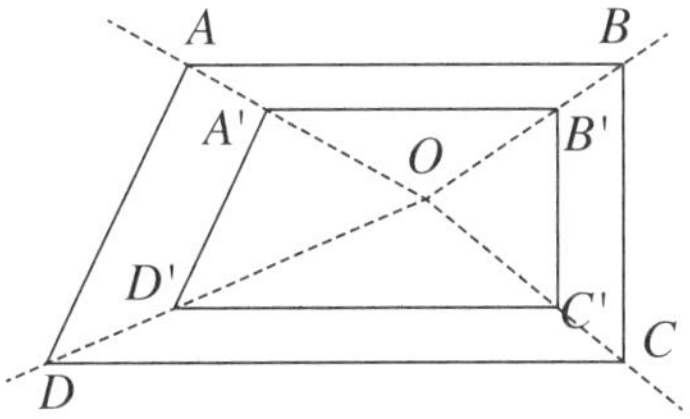

a Find the length (in millimetres) of:

i AB ________ **ii** BC ________

iii CD ________ **iv** DA ________

v $A'B'$ ________ **vi** $B'C'$ ________ **vii** $C'D'$ ________ **viii** $D'A'$ ________

b Find in simplest form.

i $\frac{A'B'}{AB}$ = ________ **ii** $\frac{B'C'}{BC}$ = ________ **iii** $\frac{C'D'}{CD}$ = ________ **iv** $\frac{D'A'}{DA}$ = ________

v $\frac{O'A'}{OA}$ = ________ **vi** $\frac{O'B'}{OB}$ = ________ **vii** $\frac{O'C'}{OC}$ = ________ **viii** $\frac{O'D'}{OD}$ = ________

c What is the enlargement factor? ____________

QUESTION 3 Briefly explain what it means if an enlargement factor is between 0 and 1.

____

QUESTION 4 Complete each enlargement, given the centre of enlargement, O, if the enlargement factor is:

a $\frac{1}{3}$

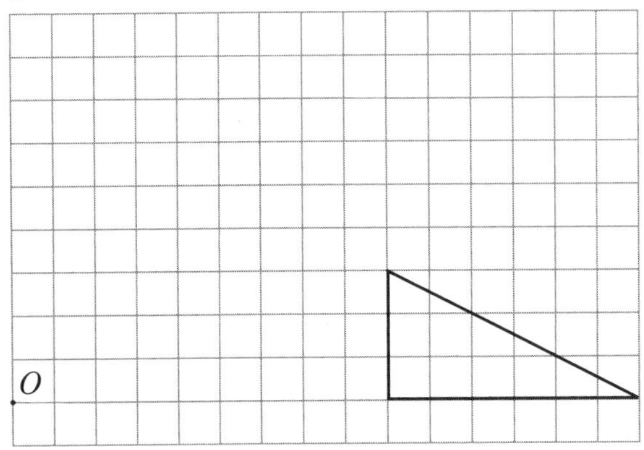

b 3

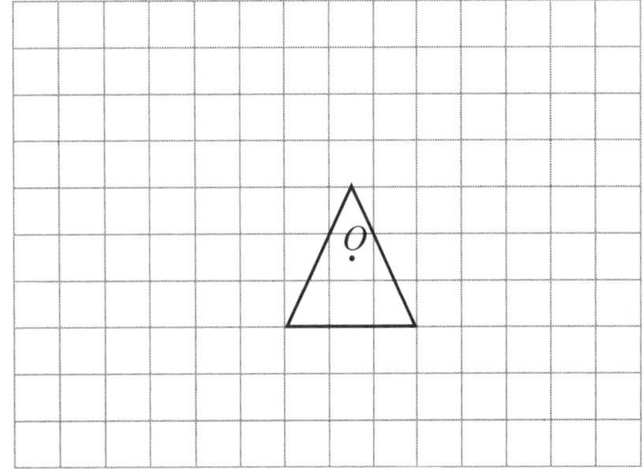

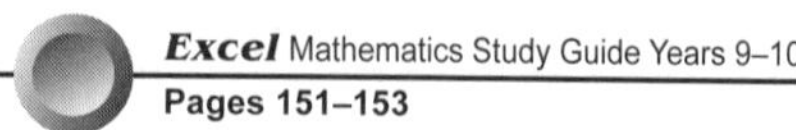

UNIT 2: Further enlargement factors

QUESTION 1 A diagram that was 6 cm long and 4.5 cm wide, has been enlarged by a factor of 2. What are its new dimensions?

QUESTION 2 A drawing was 18 cm long and 13.2 cm wide. If it was reduced by a factor of 3, what will be its new length and width?

QUESTION 3 A diagram was not thought to be large enough and so was enlarged by a factor of 4. If it is now 26 cm long and 18 cm high, what were its original dimensions?

QUESTION 4 Two triangles are congruent. The first triangle has a base of length 19 cm and a height of 13 cm. For the second triangle, what is:

a the length of its base?

b its height?

QUESTION 5 Two rectangles are similar. The first rectangle is 9 cm long and 4 cm high. The second rectangle is 45 cm long.

a What is the enlargement factor?

b How high is the second rectangle?

QUESTION 6 Each side of a regular hexagon is 6 cm long. If the hexagon is enlarged by a factor of 4 and then reduced by a factor of 3, how long will each side be?

QUESTION 7 A triangle has sides of length 30 cm, 72 cm and 78 cm. It is reduced to $\frac{2}{3}$ the size. For the reduced triangle, what is the length of:

a the shortest side?

b the longest side?

QUESTION 8 A design is 27.6 cm long and 15.6 cm wide. The design is too large and is reduced so that the length is 20.7 cm.

a What is the reduction factor?

b What is the width of the reduced design?

Excel Mathematics Study Guide Years 9–10
Pages 151–153

UNIT 3: Properties of similar figures (1)

QUESTION 1 $ABCD$ and $EFGH$ are similar figures.

a What is the enlargement factor? ____________

b Measure the size of each angle to the nearest degree.

i $\angle ABC =$ ______ **ii** $\angle BCD =$ ______

iii $\angle CDA =$ ______ **iv** $\angle DAB =$ ______

v $\angle EFG =$ ______ **vi** $\angle FGH =$ ______

vii $\angle GHE =$ ______ **viii** $\angle HEF =$ ______

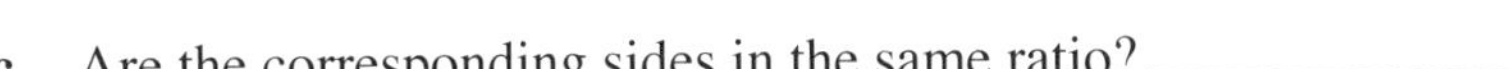

c Are the corresponding sides in the same ratio? ____________

d Are the corresponding angles equal? ____________

e What side of $EFGH$ corresponds to side AD of $ABCD$? ____________

f Which angle of $ABCD$ corresponds to $\angle FGH$ of $EFGH$? ____________

QUESTION 2 These two triangles are similar.

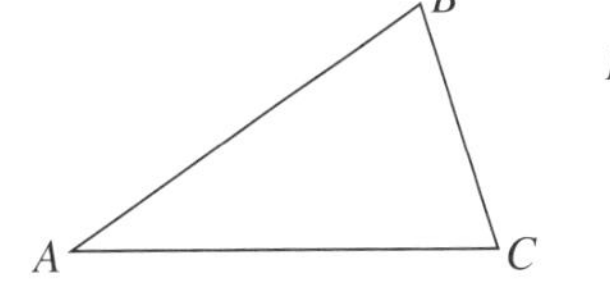

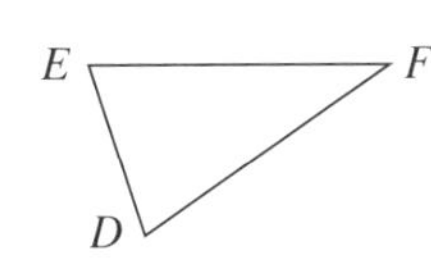

a Measure the size of each angle and show the size on the diagram.

b Which angle of ΔDEF corresponds to:

i $\angle ABC$? ______ **ii** $\angle CAB$? ______ **iii** $\angle BCA$? ______

c Which side of ΔABC corresponds to:

i DE? ______ **ii** DF? ______ **iii** EF? ______

QUESTION 3 This pair of figures is similar.

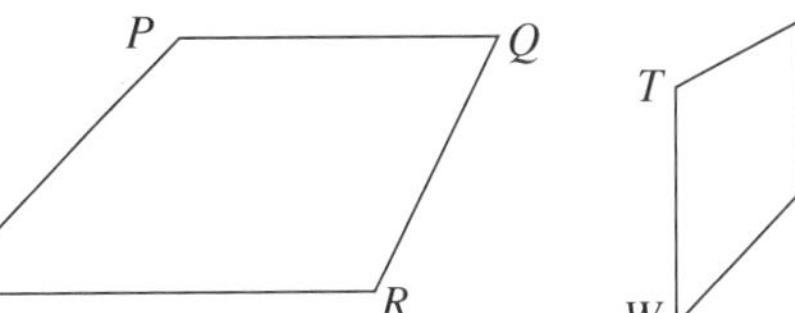

a List the corresponding angles (showing their size).

____________________ ____________________

____________________ ____________________

____________________ ____________________

____________________ ____________________

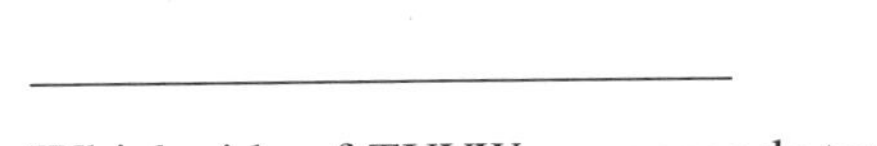

b Which side of $TUVW$ corresponds to:

i PQ? ______ **ii** QR? ______ **iii** RS? ______ **iv** SP? ______

QUESTION 4 Complete the following.

a If two figures are similar then the ____________ angles are ____________.

b If two figures are similar then the lengths of the ____________ sides are in the same ____________.

Similarity

UNIT 4: Properties of similar figures (2)

QUESTION 1 State whether the following statements are true or false.

a If two figures are similar, they are the same shape. ______________

b If two figures are similar, they are the same size. ______________

c If two figures are similar, the corresponding angles must be equal. ______________

d If two figures are similar, the corresponding sides must be equal. ______________

e If two similar figures have a scale factor of 2, then each side of the second figure is twice as long as the corresponding side of the first figure. ______________

f If two similar figures have a scale factor of 3, then each side of the second figure is three units longer than the corresponding side of the first figure. ______________

g If two similar figures have a scale factor of 1, they are congruent. ______________

h If two figures are congruent they are the same shape and the same size. ______________

i An enlargement factor of $\frac{1}{2}$ is the same as a reduction factor of 2. ______________

QUESTION 2 Darren drew this design. 'It makes use of similar figures,' he commented. Do you agree? Briefly comment.

____

____

QUESTION 3 List some of the similar figures that appear in the design of this building.

____

____

____

____

____

____

____

____

____

____

QUESTION 4 List a few places where you might see similar figures in everyday life.

____

____

____

____

Similarity

UNIT 5: Similar figures

QUESTION 1 For the following similar figures, list the pairs of corresponding sides.

a

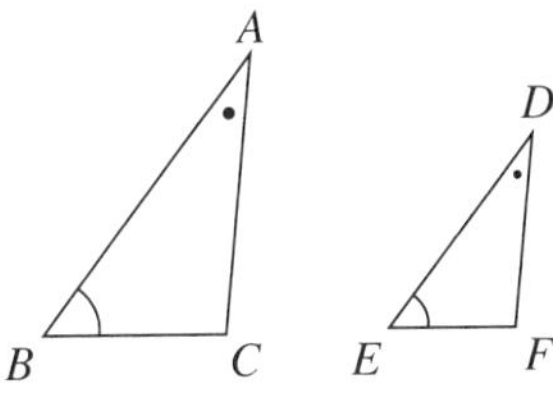

b

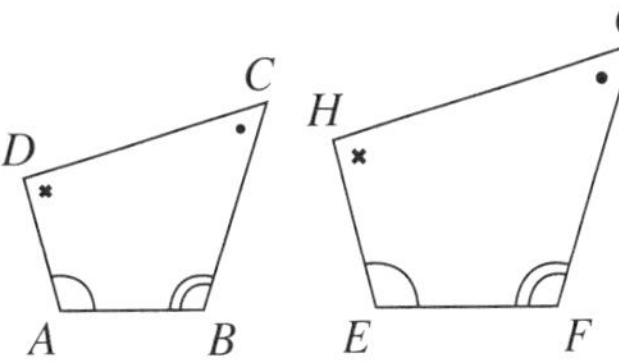

QUESTION 2 For the following similar figures, write the proportion statements.

a

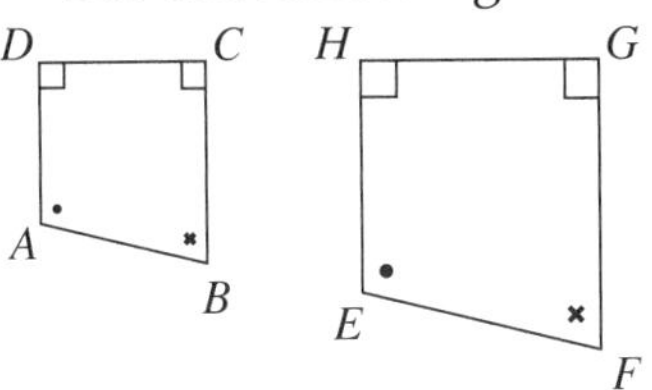

b

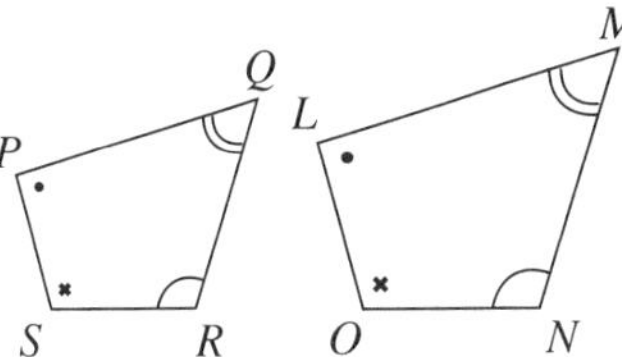

QUESTION 3 ΔABC ||| ΔPRQ. Which angle of ΔPRQ is equal to:

a $\angle ABC$? ____________ **b** $\angle BAC$? ____________ **c** $\angle ACB$? ____________

QUESTION 4 Quadrilateral $ABCD$ ||| quadrilateral $XZYW$. Which diagonal or side of $ABCD$ corresponds to:

a XY? __________ **b** YZ? __________ **c** XW? __________ **d** WZ? __________

QUESTION 5 Complete.

a Two figures are similar if an enlargement of one is ____________ to the other.

b If two figures are similar they are the same ____________ but not necessarily the same ____________.

c If two similar figures are the same size they are ____________.

QUESTION 6

a Are any two squares similar? Explain why or why not.

b Are any two circles similar? Explain why or why not.

c Are any two rectangles similar? Explain why or why not.

d Are any two equilateral triangles similar? Explain why or why not.

e Are any two parallelogram similar? Explain why or why not.

Similarity

UNIT 6: Similar triangles (1)

QUESTION 1

a In $\triangle ABC$ and $\triangle DEF$, write the matching angles.

$\angle A =$ ______________, $\angle B =$ ______________, $\angle C =$ ______________

b Write the matching sides

$AB =$ ______________, $BC =$ ______________, $AC =$ ______________

c Complete: $\triangle ABC$ ||| $\triangle$______________

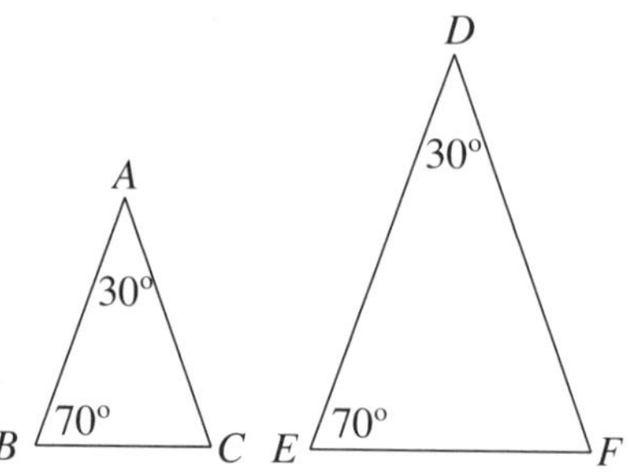

QUESTION 2 In the following pair of triangles, write the ratio of the matching sides.

a **i** $\frac{LM}{PQ} =$ ______________

ii $\frac{MN}{QR} =$ ______________

iii $\frac{NL}{RP} =$ ______________

b Complete: $\triangle LNM$ ||| $\triangle$______________

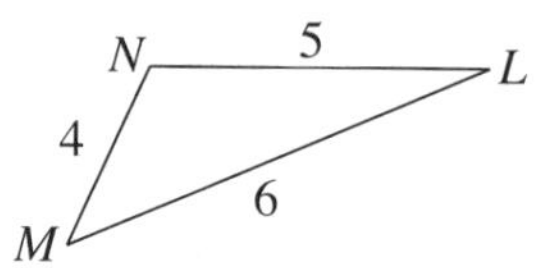

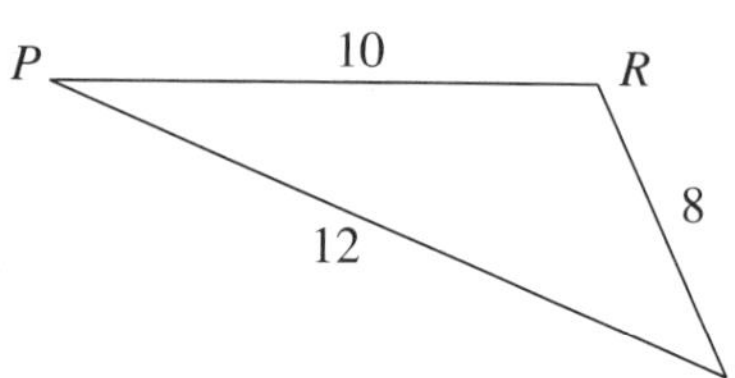

QUESTION 3 Complete.

a In simplest form $\frac{AB}{FE} =$ ______________

b In simplest form $\frac{AC}{FD} =$ ______________

c $\angle BAC = \angle$______________

d $\triangle ABC$ ||| $\triangle$______________

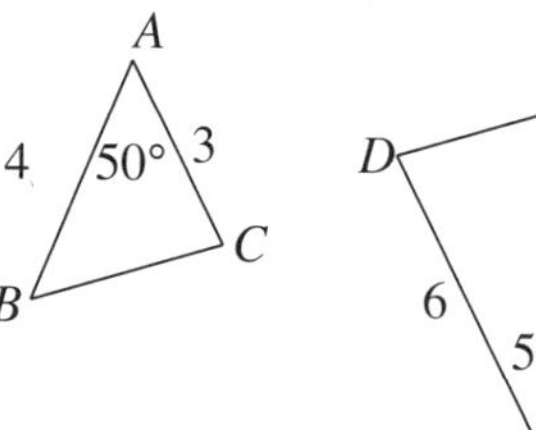

QUESTION 4 Complete.

a In simplest form $\frac{PQ}{ZY} =$ ______________

b In simplest form $\frac{PR}{ZX} =$ ______________

c $\angle PQR = \angle$______________ = ______________

d $\triangle PRQ$ ||| $\triangle$______________

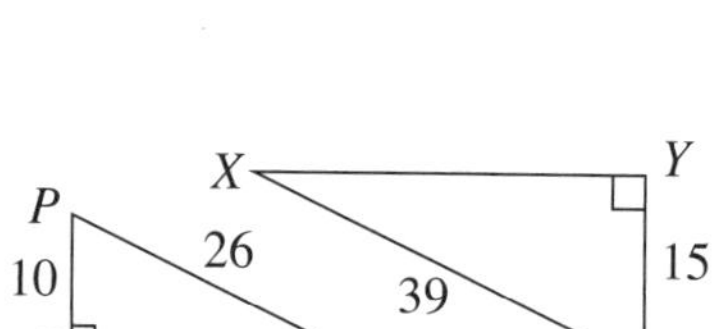

QUESTION 5 Complete the following statements.

a Two triangles are similar if two angles of one triangle are equal to ______________ of the other triangle.

b Two triangles are similar if their matching sides are in the ______________.

c Two triangles are similar if one angle of one triangle is equal to ______________ of the other and the lengths of the sides that form the angle are in the ______________.

d Two right-angled triangles are similar if the hypotenuse and a second side of one triangle ______________.

Excel Mathematics Study Guide Years 9–10
Pages 151–153

UNIT 7: Similar triangles (2)

QUESTION 1 Determine whether or not the pair of triangles is necessarily similar. If they are similar write the similarity statement and state the test.

a

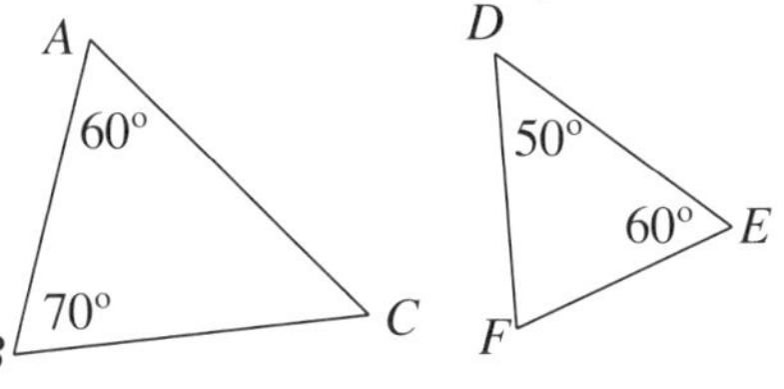

b

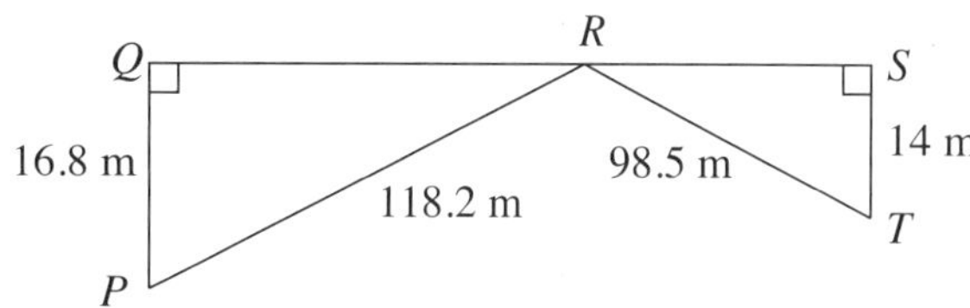

c

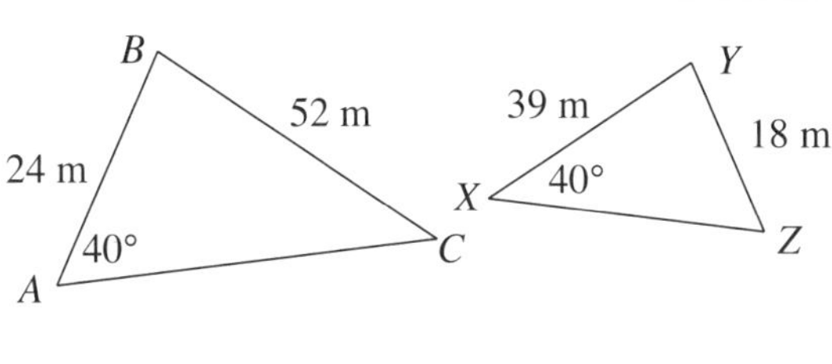

d

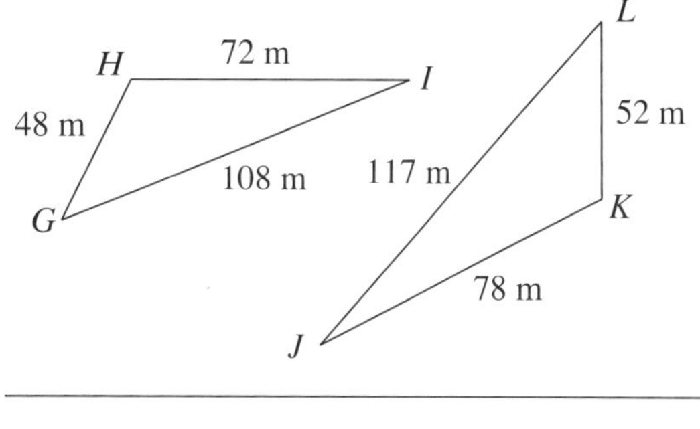

QUESTION 2 In each diagram, state the test that would be used to prove that the triangles are similar and write the similarity statement.

a

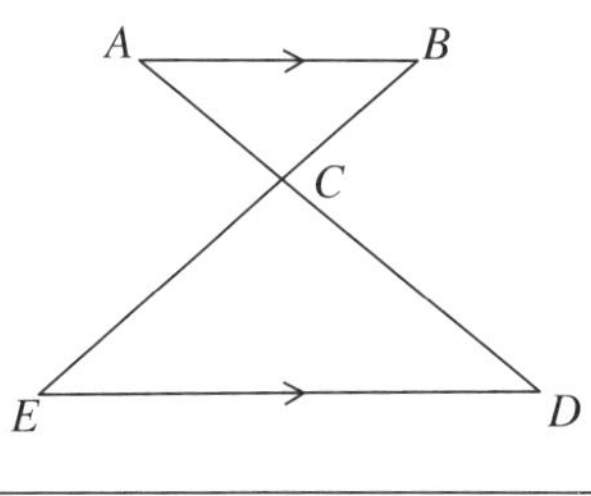

b

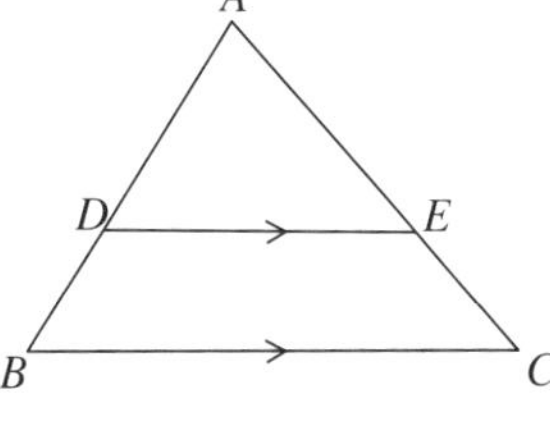

QUESTION 3 $PQ \parallel BC$ in the diagram.

a Show that $\Delta APQ \mid\mid\mid \Delta ABC$

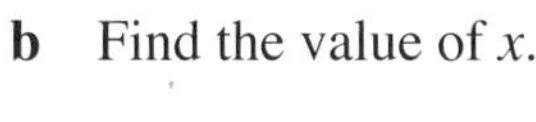

b Find the value of x.

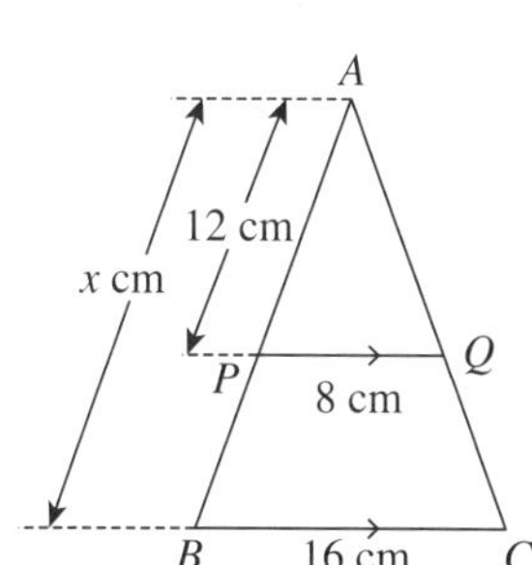

Similarity

UNIT 8: Further similar figures

Excel Mathematics Study Guide Years 9–10
Pages 151–153

QUESTION **1** Complete the following sentences.

a Two triangles are similar if their corresponding sides are in the ______________________.

b Two triangles are similar if an angle of one triangle is equal to ______________________ of the other and the lengths of the sides that form the angle are in the ______________________.

c The symbol for similar triangles is ______________________.

QUESTION **2** In ΔPQR, ST is drawn parallel to QR.

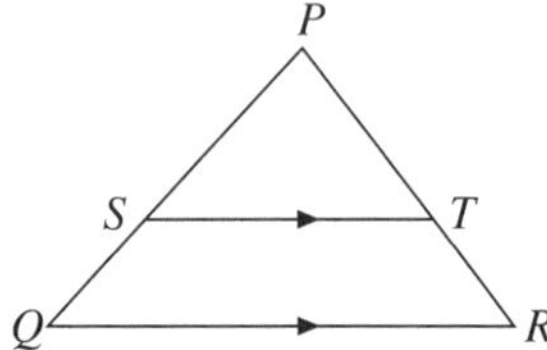

a Name two similar triangles.

____

b Complete: $\frac{\quad}{PS} = \frac{QR}{ST}$

c $PT = 8$ cm and $TR = 4$ cm. What is the enlargement factor between the two triangles?

____

____

d If $ST = 6$ cm find the length of QR.

____

____

QUESTION **3**

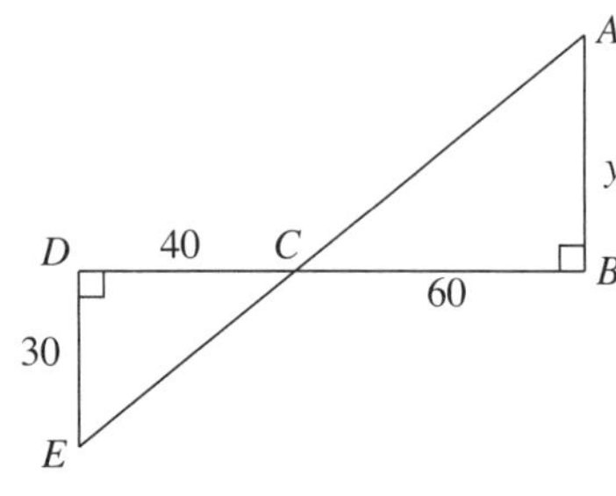

a Briefly explain why the two triangles are similar.

____

____

b Complete: ΔABC ||| Δ______________

c Find the value of y.

____

____

QUESTION **4** In ΔABC, DE is parallel to BC.

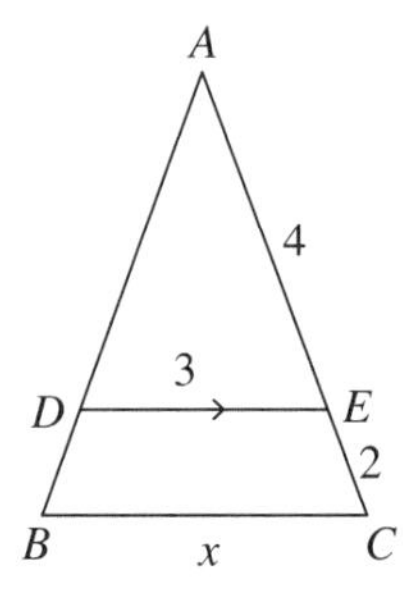

a Name the test used for similar triangles.

____

b Write the ratio of the corresponding sides.

____

c Find the value of x.

____

____

Excel Mathematics Study Guide Years 9–10
Pages 151–153

UNIT 9: Using similar triangles to find lengths and angles

QUESTION **1** In each diagram, use a test of similarity to find lengths and angles.

a

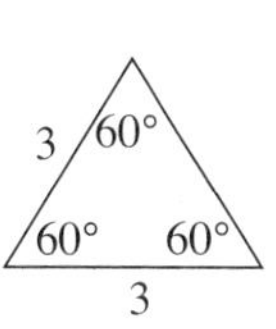

b

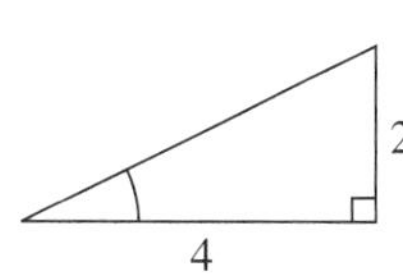

c

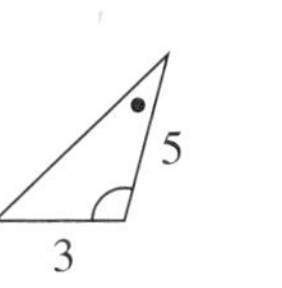

d

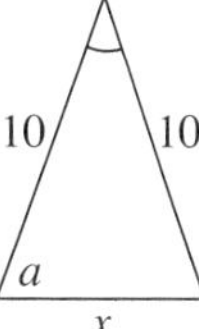

QUESTION **2** In each diagram, use a test of similarity to find the value of the pronumeral.

a

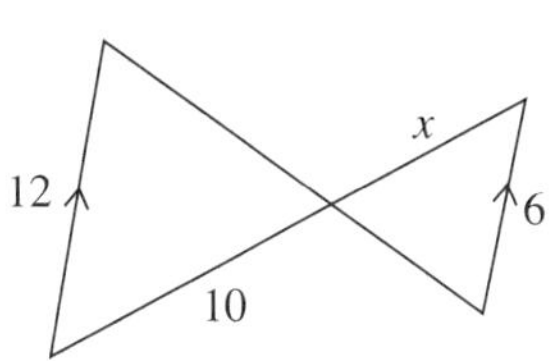

b

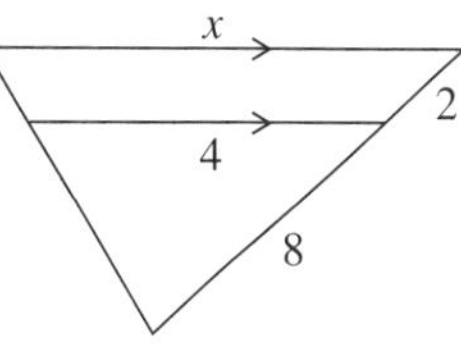

c

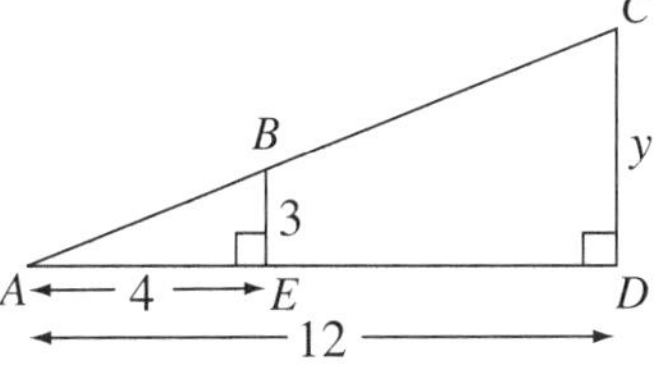

d

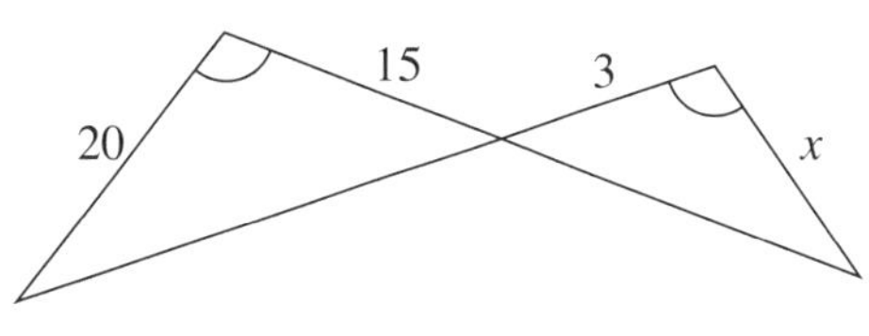

Excel Mathematics Study Guide Years 9–10
Pages 151–153

UNIT 10: Miscellaneous questions

QUESTION 1 Find the value of the pronumerals in each pair of similar triangles.

a

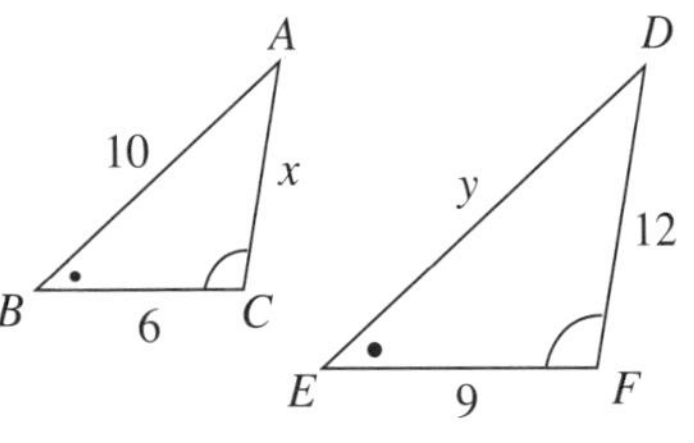

b

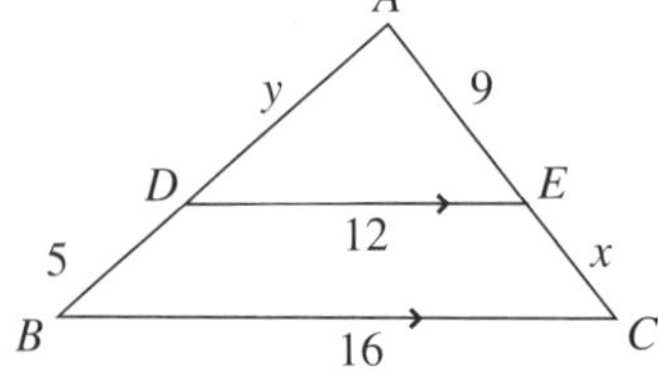

c

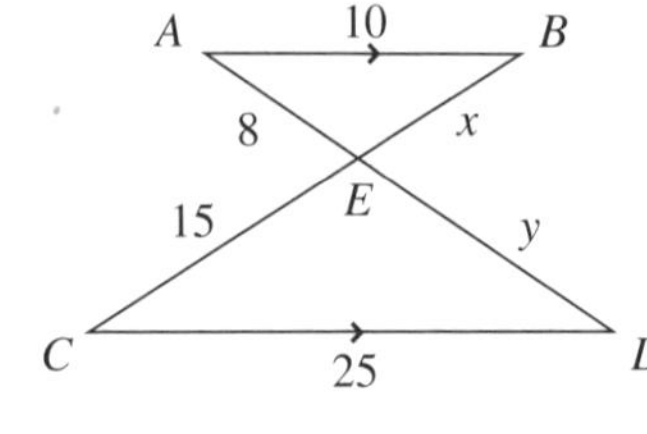

d

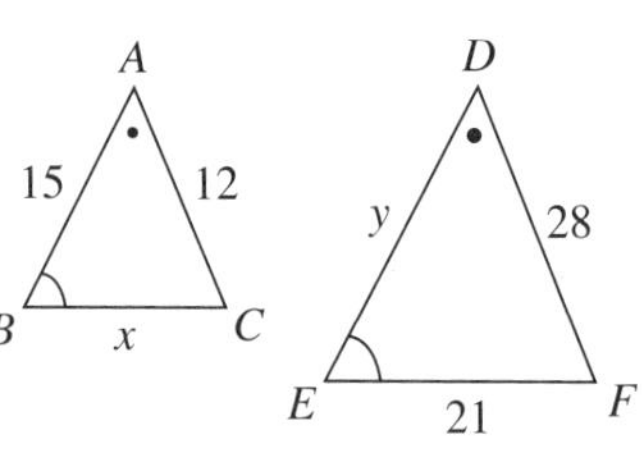

e

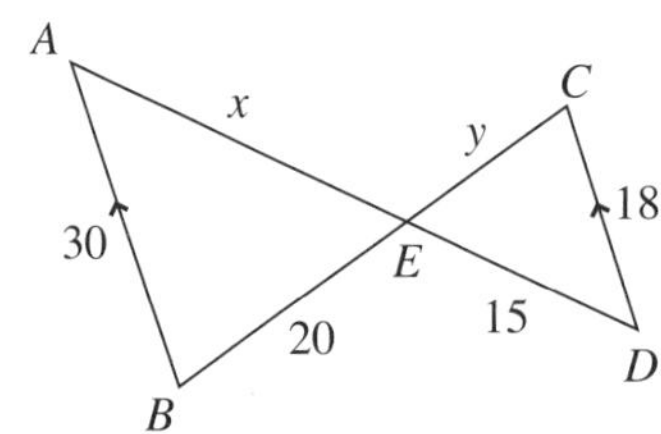

f

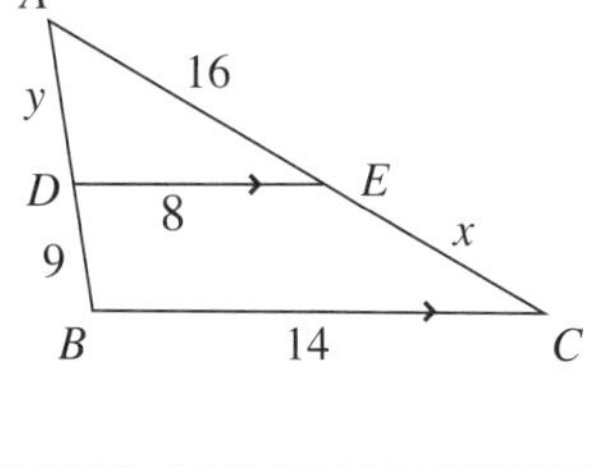

QUESTION 2 A post 1 m high casts a shadow, on level ground, that is 1.3 m long. At the same time a tree casts a shadow 71.5 m long. Use similar triangles to find the height of the tree.

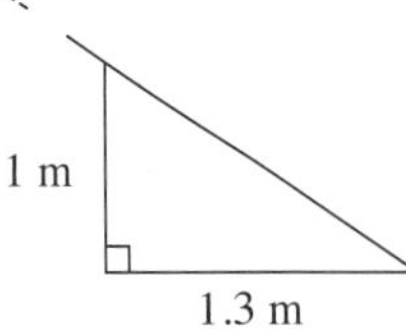

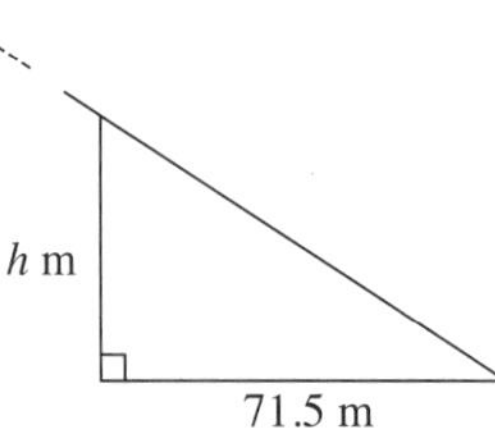

Excel Advanced Mathematics Study Guide Years 9–10
Pages 174–176

UNIT 11: Areas of similar figures

QUESTION 1 For each of the following similar figures, find the ratio of the smaller area to the larger area. All measurements are in centimetres.

a

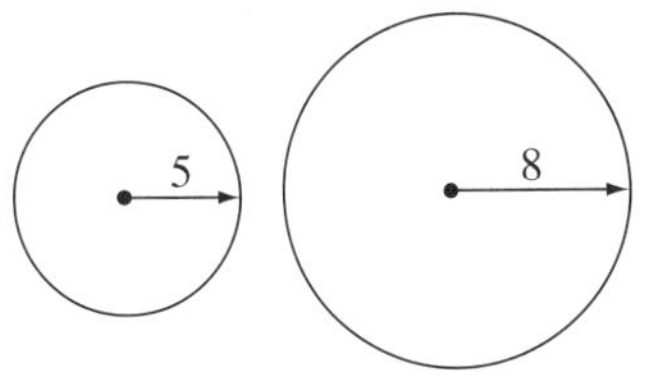

b

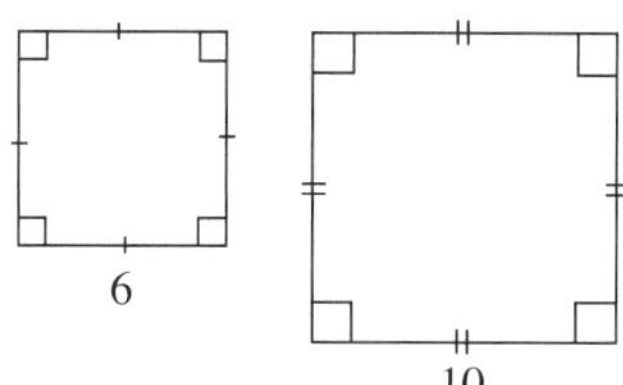

c

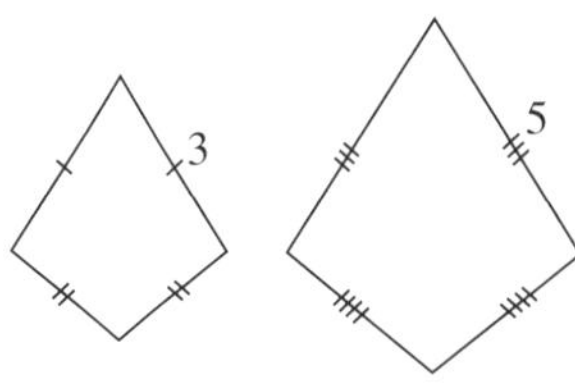

QUESTION 2 For each of the following similar figures, the ratio of the sides is given and one area is also given. Find the other area. All measurements are in centimetres.

a

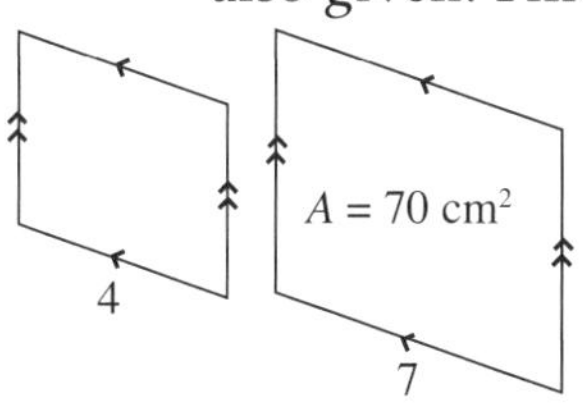

b

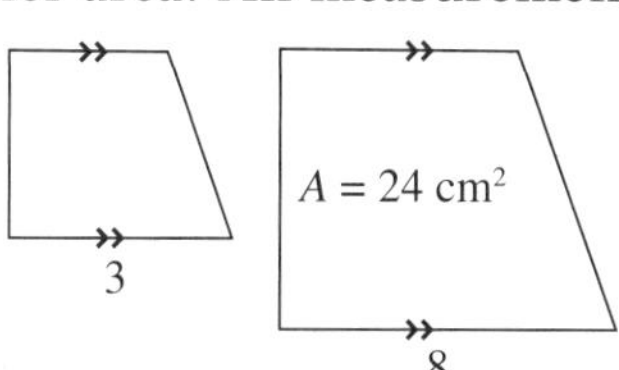

c

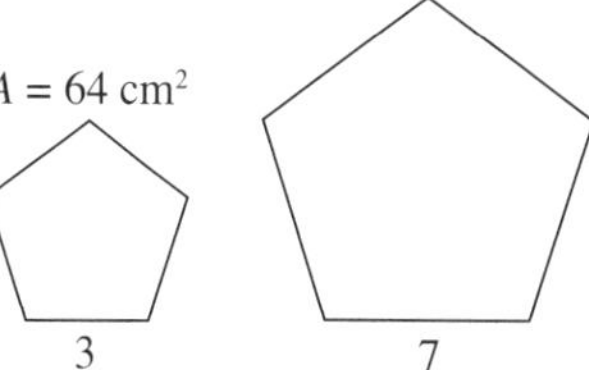

QUESTION 3

a If the sides of a rectangle are doubled, how much larger is the area now?

b The area of one square is nine times the area of another square. What is the ratio of the lengths of their sides?

c What can we say in regard to similarity of any two circles?

d What is the ratio of the area of a circle of radius 2 cm and a circle of radius 3 cm?

Excel Advanced Mathematics Study Guide Years 9–10
Pages 174–176

UNIT 12: Volumes of similar solids

QUESTION 1 For each of the following similar figures, find the ratio of the smaller volume to the larger volume. All measurements are in centimetres.

a

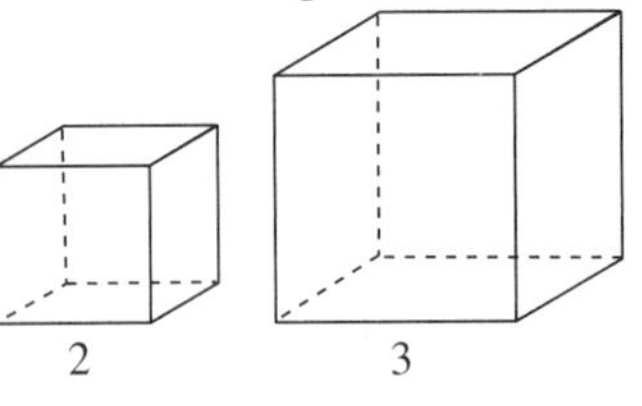

b

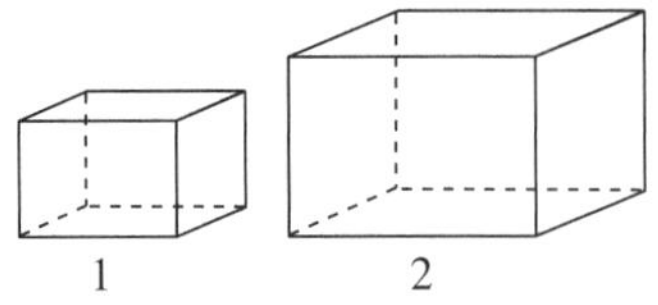

c

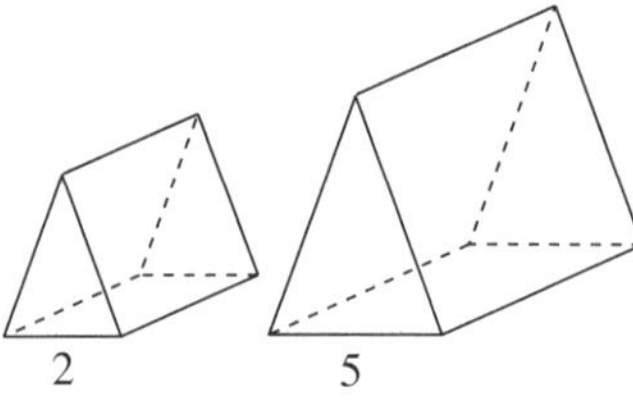

QUESTION 2 For each of the following similar figures, the ratio of the sides and one volume are given. Find the other volume. All measurements are in centimetres.

a

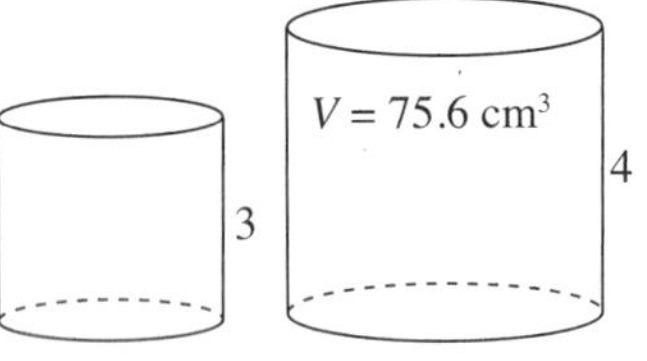

b

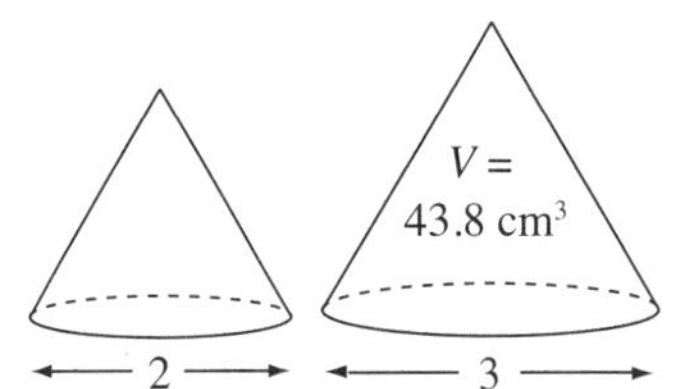

c

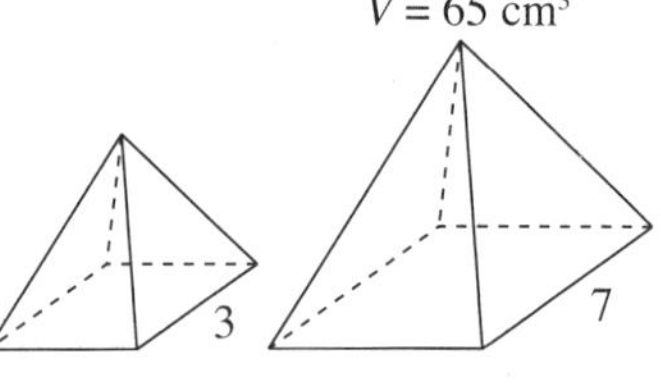

QUESTION 3

a The side lengths of two cubes are in the ratio 5 : 4. Find the ratio of their volumes.

b The surface areas of two spheres are in the ratio 64 : 49.

i What is the ratio of their radii?

ii Find the ratio of their volumes.

c The side lengths of two rectangular prisms are in the ratio 1 : 2. If the smaller prism has a volume of 68 cm^3, what is the volume of the larger prism?

Similarity

TOPIC TEST **PART A**

Instructions
- This part consists of 10 multiple-choice questions.
- Fill in only ONE CIRCLE for each question.
- Each question is worth 1 mark.

Time allowed: 10 minutes **Total marks: 10**

Marks

1 The symbol used to show that shapes are similar is
(A) || (B) = (C) ≡ (D) ||| 1

2 All similar triangles are
(A) equilateral. (B) equiangular. (C) different. (D) congruent. 1

3 Similar figures must be
(A) the same shape but not necessarily the same size. (B) the same shape but not the same size.
(C) the same shape and size. (D) neither the same shape nor size. 1

4 In similar figures the lengths of the corresponding sides must be
(A) the same. (B) different. (C) in proportion. (D) none of these. 1

5 A photo 12 cm long and 8 cm wide is enlarged. If it is now 96 cm wide, how long is it?
(A) 100 cm (B) 120 cm (C) 144 cm (D) 64 cm 1

6 Which statement is NOT correct?
(A) Any two equilateral triangles are similar. (B) Any two circles are similar.
(C) Any two squares are similar. (D) Any two parallelograms are similar. 1

7 The diagram shows ΔPQR. ΔKJL ||| ΔPQR. What is the size of $\angle JLK$?
(A) 30° (B) 60°
(C) 90° (D) There is not enough information. 1

Q
P 60° 30° R

8 A triangle with sides 9 cm, 7 cm and 12 cm has been enlarged so that its perimeter is now 49 cm. What is the enlargement factor?
(A) 1.25 (B) 1.5 (C) 1.75 (D) 2.25 1

9 This pair of triangles must be
(A) neither similar nor congruent.
(B) similar but not congruent.
(C) congruent but not similar.
(D) both similar and congruent. 1

50° 70° 8 cm
8 cm 70° 60°

10 What is the length of DF?
(A) 4.5 cm (B) 7.5 cm
(C) 8 cm (D) 10 cm 1

B 90° 12 cm 9 cm A 37° 15 cm C
D 53° 37° E 6 cm F

Total marks achieved for PART A /10

Similarity

TOPIC TEST — PART B

Time allowed: 20 minutes — **Total marks: 15**

Marks

1 A diagram was 27 cm long and 18 cm high. It was reduced by a factor of 1.5. What is its new length? ________________ 1

2 In the diagram shown:

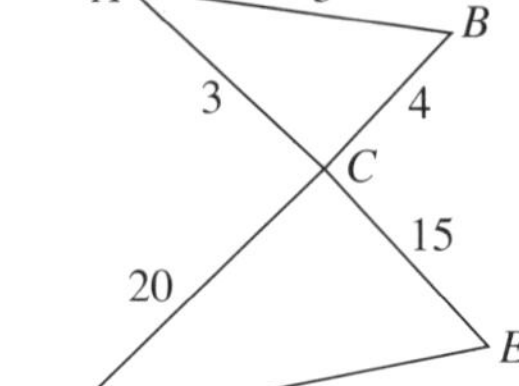

a briefly explain why the triangles are similar.

____

b Complete $\triangle ABC \;|||\; \triangle$________

c $\angle ABC = 37°$ and $\angle ACB = 90°$, what is the size of $\angle CDE$?________ 3

3 a Briefly explain why the triangles are similar.

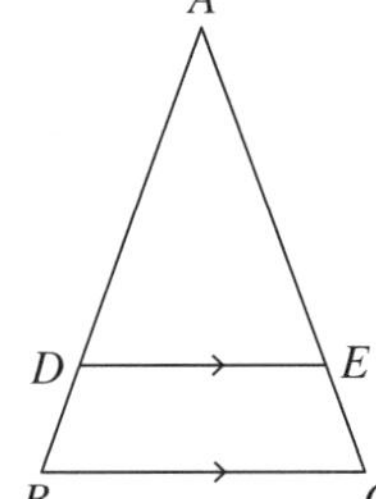

____

____

____

b Complete $\triangle ADE \;|||\; \triangle$________

c If $AE = 16$ cm, $EC = 2$ cm and $DE = 12$ cm, find the length of BC.

____

________________ 3

4 a Find the value of x **b** Find the value of y

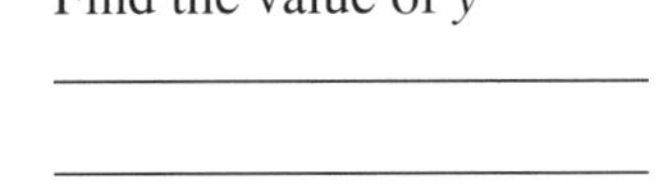

________ ________

________ ________

________ ________

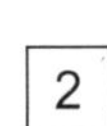

5 a Find the value of x **b** Find the value of y

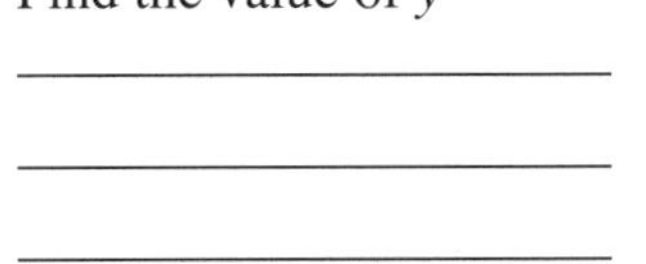

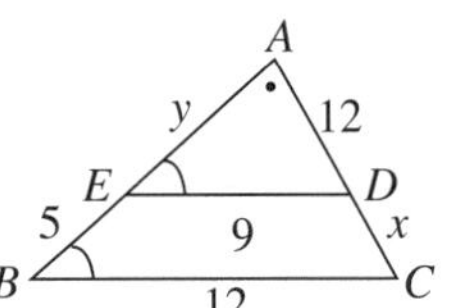

________ ________

________ ________

________ ________ 2

6 The edges of two cubes are in the ratio 2:5. Find the ratio of their:

a surface areas **b** volumes

________ ________ 2

7 The surface areas of two cylinders are in the ratio 9:49. Find the ratio of their:

a heights **b** volumes

________ ________ 2

Total marks achieved for PART B

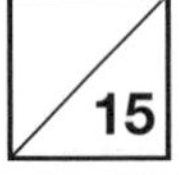

CHAPTER 9
Trigonometry

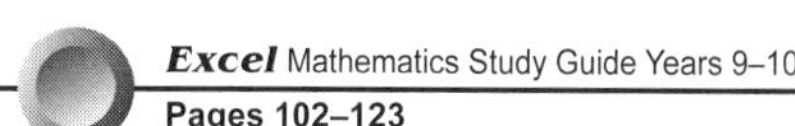

UNIT 1: Naming the sides of a right-angled triangle

QUESTION 1 In each of the following triangles, state whether x, y and z are the opposite side, adjacent side or hypotenuse with reference to the angle marked.

a

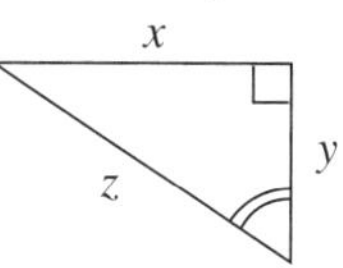

$x =$ ______________

$y =$ ______________

$z =$ ______________

b

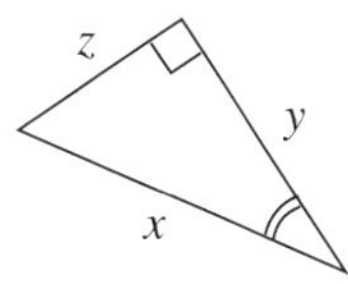

$x =$ ______________

$y =$ ______________

$z =$ ______________

c

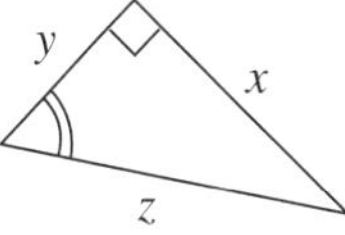

$x =$ ______________

$y =$ ______________

$z =$ ______________

d

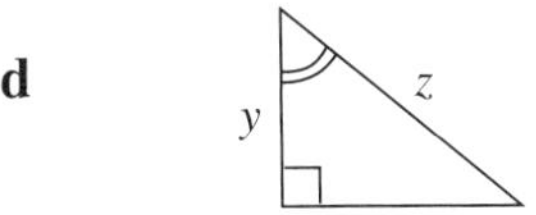

$x =$ ______________

$y =$ ______________

$z =$ ______________

e

$x =$ ______________

$y =$ ______________

$z =$ ______________

f

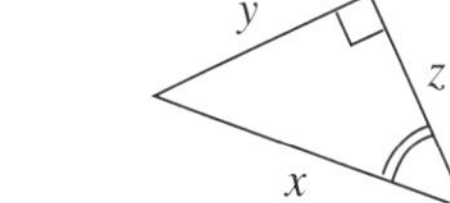

$x =$ ______________

$y =$ ______________

$z =$ ______________

QUESTION 2 Name the sides in the following right-angled triangles with reference to the angle marked.

a

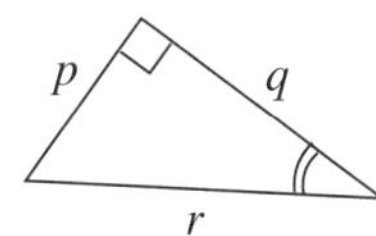

$p =$ ______________

$q =$ ______________

$r =$ ______________

b

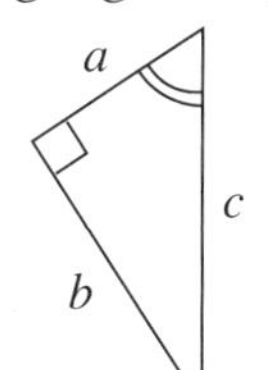

$a =$ ______________

$b =$ ______________

$c =$ ______________

c

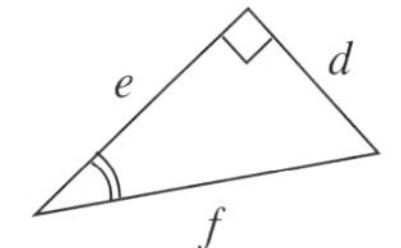

$d =$ ______________

$e =$ ______________

$f =$ ______________

d

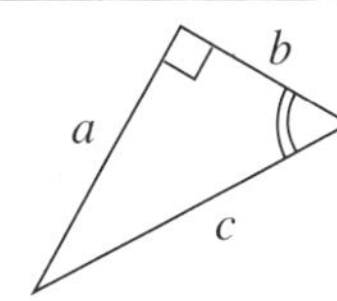

$a =$ ______________

$b =$ ______________

$c =$ ______________

e

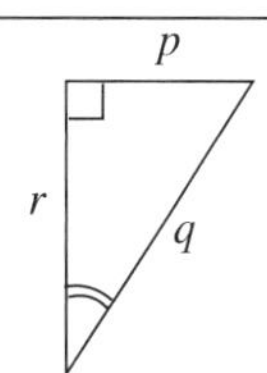

$p =$ ______________

$q =$ ______________

$r =$ ______________

f

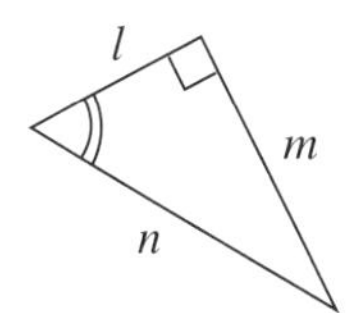

$l =$ ______________

$m =$ ______________

$n =$ ______________

QUESTION 3 Name the hypotenuse in each triangle given below.

a

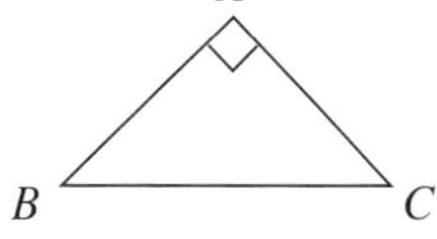

____

b

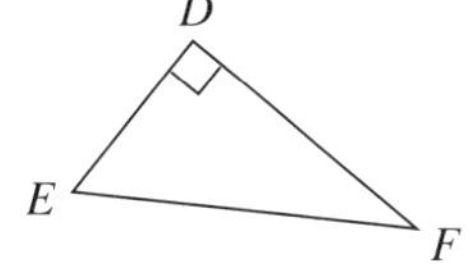

____

c

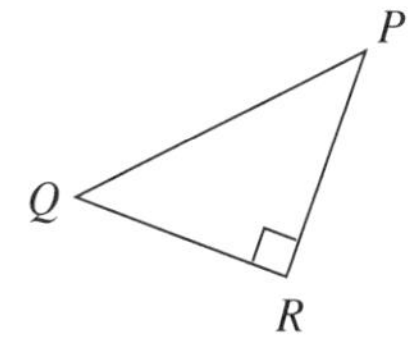

____

Trigonometry

UNIT 2: The trigonometric ratios

QUESTION 1 Write the trigonometric ratios for the following triangles.

a

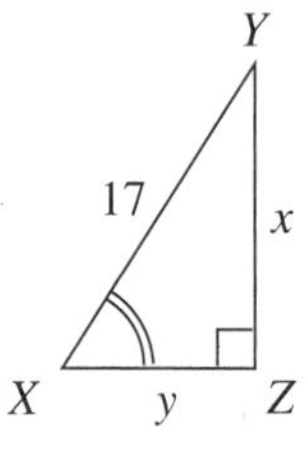

sin X = ________

cos X = ________

tan X = ________

b

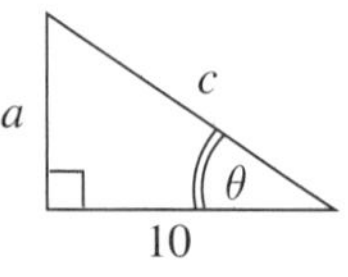

sin □= ________

cos □= ________

tan □= ________

c

30°
p
m
8

sin 30° = ________

cos 30° = ________

tan 30° = ________

d

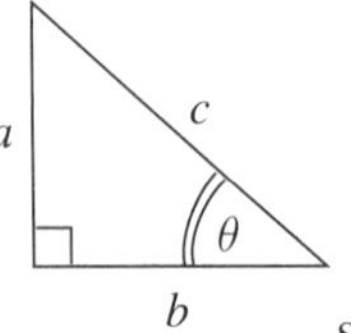

sin □= ________

cos □= ________

tan □= ________

e

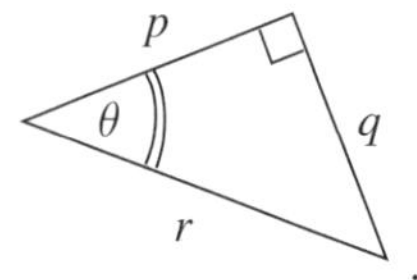

sin □= ________

cos □= ________

tan □= ________

f

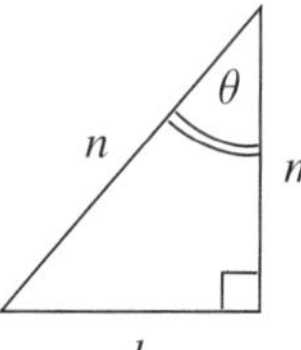

sin □= ________

cos □= ________

tan □= ________

QUESTION 2 Find sin θ, cos θ and tan θ in the following triangles.

a

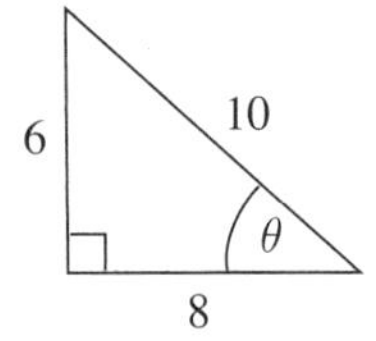

sin □= ________

cos □= ________

tan □= ________

b

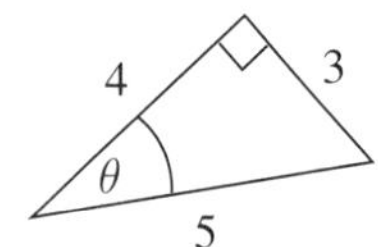

sin □= ________

cos □= ________

tan □= ________

c

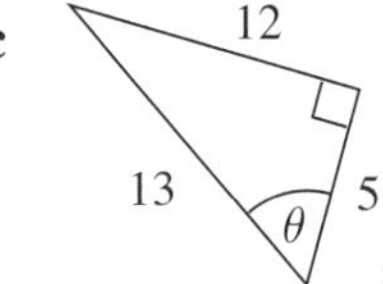

sin □= ________

cos □= ________

tan □= ________

d

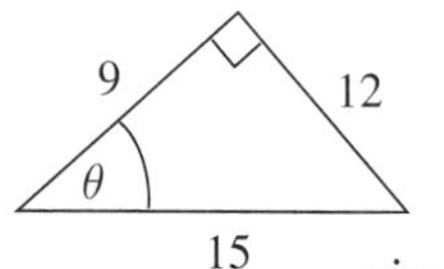

sin □= ________

cos □= ________

tan □= ________

e

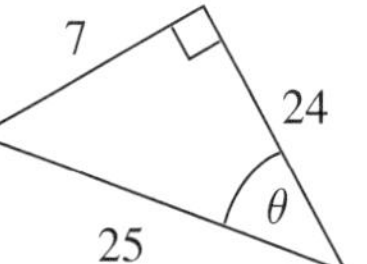

sin □= ________

cos □= ________

tan □= ________

f

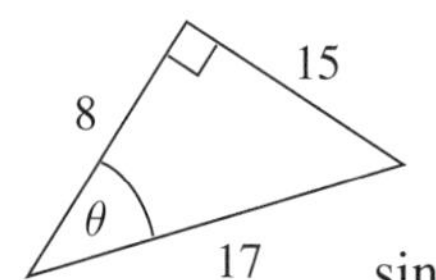

sin □= ________

cos □= ________

tan □= ________

QUESTION 3 Which ratio, sin, cos or tan links the given sides of the triangles?

a

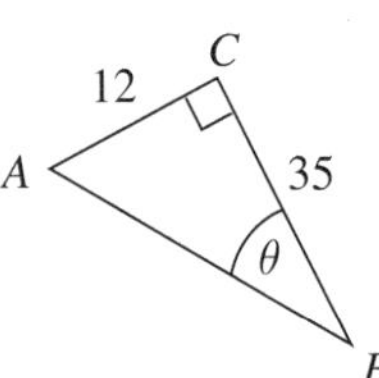

____

b

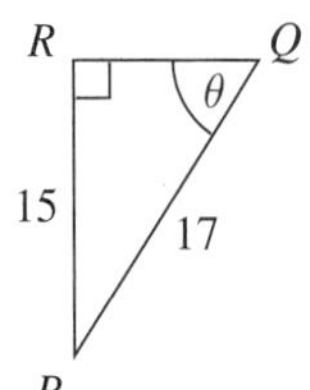

____

c

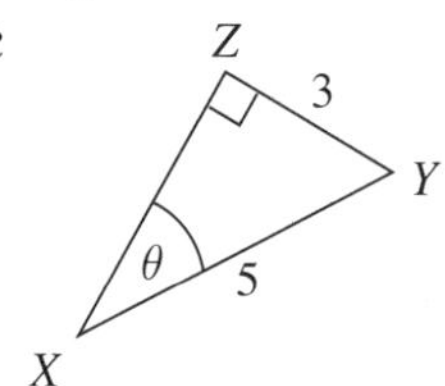

____

Trigonometry

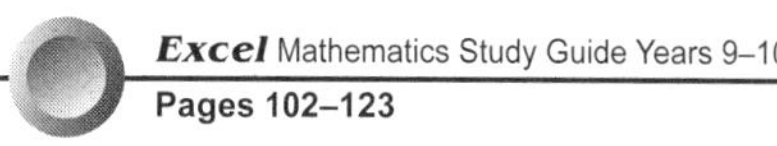

UNIT 3: Trigonometric ratios and the calculator

QUESTION 1 Find the value of the following correct to 3 decimal places.

a $\sin 69° =$ ________ **b** $\cos 70° =$ ________ **c** $\tan 23° =$ ________

d $\cos 83° =$ ________ **e** $\tan 21° =$ ________ **f** $\sin 75° =$ ________

g $\tan 48° =$ ________ **h** $\sin 36° =$ ________ **i** $\cos 48° =$ ________

QUESTION 2 Find the value of the following correct to 3 significant figures.

a $3.8 \sin 56° =$ ________ **b** $\tan 63°8' =$ ________ **c** $\sin 43°19' =$ ________

d $9 \cos 29° =$ ________ **e** $\sin 68°31' =$ ________ **f** $\cos 65°34' =$ ________

g $\sin 64°35' =$ ________ **h** $53.7 \cos 68°14' =$ ________ **i** $\tan 24°45' =$ ________

QUESTION 3 Find the value of the following correct to 2 decimal places.

a $\frac{\tan 65°}{7} =$ ________ **b** $\frac{\cos 75°}{6} =$ ________ **c** $\frac{18.6}{\sin 55°} =$ ________

d $\frac{\sin 28°43'}{5.9} =$ ________ **e** $\frac{\sin 58°36'}{5.9} =$ ________ **f** $\frac{23.8}{\cos 34°24'} =$ ________

g $\frac{\tan 27°58'}{10.35} =$ ________ **h** $\frac{\tan 48°33'}{7.5} =$ ________ **i** $\frac{864}{\tan 85°38'} =$ ________

QUESTION 4 A is an acute angle. Find its size to the nearest degree.

a $\sin A = 0.4356$ **b** $\tan A = 0.7885$ **c** $\cos A = 0.5463$ **d** $\cos A = 0.4963$

e $\tan A = 1.635$ **f** $\tan A = 1.4885$ **g** $\cos A = 0.3149$ **h** $\sin A = 0.8939$

i $\cos A = \frac{1}{3}$ **j** $\sin A = \frac{15}{19}$ **k** $\tan A = \frac{18.5}{13.63}$ **l** $\tan A = \frac{17}{23}$

QUESTION 5 A is an acute angle. Find its size in degrees and minutes.

a $\sin A = 0.6$ **b** $\cos A = 0.4831$ **c** $\tan A = 2.356$ **d** $\cos A = 0.3985$

e $\tan A = 0.8657$ **f** $\sin A = 0.4823$ **g** $\cos A = \frac{7.5}{12.3}$ **h** $\sin A = \frac{1}{4}$

Trigonometry

UNIT 4: Finding the length of a side (1)

QUESTION 1 Use the tangent ratio to find the value of x to 1 decimal place.

a

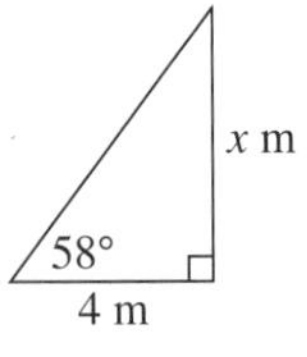

b

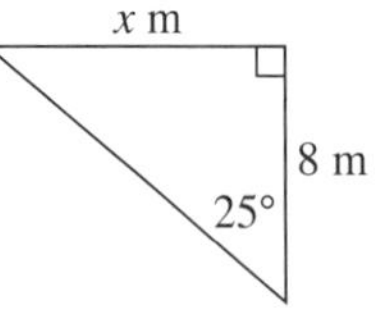

c

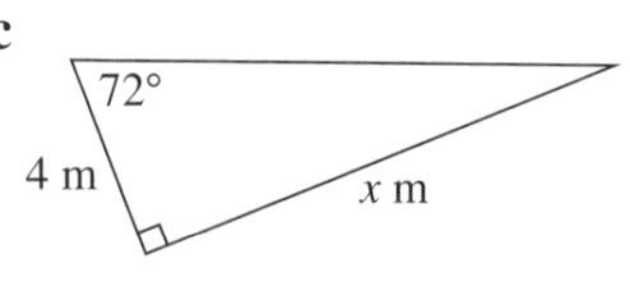

d

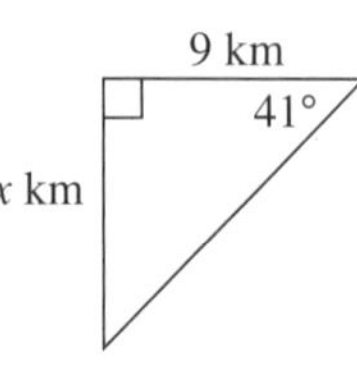

QUESTION 2 Use the sine ratio to find the value of x to 1 decimal place.

a

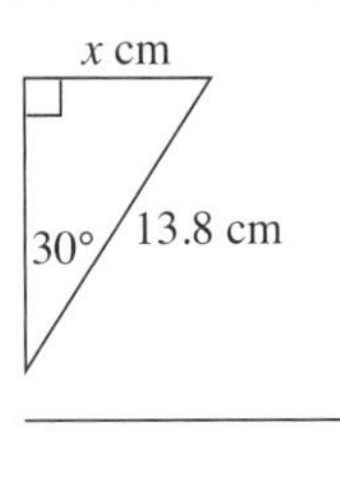

b

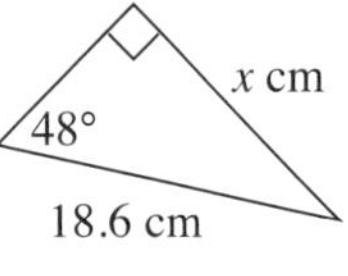

c

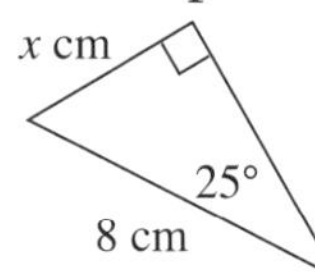

d

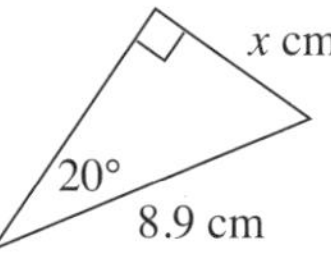

QUESTION 3 Use the cosine ratio to find the value of x to 1 decimal place.

a

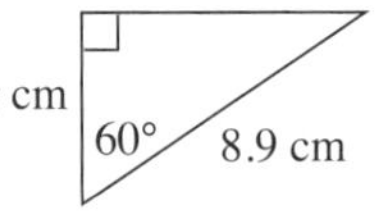

b

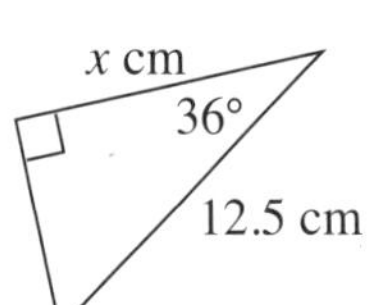

c

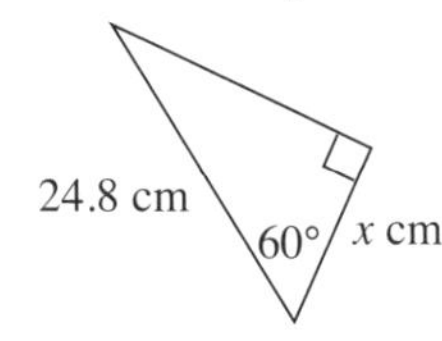

d

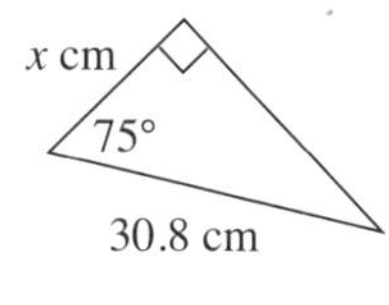

QUESTION 4 Find the value of x to 2 decimal places.

a

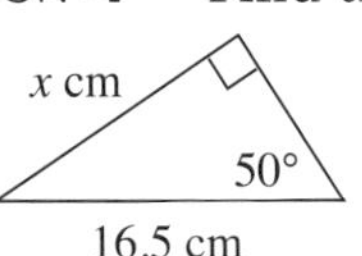

b

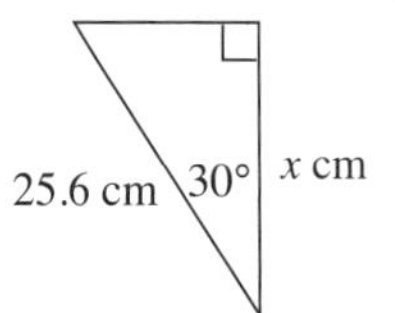

c

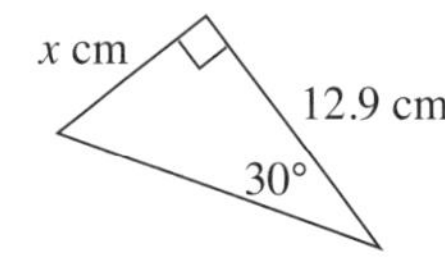

d

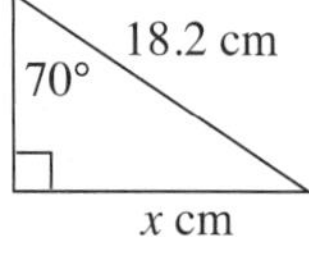

Trigonometry

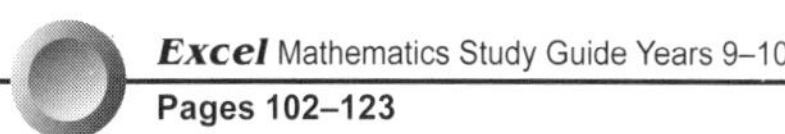

Excel Mathematics Study Guide Years 9–10
Pages 102–123

UNIT 5: Finding the length of a side (2)

QUESTION 1 Find the length of the unknown side and give the answer correct to one decimal place.

a

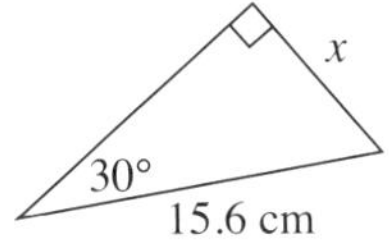

b

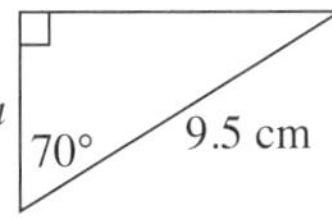

c

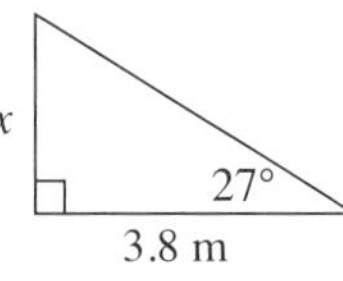

d

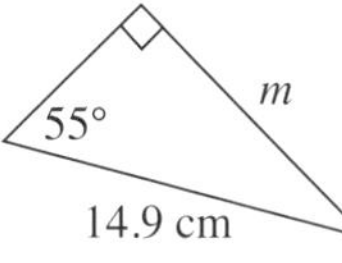

e

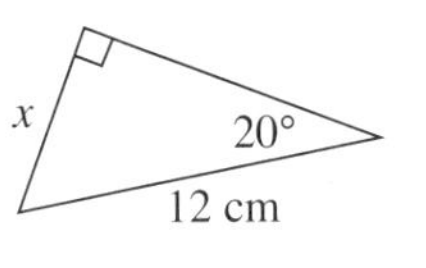

f

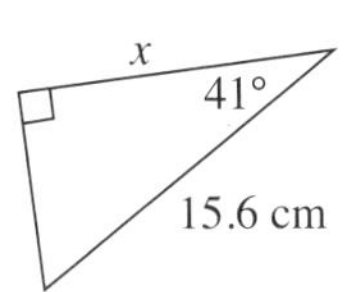

g

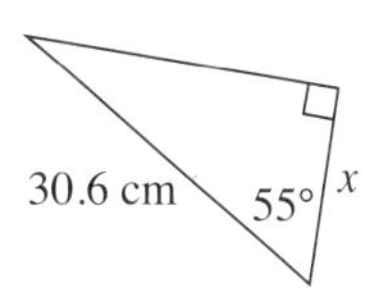

h

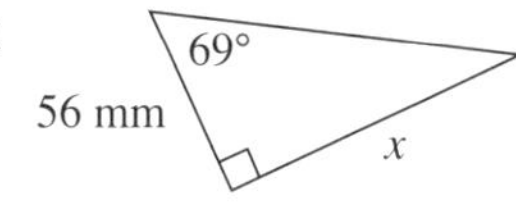

QUESTION 2 Find the value of the pronumeral to two decimal places.

a

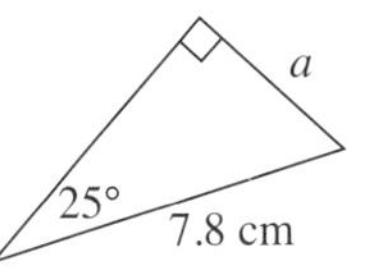

b

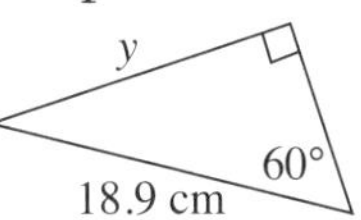

c

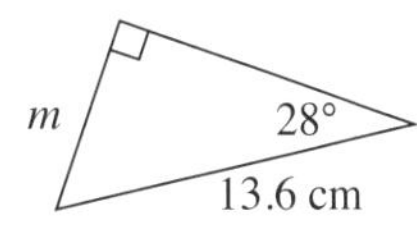

d

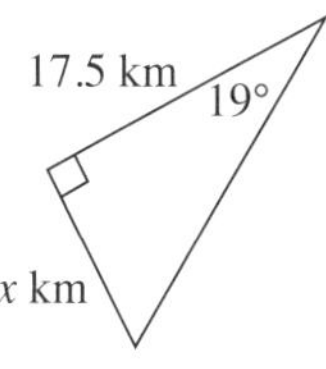

e

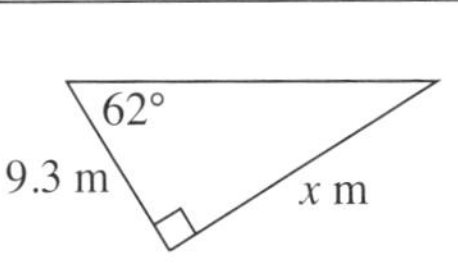

f

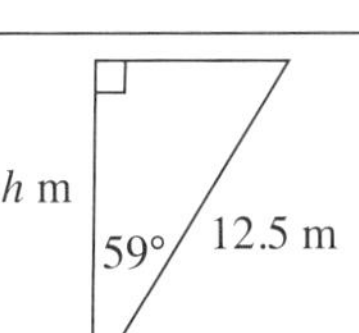

g

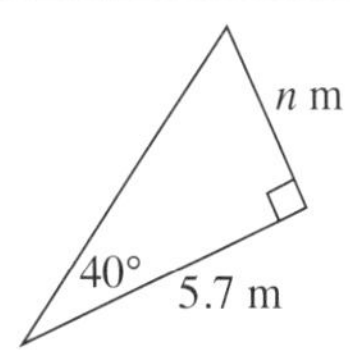

h

13.2 km
x km
51°

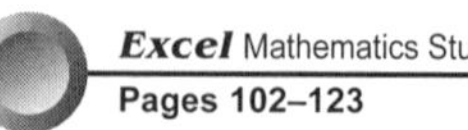

UNIT 6: Finding the length of a side (3)

QUESTION 1 Find the length of the hypotenuse correct to one decimal place.

a

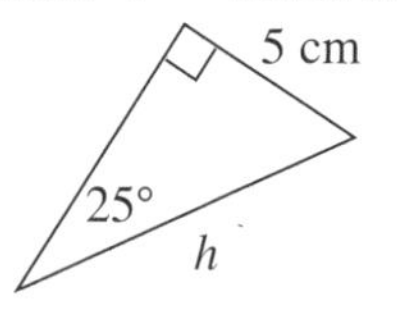

b

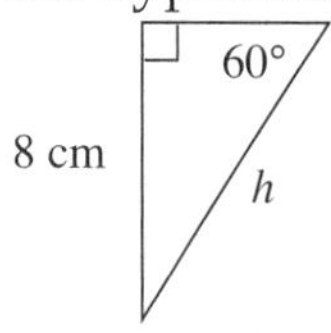

c

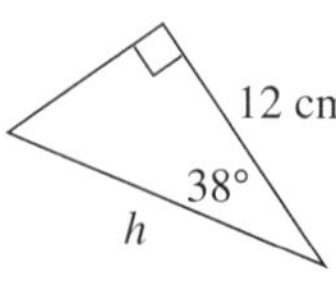

d

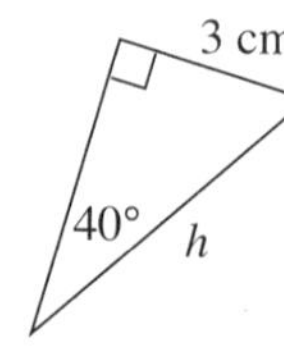

e

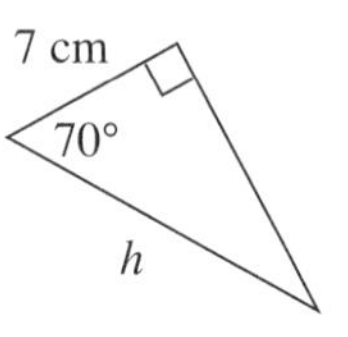

f

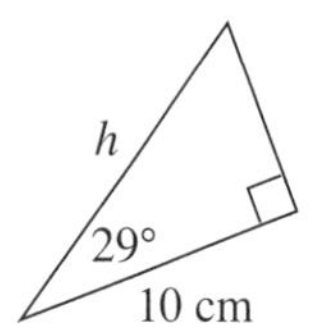

g

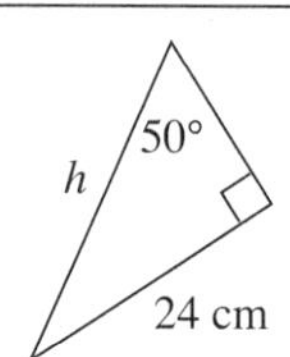

h

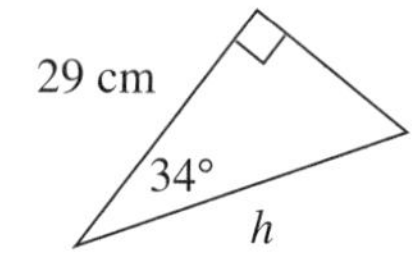

i

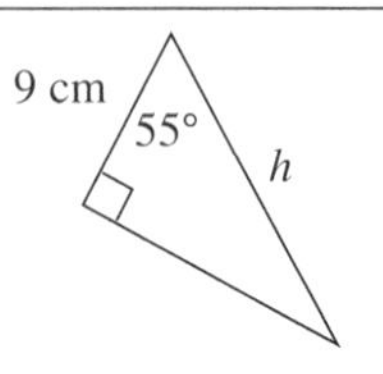

j

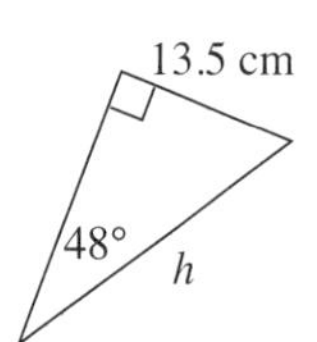

k

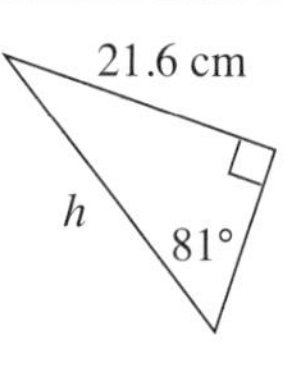

l

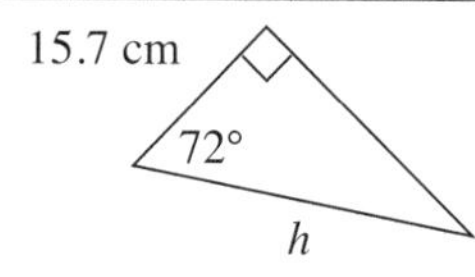

QUESTION 2 Find the value of x correct to one decimal place.

a

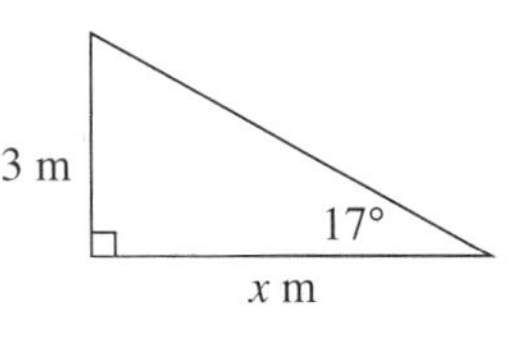

b

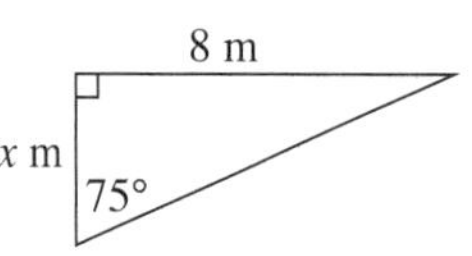

c

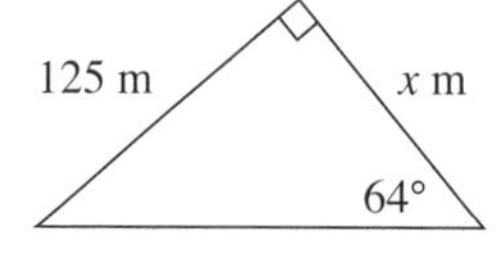

d

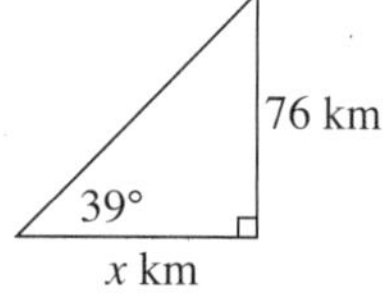

Trigonometry

UNIT 7: Finding the unknown angle (1)

QUESTION 1 Find θ. Give the answer to the nearest whole degree.

a

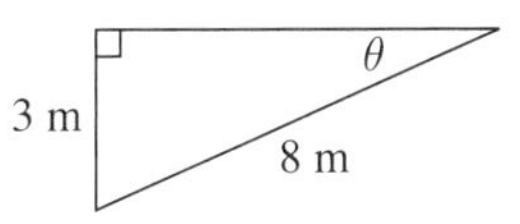

b

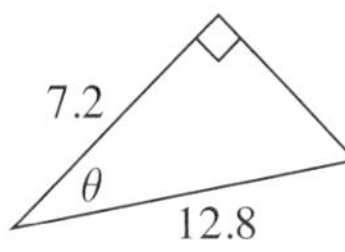

c

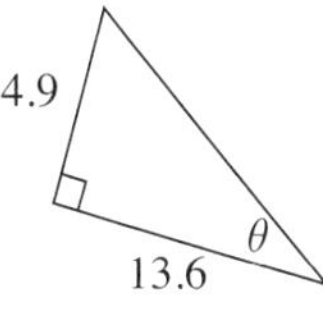

QUESTION 2 Find the size of angle A. Give the answer to the nearest degree.

a

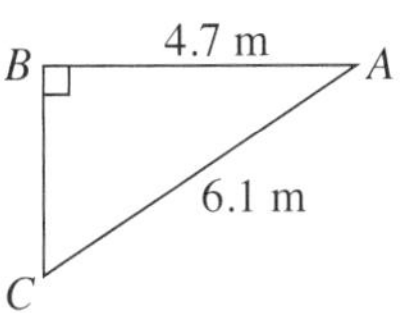

b

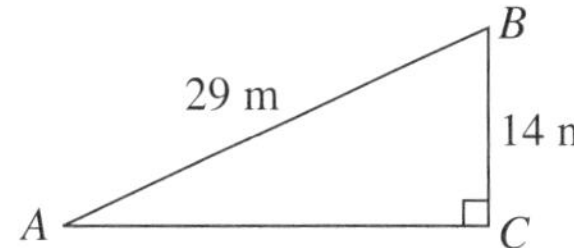

c

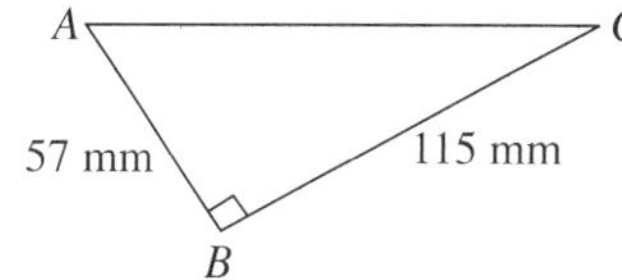

QUESTION 3 Find the size of the marked angle, to the nearest minute.

a

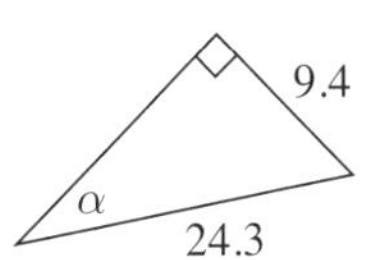

b

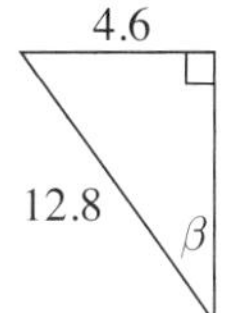

c

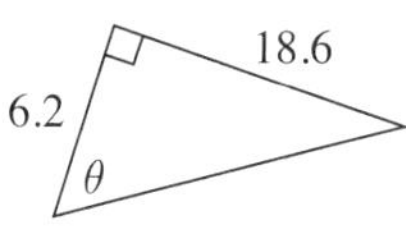

d

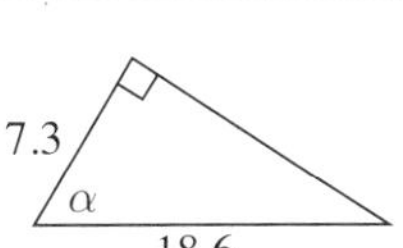

e

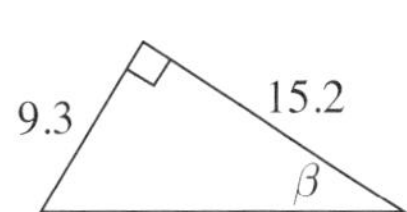

f

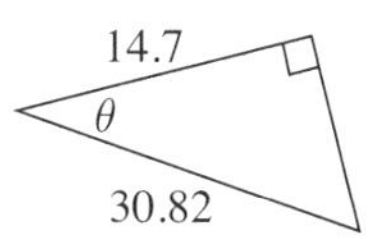

Excel Mathematics Study Guide Years 9–10
Pages 102–123

UNIT 8: Finding the unknown angle (2)

QUESTION 1 Find the size of the angle marked to the nearest minute.

a

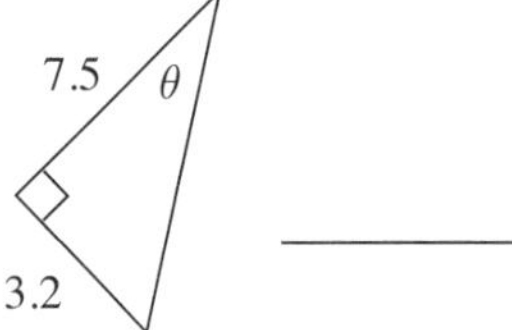

b

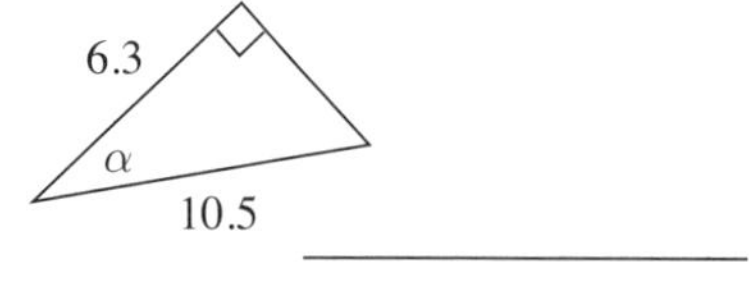

c

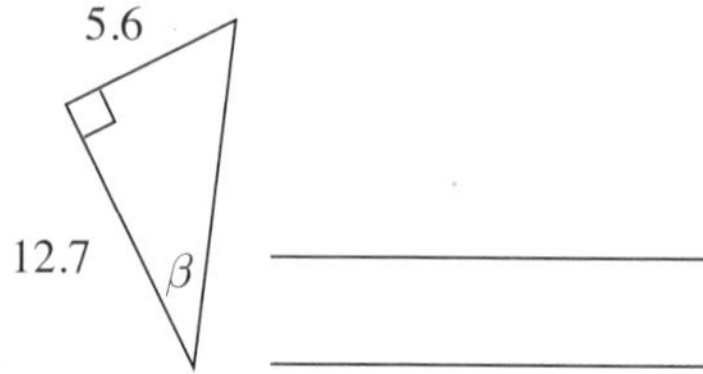

d

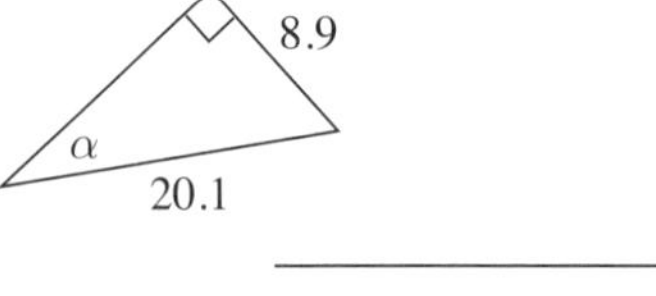

e

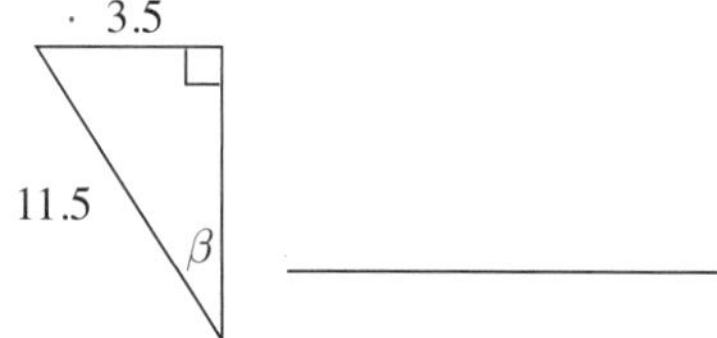

f

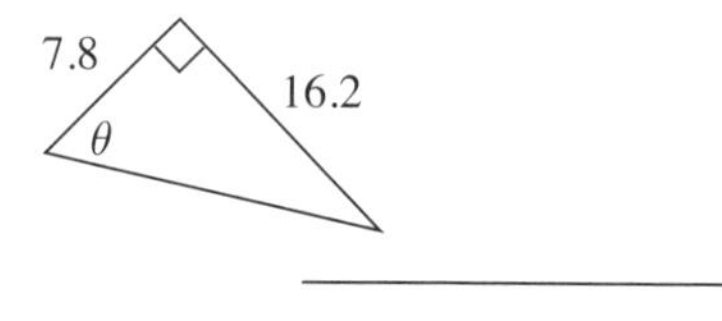

g

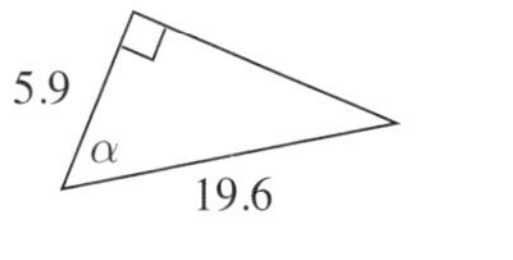

h

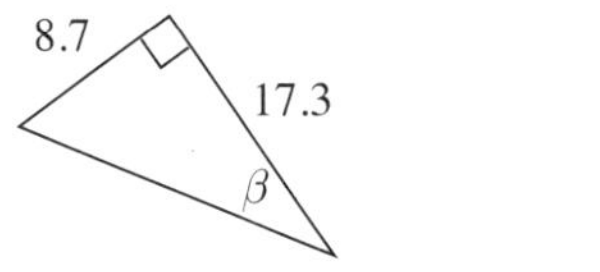

i

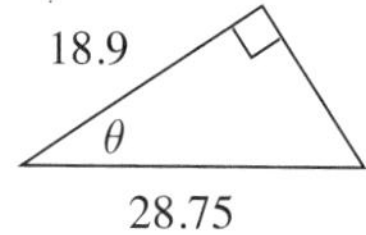

j

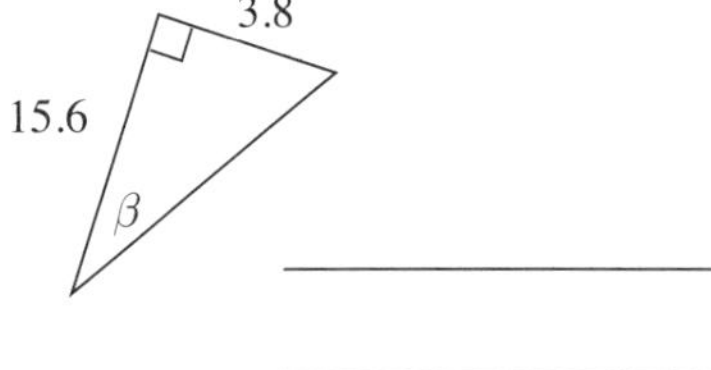

k

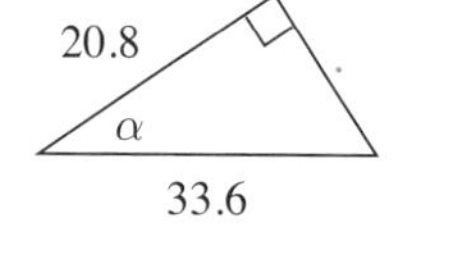

l

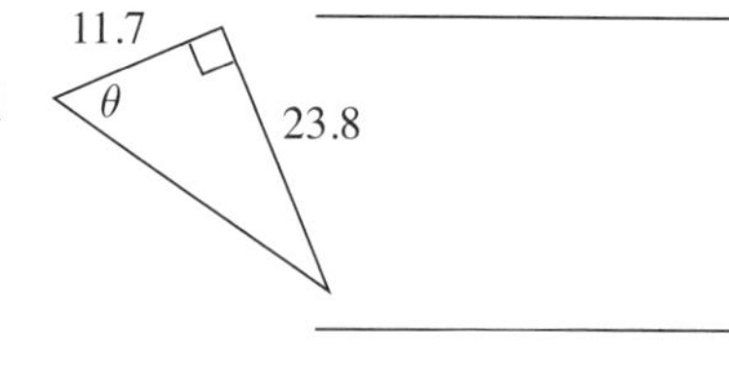

QUESTION 2 An 18 m ladder standing on level ground reaches 14 m up a vertical wall. Find the angle that the ladder makes with the ground (give your answer to the nearest degree).

QUESTION 3 $ABCD$ is a rectangle with $AC = 25$ cm and $AD = 14$ cm. Find $\angle ACD$ correct to the nearest degree.

Trigonometry

UNIT 9: Mixed exercises

QUESTION 1 Find the value of x. Give the answer correct to one decimal place.

a

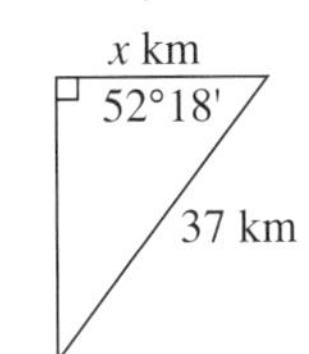

b

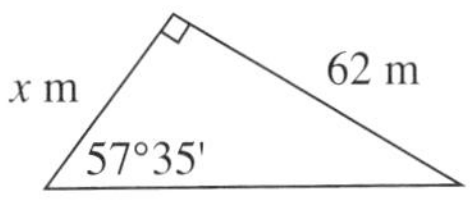

c

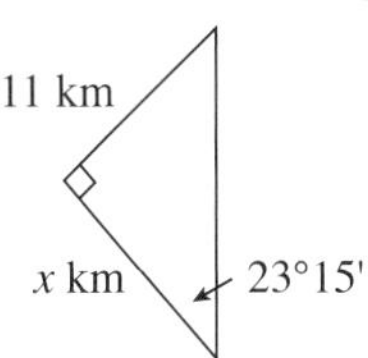

d

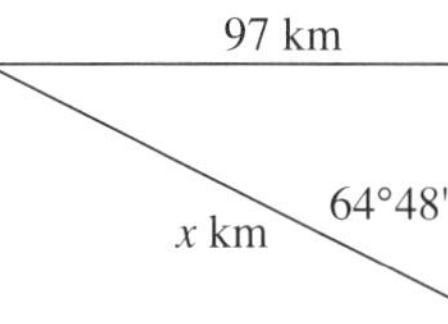

QUESTION 2 Find the value of θ to the nearest degree.

a

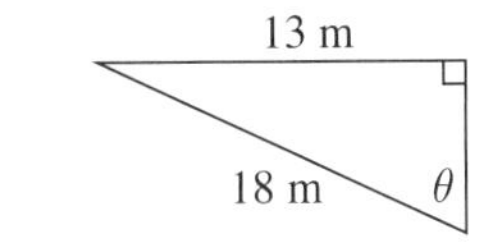

b

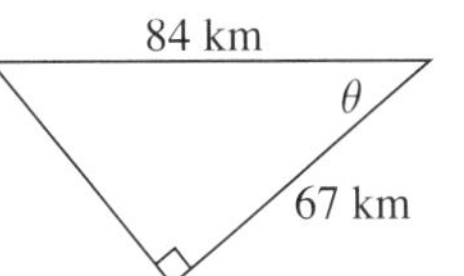

c

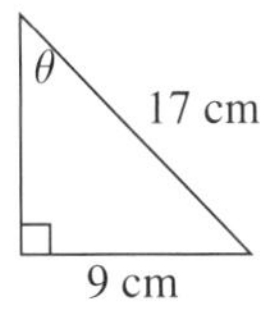

d

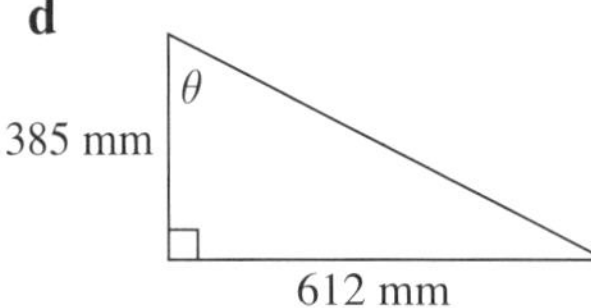

QUESTION 3 The diagram shows a trapezium with DC parallel to AB. Find the length, to one decimal place, of:

a BD

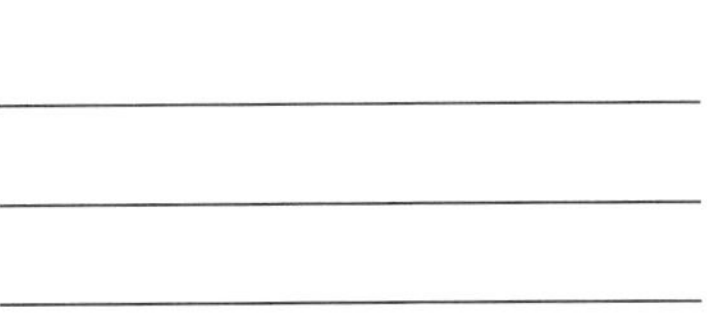

b AB

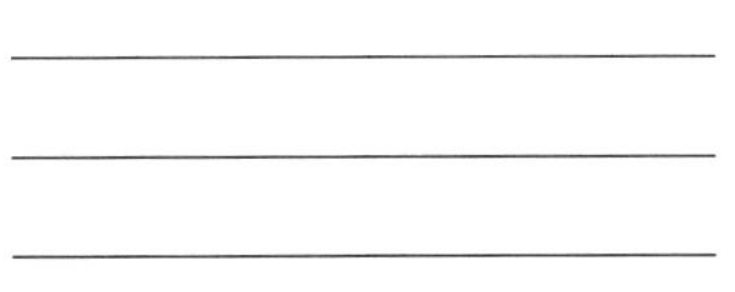

D 10 cm C

60°

A B

QUESTION 4 In the diagram, $DC = 100$ m, $CB = 150$ m and $\angle ABD = 25°$. Find.

a AD to the nearest metre

b $\angle ACD$ to the nearest degree

A

25°

D 100 m C 150 m B

Excel Mathematics Study Guide Years 9–10
Pages 102–123

UNIT 10: Problem solving

QUESTION 1 Michelle is flying a kite on a 55 metre string that makes an angle of 56° with the horizontal. Calculate the height of the kite to the nearest metre.

QUESTION 2 Find the length of the diagonal of a rectangle if the length of the rectangle is 10.7 cm and the diagonal makes an angle of 30° with the longer side.

QUESTION 3 A 3 metre ladder leans against a building with its top reaching a height of 2.6 metres. What angle, to the nearest degree, does the ladder make with the wall?

QUESTION 4 In ΔABC, $\angle C = 90°$, $\angle B = 34.5°$ and $AC = 3.6$ cm. Find AB.

QUESTION 5 In the triangle ABC, the angle B is 90°, AB is 4 m and AC is 5 m. Find the size of angle A correct to the nearest degree.

QUESTION 6 Point P is 18 m due south of point Q. Point R is due east of point P. If $\angle PQR = 68°$, find the distance from P to R to the nearest metre.

Trigonometry

TOPIC TEST — PART A

Instructions
- This part consists of 10 multiple-choice questions.
- Fill in only ONE CIRCLE for each question.
- Each question is worth 1 mark.

Time allowed: 10 minutes **Total marks: 10**

Marks

1 Use your calculator to find sin 36° correct to two decimal places.

(A) 0.58 (B) 0.57 (C) 0.59 (D) 0.81 — 1

2 Evaluate 12 sin 85° correct to two decimal places.

(A) 12.05 (B) 11.95 (C) 1.05 (D) 137.16 — 1

3 Find the value of $\frac{\sin 38^\circ - \cos 55^\circ}{\tan 36^\circ}$ correct to one decimal place.

(A) 0.2 (B) 0.5 (C) 0.05 (D) 0.1 — 1

4 If sin □ = $\frac{4}{7}$, calculate the size of the angle □ to the nearest degree.

(A) 55° (B) 30° (C) 35° (D) 45° — 1

5 If cos □ = $\frac{1}{2}$, find the size of angle □

(A) 30° (B) 60° (C) 45° (D) 55° — 1

6 Which is NOT correct?

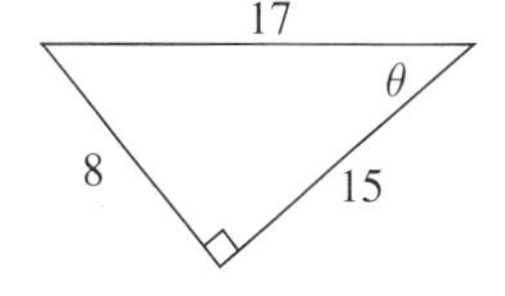

(A) $\tan\theta = \frac{8}{15}$ (B) $\sin\theta = \frac{8}{17}$

(C) $\cos\theta = \frac{15}{17}$ (D) $\cos\theta = \frac{15}{8}$ — 1

7 Which expression gives the value of x?

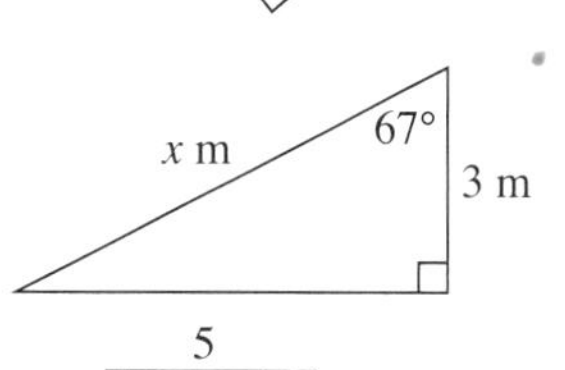

(A) 3 sin 67° (B) 3 cos 67°

(C) $\frac{3}{\sin 67^\circ}$ (D) $\frac{3}{\cos 67^\circ}$ — 1

8 What is the value of θ to the nearest degree?

5, θ, 8

(A) 39° (B) 51° (C) 32° (D) 58°

9 Which expression is correct?

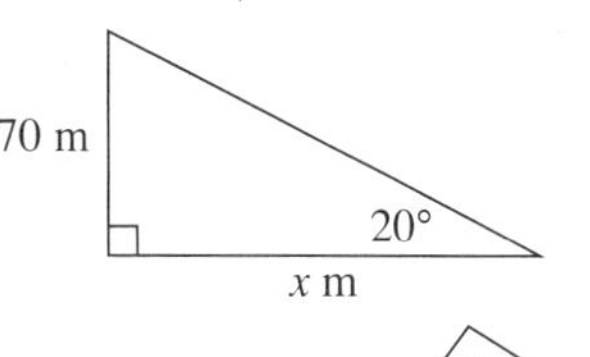

(A) $x = 70 \tan 20^\circ$ (B) $x = \frac{70}{\tan 70^\circ}$

(C) $x = \frac{\tan 20^\circ}{70}$ (D) $x = 70 \tan 70^\circ$

10 What is the value of x to one decimal place?

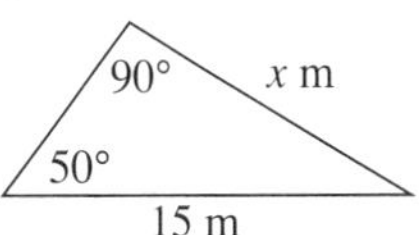

(A) 11.5 (B) 9.6 (C) 19.6 (D) 17.9 — 1

Total marks achieved for PART A /10

Trigonometry

TOPIC TEST — PART B

Time allowed: 20 minutes — **Total marks: 15**

Marks

1 Find, to 3 decimal places, the value of:

a tan 58° 25’ **b** 19.7 cos 78° **c** $\dfrac{\sin 46^\circ}{28.67}$ 3

2 Find the value of □ to the nearest degree.

a $\cos \square = \dfrac{4}{5}$ **b** $\tan \square = 0.6781$ 2

3 Find the value of □ to the nearest minute if $\sin \square = \dfrac{12}{13}$ 1

4 Find the length of side *AC*. Give the answer correct to one decimal place.

a

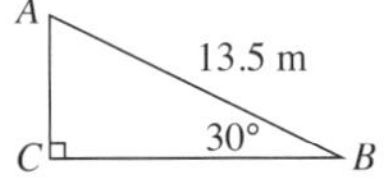

b

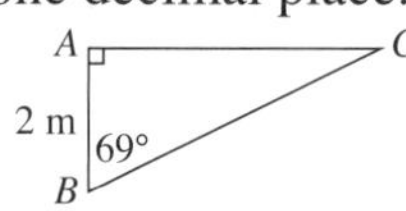

c

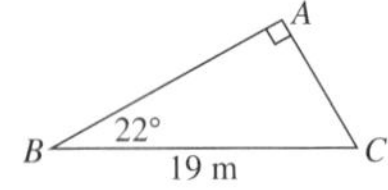

d

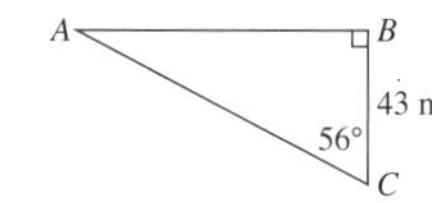

4

Find the value of □ to the nearest minute.

a

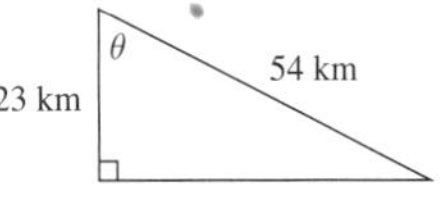

b

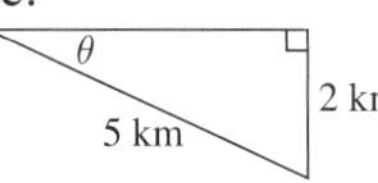

c

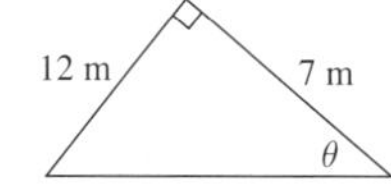

3

6 Jane is flying a kite on a 100 m string that makes an angle of 48° with the horizontal. How high is the kite above Jane’s hand? Give your answer correct to the nearest metre.

7 The diagonal of a rectangle makes an angle of 42° with one of the longer sides. If the length of the rectangle is 12 cm, find the length of the diagonal correct to one decimal place.

2

Total marks achieved for PART B

CHAPTER 10

Probability

UNIT 1: Review of basic probability

QUESTION 1 A die is thrown once. Find the probability that it shows:

a a six ____________ **b** a four ____________ **c** a seven ____________

d an even number ____________ **e** a number less than 4 ____________ **f** 5 or higher ____________

QUESTION 2 A bag contains 4 red balls, 5 blue balls and 1 white ball. If a ball is drawn at random, find the probability that it is:

a white ____________ **b** red ____________ **c** blue ____________

d not white ____________ **e** yellow ____________ **f** either blue or white ____________

QUESTION 3 From the letters of the word MATHEMATICS, one letter is selected at random. What is the probability that the letter is:

a a vowel? ____________ **b** a consonant? ____________ **c** the letter M? ____________

d the letter T? ____________ **e** the letter M or T? ____________ **f** the letter S? ____________

QUESTION 4 The numbers from 1 to 5 are written on separate cards. One card is chosen at random. What is the probability that the number is:

a odd? ____________ **b** zero? ____________ **c** even? ____________

d 5? ____________ **e** divisible by 3? ____________ **f** a prime number? ____________

QUESTION 5 A card is drawn at random from a normal pack of 52 cards. Find the probability that the card is:

a a spade ____________ **b** a black card ____________ **c** a queen ____________

d not a diamond ____________ **e** a red ten ____________ **f** a jack or king ____________

QUESTION 6 A bag holds 9 blue, 6 red and 3 yellow golf tees. A tee is randomly selected from the bag. What is the probability, as a fraction in simplest form, that the tee is:

a blue? ________ **b** red? ________ **c** yellow? ________

d red or blue? ________ **e** green? ________ **f** red, yellow or blue? ________

QUESTION 7 Complete.

The probability of any event is always in the range from ____________ to ____________.

Probability

Excel Mathematics Study Guide Years 9–10
Pages 186–203

UNIT 2: Relative frequency

QUESTION 1 Write the relative frequency of the score '3', as a fraction, in the following number sets.

a 8, 4, 2, 8, 4, 5, 3, 3 ______

b 1, 8, 10, 7, 1, 12, 8, 7, 3, 3 ______

c 3, 7, 6, 7, 7, 5, 6, 7, 3, 3, 3 ______

d 6, 9, 8, 9, 7, 9, 6, 5, 3, 3 ______

e 2, 9, 5, 9, 3, 9, 6 ______

f 7, 7, 5, 7, 5, 7, 3, 5, 7 ______

g 1, 8, 6, 8, 4, 3, 8, 2, 1 ______

h 3, 4, 4, 5, 6, 8, 5, 7, 5, 5, 4, 5, 3, 3 ______

i 4, 8, 3, 2, 5, 4, 8, 5 ______

j 6, 5, 6, 6, 7, 8, 10, 12, 6, 3, 3, 3 ______

k 3, 7, 9, 11, 12, 15, 9, 7, 7, 7, 9, 3, 3 ______

l 6, 8, 10, 12, 10, 11, 10, 10, 11, 3, 3 ______

m 2, 3, 3, 2, 4, 3, 3, 4, 2, 4 ______

n 3, 3, 4, 5, 4, 5, 4, 4, 4, 5 ______

o 3, 8, 7, 6, 8, 7, 8, 8, 6, 8, 8, 7 ______

p 5, 4, 9, 8, 7, 8, 8, 8, 7, 8, 8, 7, 3, 3, 3 ______

QUESTION 2 Complete the relative frequency column for the following tables. Give each answer as a decimal, correct to two decimal places.

a

Score (x)	Frequency (f)	Relative frequency
1	2	
2	4	
3	3	
4	2	
5	8	
6	4	
7	7	

b

Score (x)	Frequency (f)	Relative frequency
3	3	
6	5	
9	2	
12	6	
15	3	
18	4	
21	7	

c

Score (x)	Frequency (f)	Relative frequency
2	3	
4	2	
6	3	
8	4	
10	2	
12	4	
14	2	

d

Score (x)	Frequency (f)	Relative frequency
5	3	
10	5	
15	4	
20	6	
25	7	
30	10	
35	5	

e

Score (x)	Frequency (f)	Relative frequency
10	5	
20	4	
30	5	
40	6	
50	8	
60	5	
70	7	

f

Score (x)	Frequency (f)	Relative frequency
7	6	
14	8	
21	7	
28	4	
35	8	
42	10	
49	7	

UNIT 3: Experimental and theoretical probability

QUESTION 1 Annabel tossed a coin many times and the results were tabulated.

	Heads	Tails
Frequency	59	41

a How many times did Annabel toss the coin? ______

b What is the relative frequency of tossing heads? ______

c What is the theoretical probability of tossing heads? ______

d What is the relative frequency of tossing tails? ______

e What is the theoretical probability of tossing tails? ______

f What is the sum of the relative frequencies? ______

g How many tails do you expect to get in 100 tosses of a coin? ______

QUESTION 2 Lucy rolled a die many times and recorded the results.

a Complete the table showing the relative frequencies as fractions in simplest form.

Number	Frequency	Relative frequency
1	9	
2	15	
3	18	
4	12	
5	8	
6	10	

b From Lucy's experiment, find the probability of rolling:

i a 3 ______ **ii** an odd number ______

iii a 5 or 6 ______ **iv** a 1, 2 or 3 ______

c When a die is rolled what is the theoretical probability of rolling:

i a 3? ______ **ii** an odd number? ______

iii a 5 or 6? ______ **iv** a 1, 2 or 3? ______

d Comment on any similarities or differences between the answers to parts b and c.

______

______

______

Probability

Excel Mathematics Study Guide Years 9–10
Pages 186–203

UNIT 4: Expected results

QUESTION 1 If a die is rolled 48 times, how many times would you expect to get:

a a 2? ____________ **b** an even number? ____________ **c** a number less than 5? ____________

QUESTION 2 If this spinner is spun 60 times, how many times would you expect to spin:

a 6? ____________ **b** 2? ____________

c 1? ____________ **d** 5? ____________

e 3? ____________ **f** an odd number? ____________

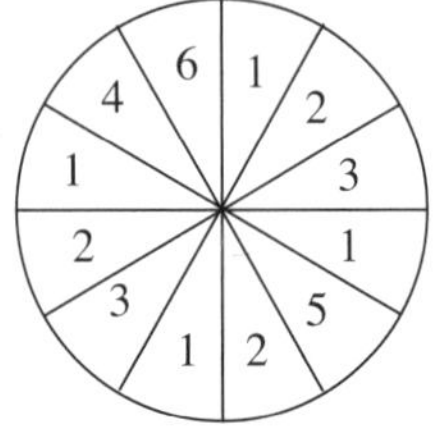

QUESTION 3 This spinner was spun 40 times. The number of times each colour was spun is shown in the table.

Colour	Red	White	Blue	Green	Yellow
Frequency	14	9	8	5	4

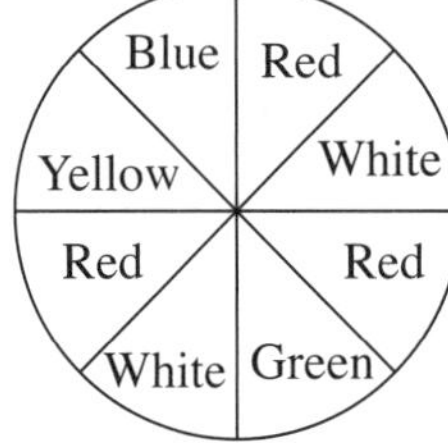

a Which colour was spun the same number of times as you would expect?

____

____

b Which colours were spun more times than you would expect?

____

QUESTION 4 Jade threw a die 100 times and recorded the results. She calculated that the relative frequency of the result 5 was 0.23. 'That is a lot higher than I would have thought' she said. Do you agree? Briefly comment, justifying your answer.

____

____

____

QUESTION 5 Jimmy believed a die was biased. He rolled the die 24 times and got these results.

Number	1	2	3	4	5	6
Frequency	3	2	6	3	9	1

a Do you agree that the die is biased? Justify your answer.

____

____

b What do you suggest that Jimmy should do to confirm his suspicions?

____

____

____

Probability

UNIT 5: Venn diagrams

QUESTION 1 The Venn diagram shows the sports played by some students.

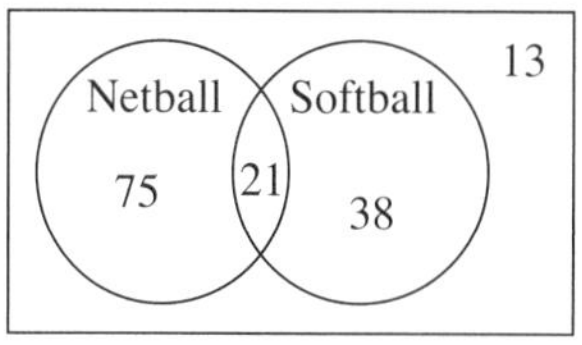

a How many students:

i were there altogether? ________________

ii played netball but not softball? ________________

iii played softball? ________________

b What is the probability that a randomly chosen student from the group plays:

i both netball and softball? ____________ **ii** neither netball nor softball? ____________

iii softball but not netball? ____________ **iv** netball or softball? ____________

v netball or softball but not both? ________________

QUESTION 2 In a class of 30 students who study either French or German or both, 18 study French and 22 study German.

a Draw a Venn diagram to represent this.

b How many study both languages? ________________

c If a student is randomly chosen from the class, what is the probability that she studies only French? ________________

QUESTION 3 A group of students were asked if they had read three books: *Fury*, *Battle Scars* and *Mannequin*. The results are shown in the Venn diagram. What is the probability that a randomly chosen student from the group has read:

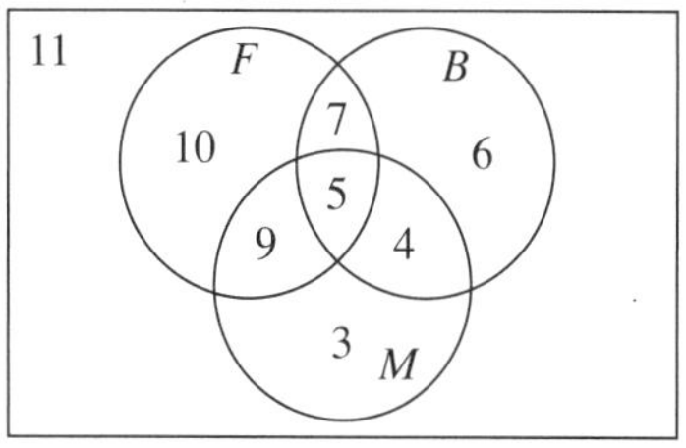

a all three books? ________________

b only *Mannequin*? ________________

c none of the books? ________________

d *Fury*? ________________

e exactly 2 of the books? ________________

f *Battle Scars* but not *Mannequin*? ________________

g *Fury* or *Battle Scars*? ________________

QUESTION 4 In a group of 35 students, 20 play tennis, 19 play hockey and 8 play both. What is the probability that a student, chosen at random from the group, plays:

a neither hockey nor tennis?

____

b tennis or hockey?

____

Probability

Excel Mathematics Study Guide Years 9–10
Pages 186–203

UNIT 6: Two-way tables

QUESTION 1 A survey was taken at a set of traffic lights. Cars were observed to see whether the driver was male or female and whether they were carrying passengers. The results are shown in the two-way table.

	Passengers	No passengers
Male driver	56	97
Female driver	32	91

a How many drivers were male? ______

b How many cars were carrying passengers? ______

c How many cars were counted altogether? ______

d What fraction of cars carrying passengers had a female driver? ______

e What percentage of cars being driven by a female had no passengers (give your answer to the nearest whole per cent)? ______

f What is the probability that a car chosen at random from the surveyed group:

i had a male driver? ______ **ii** carried passengers? ______

QUESTION 2 Liz conducted a survey of 75 men and 125 women to see whether they agreed or disagreed with a particular government policy. She drew up this table.

	Agreed	Disagreed	Total
Male	15	60	75
Female	35	90	125
Total	50	150	200

a What percentage of those who agreed were female? ______

b What percentage of females agreed? ______

c If a person is selected at random from the surveyed group, what is the probability that he or she:

i agreed with the policy? ______ **ii** is a female? ______

iii is a female who agreed with the policy? ______

QUESTION 3 300 people were tested to see if they had a particular disease. 95 tested positive and of those 58 were male. 60% of those who tested negative were female.

a Complete the two-way table showing this information.

	Positive	Negative	Total
Male			
Female			
Total			

b What is the probability that a randomly chosen:

i person tested negative? ______ **ii** male tested negative? ______

iii person who tested negative was male? ______

c Are males or females more likely to test positive? Justify your answer.

______

______

Probability

UNIT 7: Tree diagrams

QUESTION 1 Two unbiased coins are tossed at the same time.

a Complete the tree diagram and list the sample space.

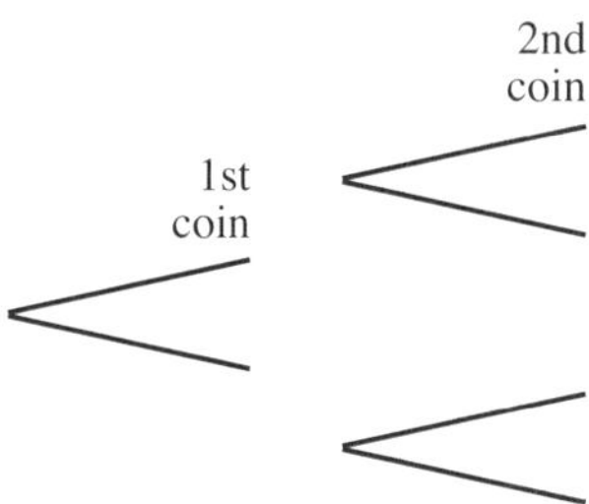

Use the tree diagram to find the probability of:

b two tails ______________________

c one tail and one head in any order ______________

d at least one head ______________________

e exactly two heads ______________

QUESTION 2 Four cards marked with the numbers 1, 2, 3 and 4 are placed in a box. Two cards are selected at random, one after the other without replacement, to form a two-digit number.

a Draw a tree diagram to show the possible outcomes.

b How many different two-digit numbers can be formed? ______________________

c What is the probability that the number formed is:

i less than 34? ______________ ______________

ii divisible by 3? ______________ ______________

iii even? ______________ ______________

Probability

Excel Mathematics Study Guide Years 9–10
Pages 186–203

UNIT 8: Two-step chance experiments

QUESTION 1 A box holds 5 blue, 4 red and 3 white marbles. One marble is taken at random from the box. It is then replaced and another marble is taken at random. What is the probability that:

a the first marble is blue? ________ **b** the second marble is blue? ________

c the first marble is red? ________ **d** the second marble is red? ________

e the first marble is white? ________ **f** the second marble is white? ________

QUESTION 2 A box holds 5 blue, 4 red and 3 white marbles. One marble is taken at random from the box. It is not replaced and another marble is taken at random. What is the probability that:

a the first marble is blue? ________ **b** the second marble is blue? ________

c the first marble is red? ________ **d** the second marble is red? ________

e the first marble is white? ________ **f** the second marble is white? ________

QUESTION 3 A die is rolled and a coin is tossed at the same time.

a List the possible outcomes.

____

____

b What is the probability of getting a head and a number greater than 4?

____

QUESTION 4 There are 6 black and 2 grey socks in a drawer. Without looking, two socks are taken from the drawer at the same time.

a If the first sock was black, what is the probability that both socks are black?

____

b What is the probability that both socks are grey?

____

QUESTION 5 Michelle has a box containing one red marble and two green marbles. She selects two marbles at random. Find the probability of her selecting:

a two green marbles if she replaces the first marble before she selects the second.

____

b one red marble if she does not replace the first marble.

____

____

Probability

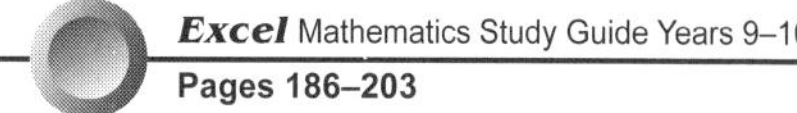

UNIT 9: Miscellaneous questions

QUESTION 1 A bag contains 3 red marbles and 5 blue marbles. Two marbles are drawn at random without replacement. Find the probability that:

a both the marbles are blue. **b** at least 1 marble is red.

QUESTION 2 John decides to play 2 games of tennis. He has an 80% chance of winning each game. What is the probability that:

a he will win the first game? **b** he will win both games?

QUESTION 3 A jar contains 6 white and 4 green jelly beans. Stacey takes a jelly bean at random and eats it. She then takes another jelly bean and eats it. What is the probability that:

a The first bean eaten is white? **b** both beans eaten are green?

QUESTION 4 There are 2 prizes in a raffle in which 50 tickets are sold. Dena buys 2 tickets. Find the probability that:

a she wins a prize in the first draw. **b** she wins both the prizes. **c** she does not win a prize.

QUESTION 5 A coin is tossed and a die is rolled. Calculate the probability of obtaining:

a a head and a 2 **b** a tail and an even number

QUESTION 6 A box contains 6 red pens and 8 blue pens. A pen is drawn at random and then replaced. A second pen is then drawn. What is he probability that:

a both pens are blue? **b** both pens are red? **c** one pen is blue and the other is red?

QUESTION 7 In an experiment, a card is drawn from a pack of playing cards and a coin is tossed. What is the probability of getting:

a an ace and a head? **b** the queen of hearts and a tail?

QUESTION 8 In another experiment, a die is thrown and a card is drawn from a pack of playing cards. What is the probability of getting:

a an even number and a king? **b** a 6 and a diamond?

Probability

TOPIC TEST — PART A

Instructions
- This part consists of 10 multiple-choice questions.
- Fill in only ONE CIRCLE for each question.
- Each question is worth 1 mark.

Time allowed: 10 minutes **Total marks: 10**

Marks

1 In a single throw of a die, the probability of obtaining a number greater than 3 is

 (A) $\frac{1}{2}$ (B) $\frac{1}{6}$ (C) $\frac{1}{3}$ (D) $\frac{2}{3}$ 1

2 A card is chosen at random from a standard pack of 52 cards. What is the probability that the card is red or a queen?

(A) $\frac{1}{4}$ (B) $\frac{1}{13}$ (C) $\frac{9}{13}$ (D) $\frac{7}{13}$ 1

3 A carton of eggs contains 3 brown eggs and 9 white eggs. Two eggs are chosen at random. What is the probability that both are brown?

(A) $\frac{9}{44}$ (B) $\frac{6}{11}$ (C) $\frac{19}{44}$ (D) $\frac{1}{22}$ 1

4 In a simultaneous tossing of 2 coins, the probability of obtaining 2 tails is

(A) $\frac{1}{2}$ (B) $\frac{3}{4}$ (C) $\frac{1}{4}$ (D) $\frac{1}{3}$ 1

5 A bag contains 2 red balls, 4 white balls and 1 green ball. Calculate the probability of selecting at random, in 2 draws without replacement, 1 red and 1 white ball.

(A) $\frac{8}{21}$ (B) $\frac{4}{21}$ (C) $\frac{7}{42}$ (D) $\frac{20}{21}$ 1

6 A jar contains 20 lollies of which 10 are red, 6 are green and 4 are yellow. Two lollies are chosen at random, the first being eaten before the second is selected. Find the probability that neither of the lollies is green.

(A) $\frac{9}{38}$ (B) $\frac{91}{190}$ (C) $\frac{2}{19}$ (D) $\frac{13}{38}$ 1

7 This spinner was spun 50 times. It showed green 20 times and blue 22 times. What is the relative frequency of red?

red
blue
green
green
blue

(A) $\frac{4}{25}$ (B) $\frac{2}{5}$ (C) $\frac{11}{25}$ (D) $\frac{10}{11}$ 1

8 The Venn diagram shows the number of students who play cricket and soccer. What is the probability that a randomly chosen student from the group plays cricket or soccer?

C
S
9
12
5
17

(A) $\frac{29}{34}$ (B) $\frac{29}{43}$ (C) $\frac{24}{43}$ (D) $\frac{34}{43}$ 1

9 There are 3 red and 2 blue balls in a bag. Two balls are taken from the bag one after the other without replacement. The first ball was red. What is the probability that the second ball was also red?

(A) $\frac{2}{5}$ (B) $\frac{1}{2}$ (C) $\frac{3}{5}$ (D) $\frac{3}{4}$ 1

10 A die was rolled many times. The table shows the results. What number occurred the number of times that was expected?

Number	1	2	3	4	5	6
Frequency	10	11	7	9	15	8

(A) 1 (B) 2 (C) 4 (D) 6 1

Total marks achieved for PART A

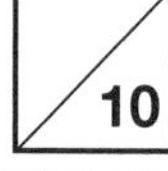
/10

Probability

TOPIC TEST — PART B

Time allowed: 20 minutes **Total marks: 15**

Marks

1 The numbers 1, 2, 3 and 4 are written on separate cards. One card is drawn at random to give the tens digit of a 2-digit number. Another card is drawn at random to give the units digit. What is the probability that the number formed is:

a odd? __________ **b** divisible by 2? __________

c greater than 32? __________ 3

2 The scores in a quiz are shown in the table on the right.

Score	4	5	6	7	8	9	10
Frequency	1	1	2	5	6	7	3

a What is the relative frequency of the score 7? __________

b Based on these results what is the percentage chance of scoring 10? __________ 2

3 In a survey some students were asked which of two options A or B they preferred. The table shows the results.

	Option A	Option B
Male	51	37
Female	44	68

a How many students were surveyed?

____

b What is the probability that a randomly chosen student from the group is a male who prefers option B? __________

c What is the probability that a male student from the group prefers option B?

____

d What is the probability that a student who prefers option B is a male?

__________ 4

4 A box holds 1 red and 2 blue pens. A pen is taken from the box, without looking, and then replaced. Another pen is then taken at random.

a Draw a tree diagram to show the possible outcomes.

b What is the probability that both pens are blue? __________

c What is the probability that one is red and one is blue? __________ 3

5 A box holds 1 red and 2 blue pens. A pen is taken from the box, without looking, and is not replaced. Another pen is then taken at random.

a Draw a tree diagram to show the possible outcomes.

b What is the probability that both pens are blue?

____

c What is the probability that one is red and one is blue?

__________ 3

Total marks achieved for PART B /15

CHAPTER 11
Data representation and analysis

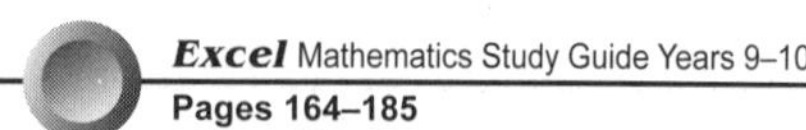

UNIT 1: Review of statistics

QUESTION 1 Fifty families were surveyed to find how many children each family had and the following set of data was obtained.

5 3 2 4 1 5 0 2 3 2 2 1 1 3 3
4 1 3 2 1 3 3 2 2 2 3 2 1 3 1
2 3 0 1 1 5 3 4 5 0 3 0 2 0 2
2 1 5 4 3

a Complete the frequency distribution table.

Score (x)	Tally	Frequency (f)	$f \times x$	Cumulative frequency
0				
1				
2				
3				
4				
5				
		$\Sigma f =$	$\Sigma fx =$	

b Draw a frequency histogram and polygon.

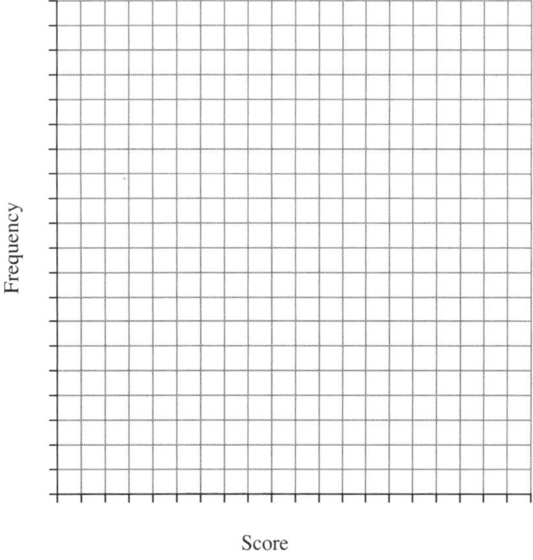

c Draw a cumulative frequency histogram and polygon.

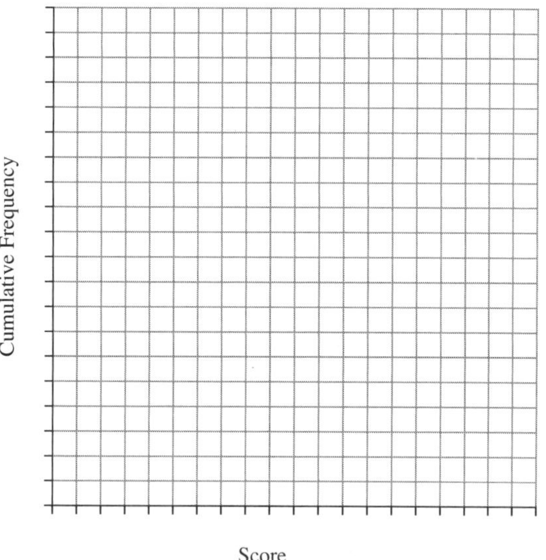

d Find:

i the mean

____

ii the mode

____

iii the median

____

iv the range

____

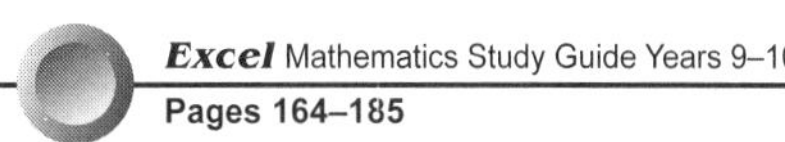

UNIT 2: Mean, mode, median and range

QUESTION 1 Find the mean, mode, median and range for each set of scores.

a 4, 5, 5, 7, 7, 8, 8, 8, 9, 9

mean = ______ mode = ______

median = ______ range = ______

b 16, 18, 15, 11, 15, 12, 17, 13, 14, 18

mean = ______ mode = ______

median = ______ range = ______

c 8, 11, 16, 13, 12, 13, 16, 11, 8, 7, 8

mean = ______ mode = ______

median = ______ range = ______

d 56, 60, 68, 49, 66, 87, 67, 56

mean = ______ mode = ______

median = ______ range = ______

e 2, 2, 3, 4, 4, 5, 5, 6, 5, 6

mean = ______ mode = ______

median = ______ range = ______

f 4, 8, 8, 9, 9, 9, 9, 9

mean = ______ mode = ______

median = ______ range = ______

g 2, 3, 3, 2, 4, 2, 5, 6, 5, 3, 3

mean = ______ mode = ______

median = ______ range = ______

h 52, 17, 18, 52, 53, 54, 52, 52, 53, 52

mean = ______ mode = ______

median = ______ range = ______

QUESTION 2 Complete the table, then find the mean (to 1 decimal place), mode, median and range.

a

Score (x)	Frequency (f)	$f \times x$	Cumulative frequency
1	3		
2	6		
3	8		
4	7		
5	5		
6	4		

mean = ______ mode = ______

median = ______ range = ______

b

Score (x)	Frequency (f)	$f \times x$	Cumulative frequency
5	12		
6	19		
7	18		
8	15		
9	10		
10	13		

mean = ______ mode = ______

median = ______ range = ______

c

Score (x)	Frequency (f)	$f \times x$	Cumulative frequency
16	8		
17	6		
18	7		
19	10		
20	5		

mean = ______ mode = ______

median = ______ range = ______

d

Score (x)	Frequency (f)	$f \times x$	Cumulative frequency
16	5		
17	7		
18	8		
19	14		
20	6		

mean = ______ mode = ______

median = ______ range = ______

Data representation and analysis

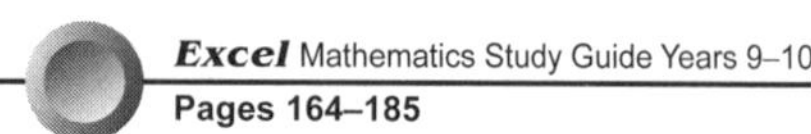

UNIT 3: Using the mean, mode and median

QUESTION **1** A foreign language class has just 6 students. The class sat for a test and the following marks resulted. 7, 93, 95, 96, 96, 99

a Find.

i the median **ii** the mean **iii** the mode

b Barry scored 93. 'I did well in the test,' Barry told his mother. 'I was way above average.' Do you agree with Barry's statement? Briefly comment.

QUESTION **2** When talking about real-estate, people in the industry and the media refer to the median house price. Why is the median a better means of describing the data than the mean or mode?

QUESTION **3** A shop sells women's clothes. The table shows the numbers of each size of dress sold over the previous month.

Size	8	10	12	14	16	18	20	22	24
Number sold	2	13	28	42	35	26	23	19	21

a Find the mean dress size.

b What is the modal dress size?

c What is the median?

d The shop owner is most interested in the modal dress size. Why do you think she would find that important?

Data representation and analysis

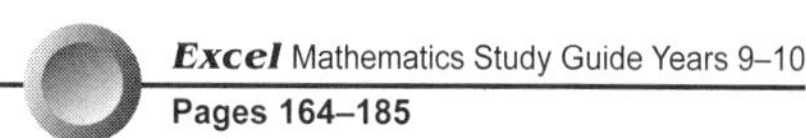

UNIT 4: Measures of location and spread

QUESTION 1

a The mean of 8 scores is 19. Another score of 11 is included with the scores. What is the new mean?

____

b After 7 tests Alice has an average (mean) mark of 78%. What would Alice need to score in her next test to increase her average to 80%?

____

QUESTION 2

The prices of 11 books have been listed below.

\$30 \$30 \$35 \$45 \$55 \$45 \$35 \$34 \$50 \$35 \$35

a What is the mean price? ______

b What is the modal price? ______

c What is the range? ______

d What is the median? ______

Another book priced at \$39 is included with the list.

e What is the new mean price? ______

f What is the new modal price? ______

g What is the new range? ______

h What is the new median? ______

i Briefly comment on any changes to the measures of location and spread.

____

QUESTION 3

These points were scored by each of the nine members of a basketball team in a game.

5, 5, 9, 10, 4, 3, 5, 12, 5

a How many points did the team score in the match? ______

b Find the mean number of points scored by each player. ______

c Find the median. ______

d What was the mode? ______

e What is the range? ______

These are the numbers of points scored by each member of the opposing team in the same game.

4, 6, 9, 7, 6, 8, 6, 10, 7

f Who won the game? Justify your answer. ______

g Which team had the lowest range? What conclusions can be drawn from this? ______

QUESTION 4

The following are a set of scores: 5, 14, 15, 18, 19, 16, 19, 17, 15

a Find the mean. ______

b Find the median. ______

c Identify the outlier. ______

d How does the outlier affect the mean? Justify your answer. ______

e Does the outlier affect the median? Justify your answer. ______

Data representation and analysis

Excel Mathematics Study Guide Years 9–10
Pages 164–185

UNIT 5: Skewed displays

QUESTION 1 For each of the following histograms, state whether the distribution is symmetrical, positively skewed, negatively skewed or none of these.

a

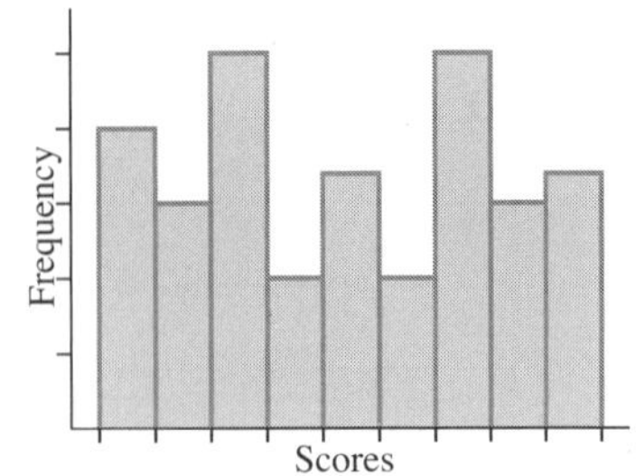

b

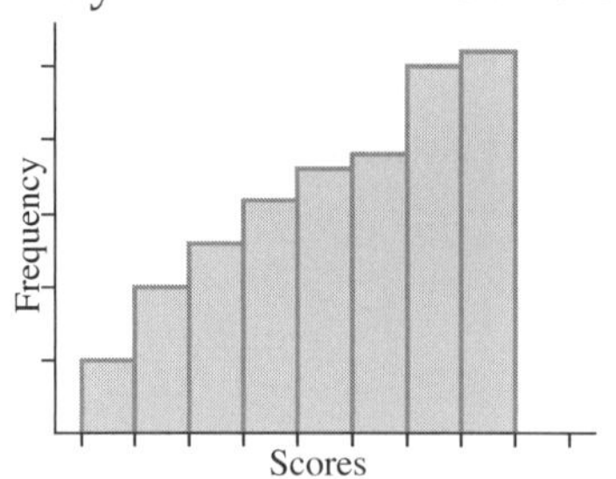

c

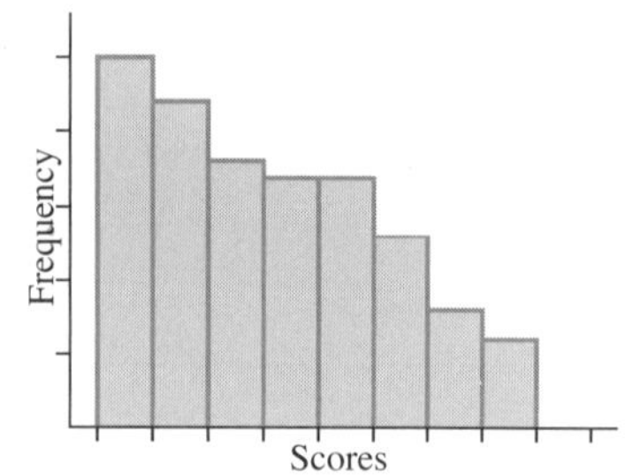

d

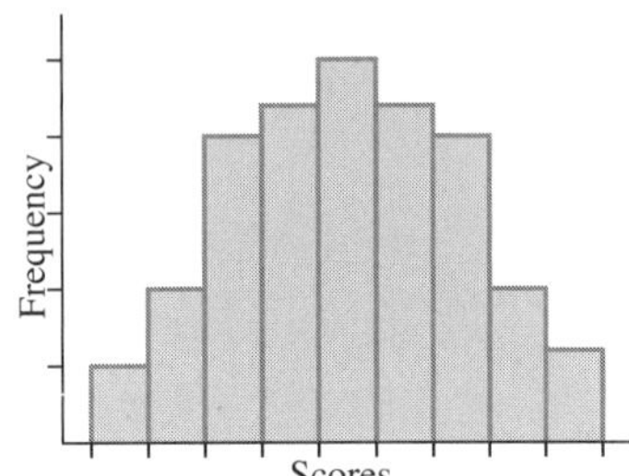

e

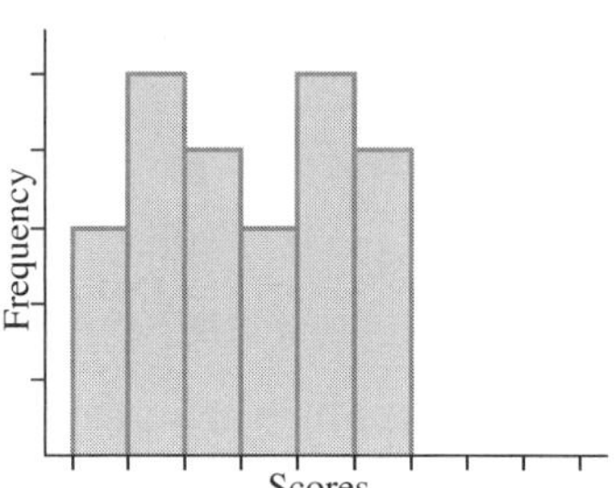

f

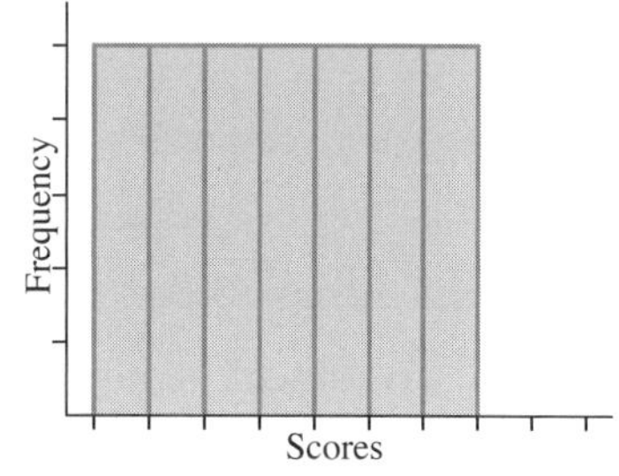

g

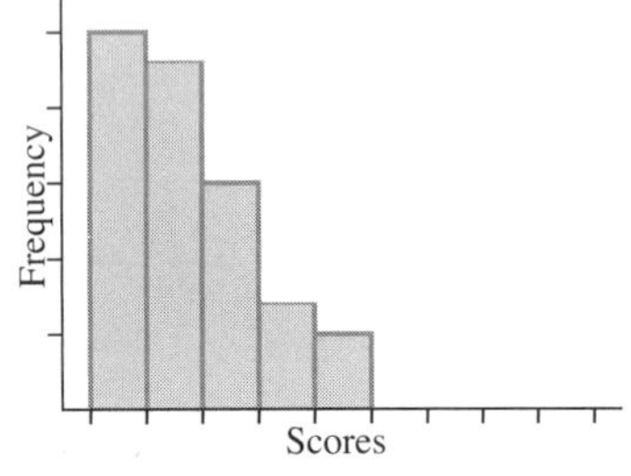

h

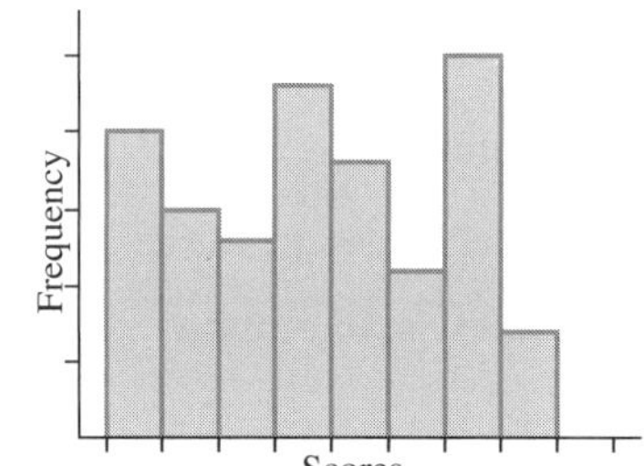

i

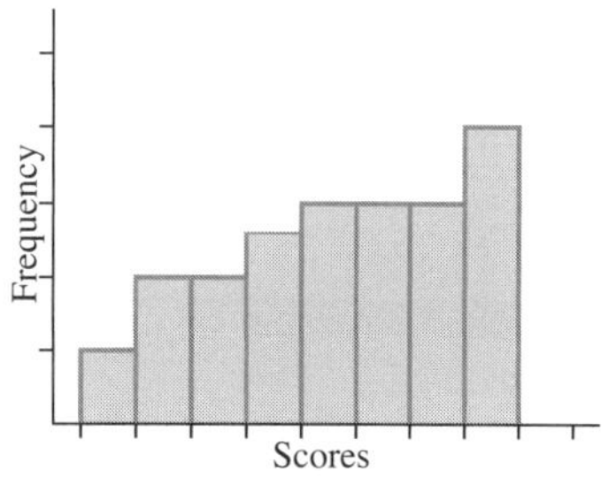

QUESTION 2 State whether the distribution data in these dot plots is symmetrical, positively skewed, negatively skewed or none of these.

a

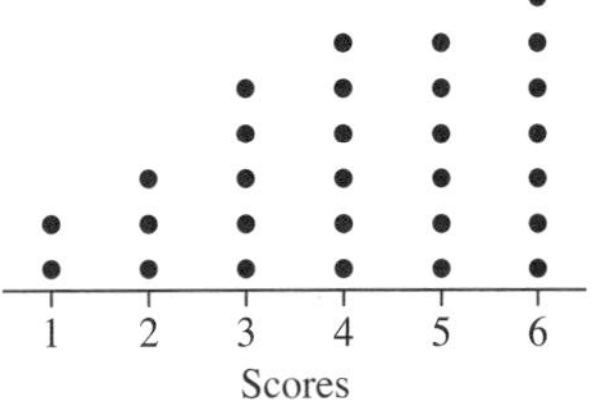

b

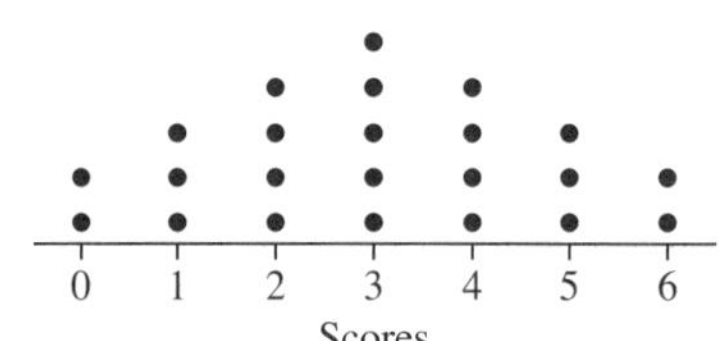

c

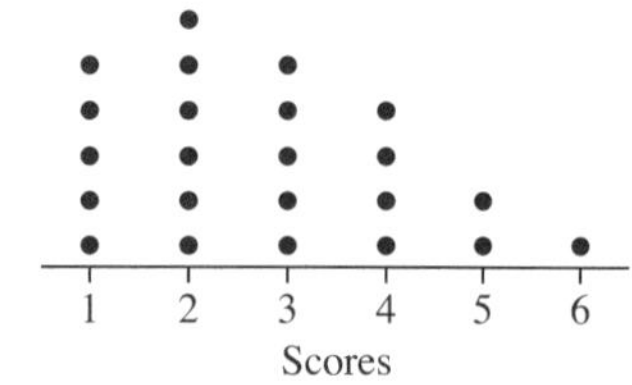

Data representation and analysis

UNIT 6: Description of data

QUESTION 1 The frequency histogram shows the scores by students in a class test.

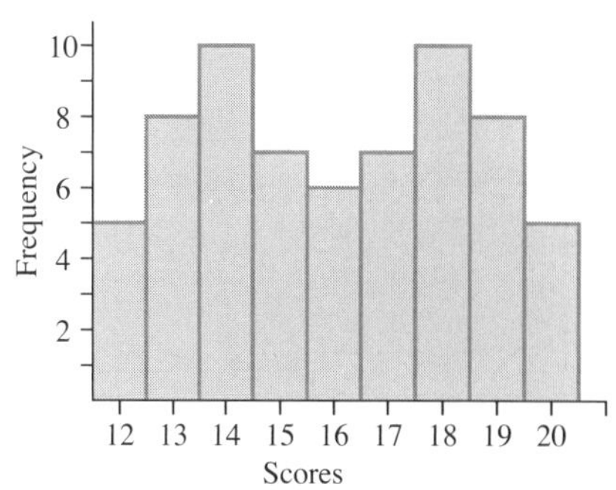

a Is the graph symmetrical? ______________________

b Find the mode(s). ______________________

c Is the data bimodal? ______________________

d Briefly explain why the mean and median can easily be seen from the histogram and give their values.

____

____

QUESTION 2 Briefly describe the data shown in each histogram. Is the display symmetrical or skewed? If skewed, is it positive or negative? Is there a single mode or is the data bimodal?

a

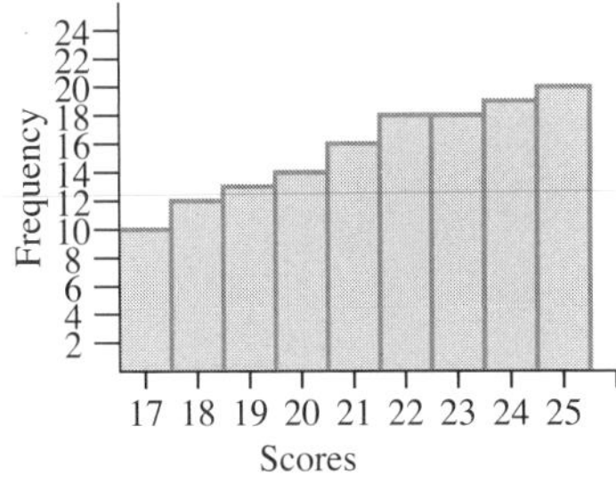

____

____

b

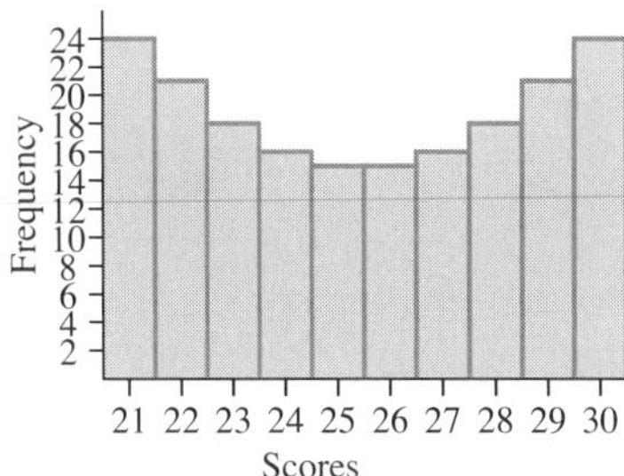

____

____

c

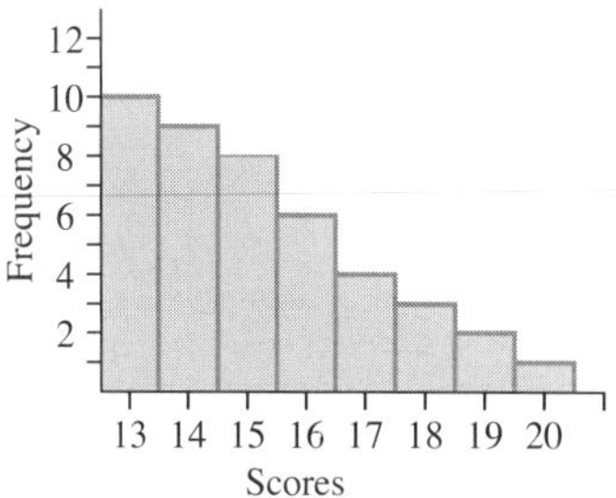

____

____

QUESTION 3 From the cumulative frequency histograms, determine whether the data is symmetrical or skewed and describe any skewness.

a

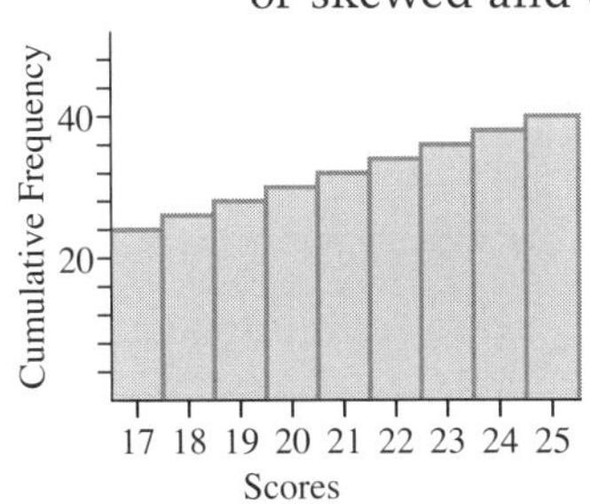

____

b

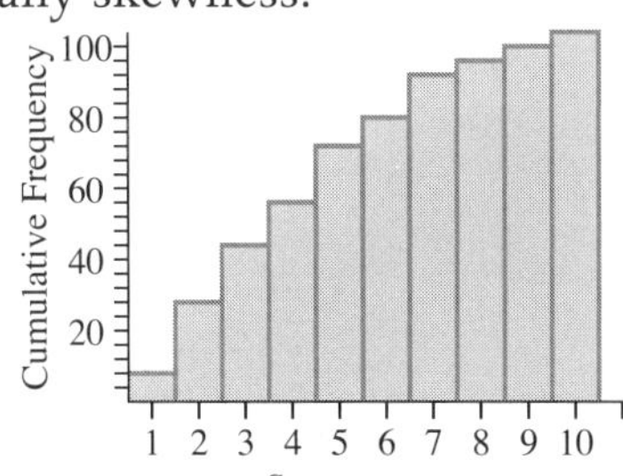

____

c

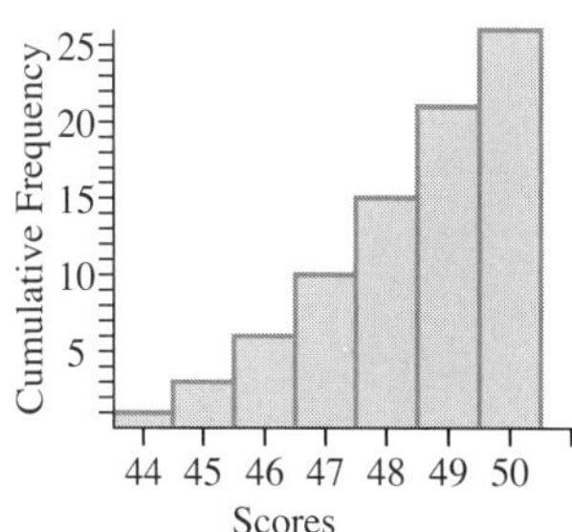

____

QUESTION 4 From these stem-and-leaf plots shown, determine whether the data is symmetrical or skewed. Describe any skewness.

a

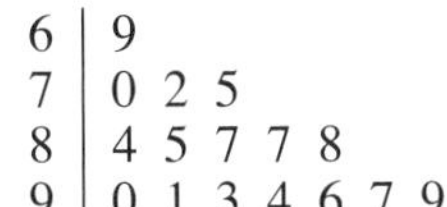

Stem	Leaf
6	9
7	0 2 5
8	4 5 7 7 8
9	0 1 3 4 6 7 9

____

b

Stem	Leaf
12	2 3
13	0 5 7 8
14	1 3 6 6 7 9
15	3 4 6 7
16	5 8

____

c

Stem	Leaf
0	1 2 4 4 7 9
1	1 1 3 6 8
2	4 5 7 9
3	0 6 8
4	3

____

Data representation and analysis

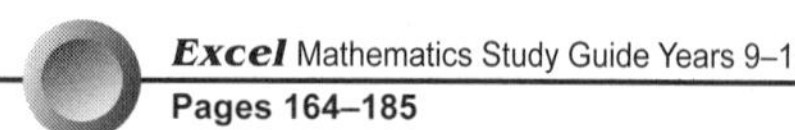

UNIT 7: The shape of a display

QUESTION 1 Using data from the frequency histograms below, how many modes are there, is it smooth, and is it skewed?

a

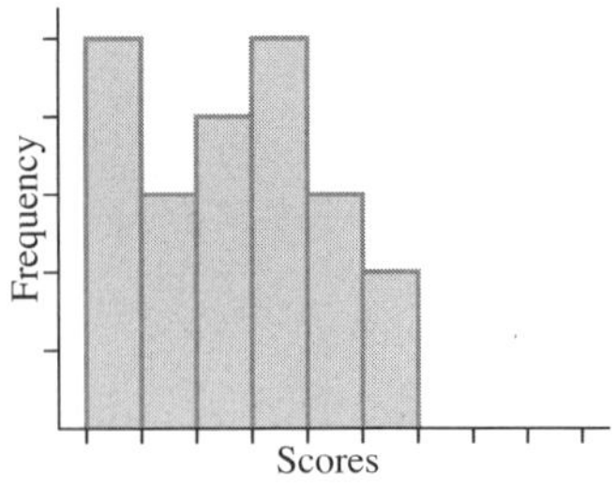

b

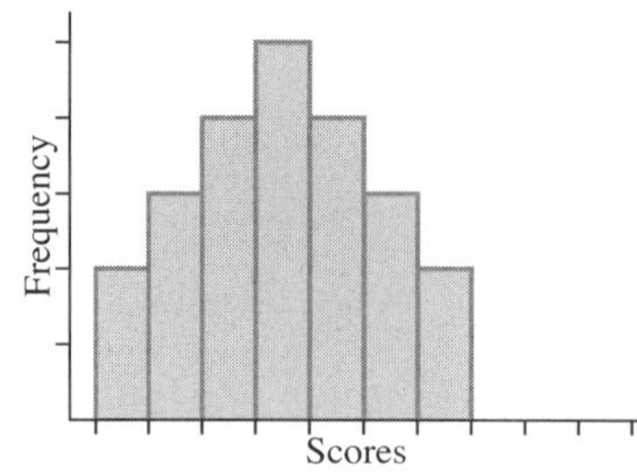

c

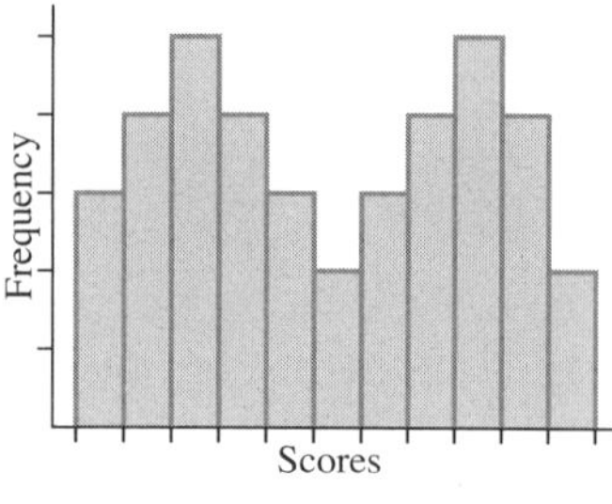

QUESTION 2 A group of 25 students was surveyed prior to an examination. The number of hours they spent studying for the exam was recorded. The results appear below.

7	20	15	36	18	12	9	24	25	40	5	32	
45	16	29	28	19	25	30	36	38	27	42	22	30

a Show the results in an ordered stem-and-leaf plot.

b What is the median? ______________

c Does the shape of the display indicate any skewness?

d What conclusions can you make about the data?

QUESTION 3 A dot plot has been drawn to show the results of a test.

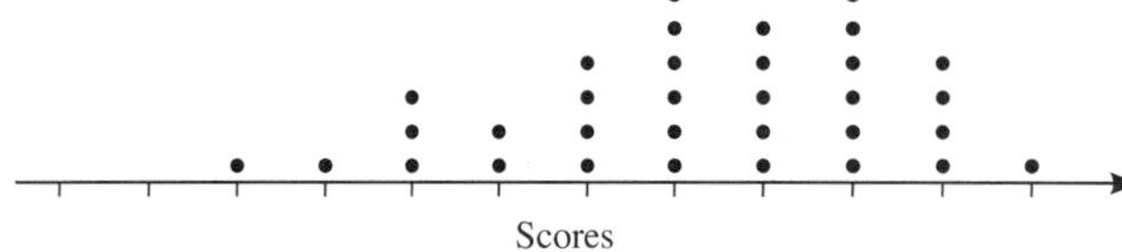

a How many modes are there?

b Is there any skewness?

c What other information, if any, can you gain about the test?

Data representation and analysis

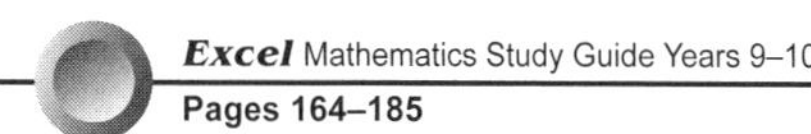

UNIT 8: Back-to-back stem-and-leaf plots

QUESTION 1 The following sets of data show the assessment marks (as percentages) for 2 different tasks given to a group of students.

Task A	8	10	15	30	32	34	35	38	43	52
	56	61	63	68	70	71	73	78	79	80
	84	84	92	92	92	92	92	98	98	99
Task B	2	6	7	12	15	21	22	23	23	25
	28	31	31	31	31	48	52	53	56	62
	69	69	69	81	85	88	93	94	99	100

a Display these sets of data on a back-to-back stem-and-leaf plot.

b Find the range of task *A* ____________________

c What is the range of task *B*? ____________________

d Find the range of both the tasks combined ____________________

e What is the total number of students? ____________________

f Find the mode for task *A* ____________________

g Find the mode for task *B* ____________________

h What is the median for task *A*? ____________________

i What is the median for task *B*? ____________________

j Which task did students find easier? ____________________

Marks (%)		
Task *A*		Task *B*

QUESTION 2 A survey was conducted into the number of magazines bought by a group of people, males and females, in 1 year. The results are displayed in the following back-to-back stem-and-leaf plot.

a What is the mode for the whole group? ____________________

b Find the range for the females ____________________

c Find the range for the males ____________________

d Which group has the higher median? Justify your answer.

____

e What is the mean number of magazines bought by males? Give the answer to 1 decimal place.

____

Magazines bought		
Males		Females
3 1 1 0	0	0 1 5 5 5 6
7 6 4 3 2	1	2 2 2 2 2 7
1 1 0	2	0 3 5
3 3 2 1	3	0 1 3 5
3	4	2 6 7
	5	1

f What is the mean number of magazines bought by females? Give the answer to 1 decimal place.

____

g Briefly comment on the location of scores for males and for females

____

h Describe differences in the shape of the distributions.

____

____

UNIT 9: Dot plots

QUESTION 1 Fifty families were surveyed to find how many children each family had. The following data was obtained.

5	3	2	4	1	5	0	2	3	2	2	1	1
3	3	4	1	3	2	1	3	3	2	2	2	3
2	1	3	1	2	3	0	1	1	5	3	4	5
0	3	0	2	0	2	2	1	5	4	3		

a Draw a dot-plot for this data.

0 1 2 3 4 5

b What information can be seen from the dot plot?

QUESTION 2 The following data shows the number of hours each of 30 students in each of two classes watched television programs in one month.

Class A:

6	8	8	7	10	6	6	7	8	12
8	7	6	6	6	9	9	8	6	9
11	9	6	9	6	8	9	12	13	6

Class B:

8	10	11	6	9	7	9	8	8	12
9	10	6	11	10	12	8	6	10	9
11	8	7	9	7	11	9	10	7	12

a Draw a dot plot for each set of data.

Class A:

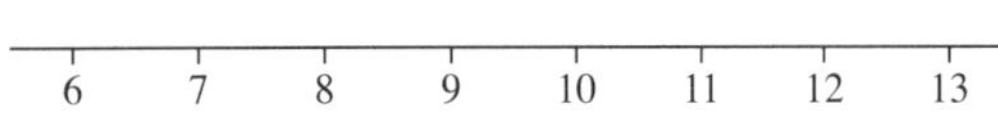

Class B:

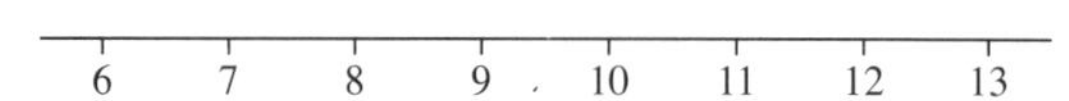

b Compare the two data sets.

QUESTION 3 The following are two sets of data.

A:

3	6	3	2	5	7	3	4	6	5
4	3	3	4	5					

B:

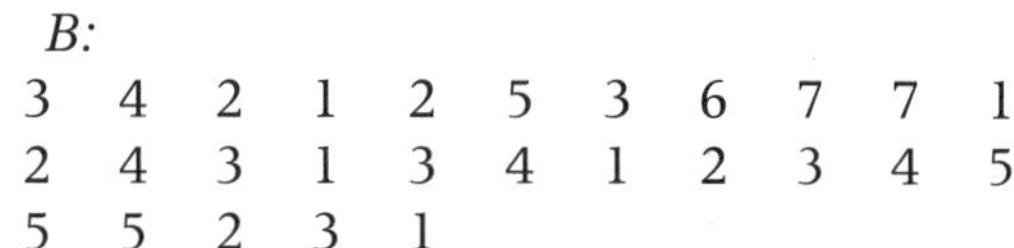

3	4	2	1	2	5	3	6	7	7	1
2	4	3	1	3	4	1	2	3	4	5
5	5	2	3	1						

a Sketch a dot plot for each data set.

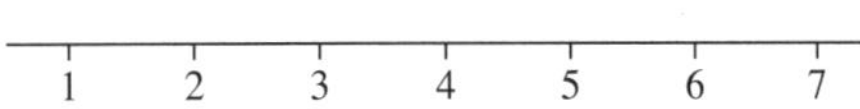

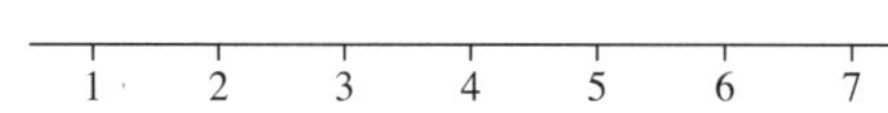

b Compare and contrast the two data sets.

Data representation and analysis

TOPIC TEST — PART A

Instructions
- This part consists of 10 multiple-choice questions.
- Fill in only ONE CIRCLE for each question.
- Each question is worth 1 mark.

Time allowed: 10 minutes **Total marks: 10**

Marks

1 The mean of the scores 7, 8, 8, 8, 9, 9, 10, 10, 12, 19 is

(A) 7 (B) 8 (C) 9 (D) 10 [1]

2 The mean of the numbers 10, 12 and x is the same as the mean of the numbers 10, 12, 14 and 16. Find the value of x.

(A) 13 (B) 15 (C) 17 (D) 30 [1]

3 What is the mode of this set of scores at right?

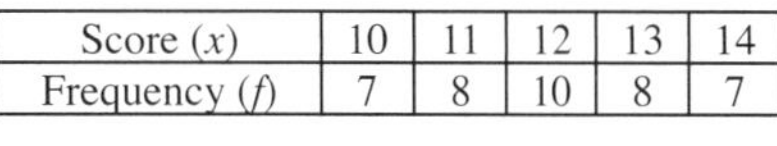

Score (x)	10	11	12	13	14
Frequency (f)	7	8	10	8	7

(A) 10 (B) 11
(C) 12 (D) 14 [1]

4 The average mass of three students is 60 kg. A fourth student of mass 72 kg joins the group. What is the average mass of the four students?

 63 (B) 62 (C) 61 (D) 60 [1]

5 The marks of 10 students in a test were as follows: 3, 4, 4, 5, 5, 5, 6, 9, 9, 10. However, the student whose mark was 6 should have obtained 7 marks. Which one of the following would have been affected by the change in this mark?

 (B) mean (C) median (D) range [1]

6 The range of scores shown in the dot plot is

 (B) 5
(C) 6 (D) 7 [1]

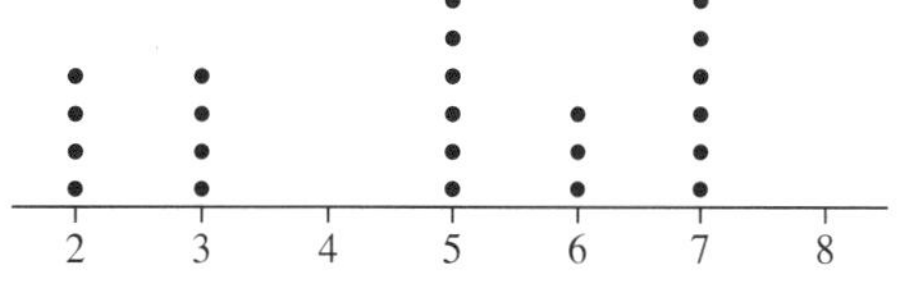

7 The median of the scores shown in the dot plot is

 (B) 5 (C) 6 (D) 7 [1]

8 Which data display is negatively skewed?

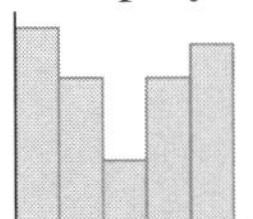
(B)
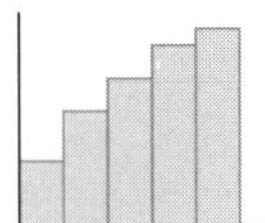
(C)
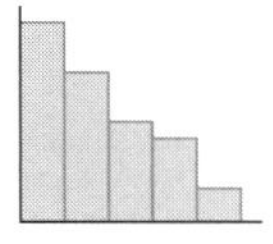
(D)
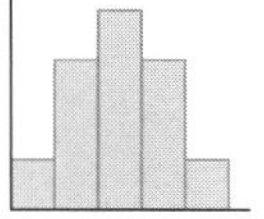
[1]

9 Which set of scores is the most consistent?

 5, 7, 9, 11, 13, 15
(B) 9, 12, 13, 13, 14, 15
(C) 11, 12, 12, 13, 13, 14
(D) 10, 11, 14, 14, 14, 15

10 Which set of data is bimodal?

(A)
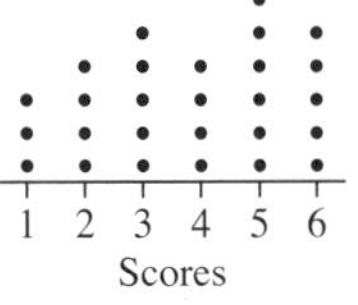

(B)
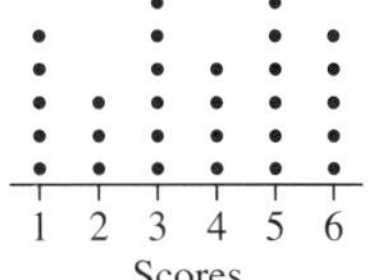

(C)
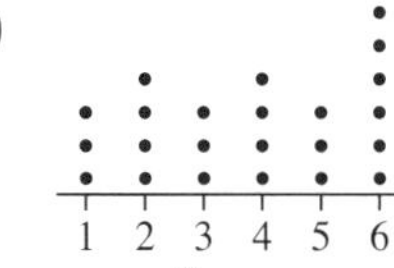

(D)
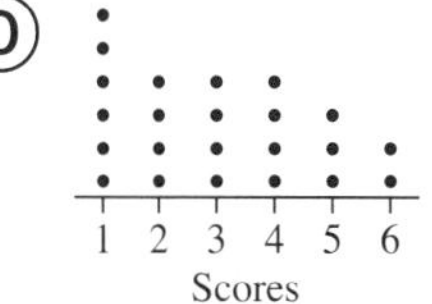

[1]

Total marks achieved for PART A

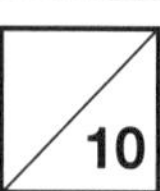

Data representation and analysis

TOPIC TEST — PART B

Time allowed: 20 minutes **Total marks: 15**

Marks

1 The results of a spelling test (out of 10) are given below:

7 6 5 4 3 5 8 10 6 3
7 4 6 4 5 8 10 9 9 6
8 7 5 7 8 7 9 6 8 4
6 5 7 4 3 8 7 7 6 5

Score (x)	Tally	Frequency (f)	$f \times x$	Cumulative frequency
3				
4				
5				
6				
7				
8				
9				
10				

a Complete the frequency distribution table for this data.

b What is the range? ____________ **c** What is the mode? ____________

d Find the median. ____________ **e** Find the mean. ____________ 5

2 The back-to-back stem-and-leaf plot shows the height (in cm) of two groups of students.

Males		Females
	13	8
	14	0 5 7 9
9 6	15	1 2 2 5 7 8
9 9 8 7 5 3 2	16	0 3 4 6 8 9
9 7 6 6 5 4 1 1	17	1 4 5
3 2 0	18	

a Which group has the greatest range and by how many?

____

b What is the median height for the females?

____

c How much taller is the median height for males than for females?

____

d What is the modal height of all the students? ______________________________

e Briefly comment on any similarities or differences between the two data sets.

______________________________ 5

3 The dot plots show the scores by the same group of students in different tests.

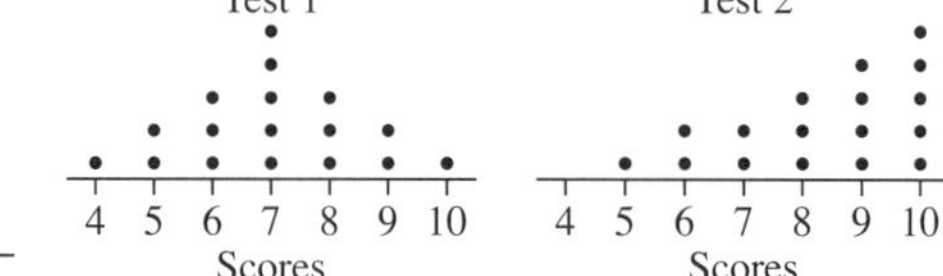

a What is the mean of Test 1?

____

b How much greater is the mean in Test 2 than Test 1?

____

c How much greater is the median in Test 2 than Test 1? ____________

d Which test shows more consistent results? Justify your answer.

____

e Briefly comment on similarities and differences between the two data sets.

______________________________ 5

Total marks achieved for PART B

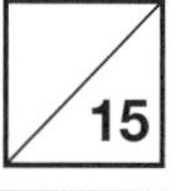

Exam Paper 1

Instructions for all parts
Time allowed: 1 hour

- Attempt all questions.
- Calculators are allowed.

Part A: Allow about 10 minutes for this part.
Part B: Allow about 20 minutes for this part.
Part C: Allow about 30 minutes for this part.

Total marks: 50

EXAM PAPER 1 PART A

Fill in only one circle for each question.

Marks

1 If $x = 3 - 2t^2$ and $t = 3$ then x is equal to **1**

Ⓐ –526 Ⓑ –33 Ⓒ –15 Ⓓ 9

2 In the diagram length x is equal to **1**

Ⓐ $a^2 + b^2$ Ⓑ $\sqrt{a + b}$ Ⓒ $\sqrt{a^2 + b^2}$ Ⓓ none of these

3 The mode of the set of scores 3, 2, 6, 1, 2, 7, 9, 2, 6 is **1**

Ⓐ 2 Ⓑ 9 Ⓒ 3 Ⓓ 4.2

4 \$500 invested for 2 years at 10% p.a. simple interest becomes **1**

Ⓐ \$600 Ⓑ \$550 Ⓒ \$625 Ⓓ \$650

5 If 516 831 is rounded off to the nearest thousand, the number is **1**

Ⓐ 51 800 Ⓑ 520 000 Ⓒ 517 000 Ⓓ 500 000

6 $2a^3 \times 3a^2 =$ **1**

Ⓐ $5a^5$ Ⓑ $5a^6$ Ⓒ $6a^5$ Ⓓ $6a^6$

7 1.25 km = **1**

Ⓐ 125 m Ⓑ 12 500 mm Ⓒ 1250 m Ⓓ 12 500 m

8 What are the coordinates of the point of intersection of the lines $x = -2$ and $y = 5$? **1**

Ⓐ (–2, 5) Ⓑ (2, –5) Ⓒ (5, –2) Ⓓ (–5, 2)

9 *SPRQ* ||| *ABCD*. What is the size of $\angle QSP$? **1**

Ⓐ 70° Ⓑ 80° Ⓒ 90° Ⓓ 120°

10 Five kittens are in a box. Three are black. Two kittens are taken from the box. The first of these kittens is black. What is the probability that both of the kittens are black? **1**

Ⓐ $\frac{1}{2}$ Ⓑ $\frac{3}{5}$ Ⓒ $\frac{2}{5}$ Ⓓ $\frac{3}{10}$

Total marks achieved for PART A /10

EXAM PAPER 1 PART B

Write only the answer in the answer column. For any working use the question column.

Questions	Answers	Marks
1 Write 0.000 406 in scientific notation.		1
2 Simplify $3a - 2ab + 5ab + 6a$		1
3 Does the point $(-3, -1)$ lie on the line $3x - 5y = 12$?		1
4 What is the simple interest on \$250 at 12% p.a. for 3 years?		1
5 What is the median of 11, 12, 10, 11, 12, 12?		1
6 What is the gradient of the line $y + \frac{2}{3}x = 3$?		1
7 Evaluate $\frac{7.8 \times 6.216^2}{\sqrt{3.5} + 2.9}$ (correct to one decimal place).		1
8 What is the value of $\tan\theta$? (Right-angled triangle with sides 5, 3, 4; θ between sides 5 and 3)		1
9 A photo that was 8 cm long and 5 cm wide has been enlarged. It is now 28 cm long. How wide is it?		1
10 Two coins are tossed at the same time. What is the probability that both show heads?		1
11 What is the equation of the circle with centre (0, 0) and radius 5 units?		1
12 A die is rolled 60 times. How many times would you expect to get 4?		1
13 Write 5^{-2} without indices.		1
14 A triangular prism has base area 18 m^2 and perpendicular height 2.5 m. What is its volume?		1
15 A train travelled at an average speed of 84 km/h. How far did it travel in $6\frac{1}{4}$ hours?		1

Total marks achieved for PART B /15

EXAM PAPER 1 PART C

Show all working for each question.

Marks

1 A straight line $y = px - 5$ passes through the point $(1, -2)$. Find the value of p. 1

2 Solve. $\frac{3a}{5} - 7 = a$ 1

3 Expand and simplify.

a $5a - 2(3a - 6)$ 1

b $x(3x + 4) - 2(3x + 4)$ 1

4 A leaking tap loses water at a rate of 3 mL/min. How many litres will leak out in one day? 1

5 Calculate the tax payable on $14 200 if $870 is payable on the first $11 000 and 30 cents for each dollar over $11 000. 1

6 The diagram shows a closed cylinder.

a Find its volume to one decimal place. 1

b Find its surface area to one decimal place. 1

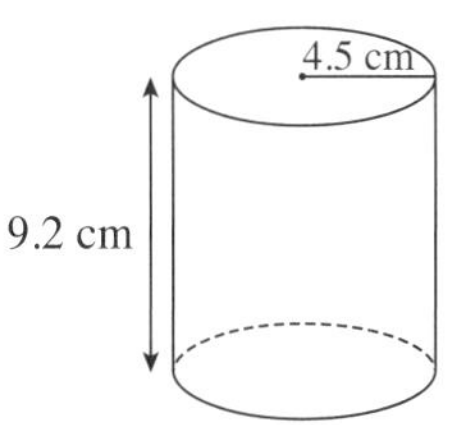

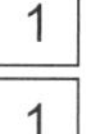

7 In the diagram $AD = 36$ cm, $AB = 24$ cm and $BC = 3$ cm.

a Explain why $\Delta ABE \mid\mid\mid \Delta ADC$ 1

b If $EB = 16$ cm, find DC. 1

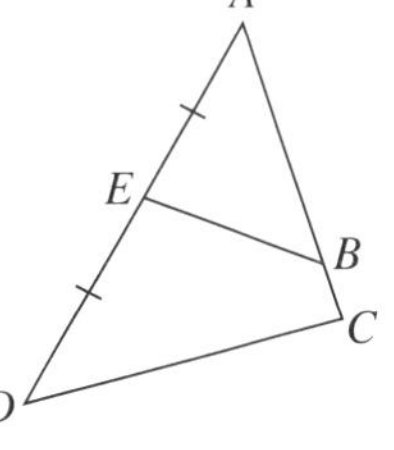

8 P is the point $(5, -2)$ and Q is the point $(-7, 7)$. Find.

a the distance PQ 1

b the midpoint of PQ 1

c the gradient of PQ 1

Continued on the next page

EXAM PAPER 1 PART C

Show all working for each question.

Marks

9 This shape is made up of a triangle and semi-circle. Find the:

a diameter of the semi-circle **b** area of the shape

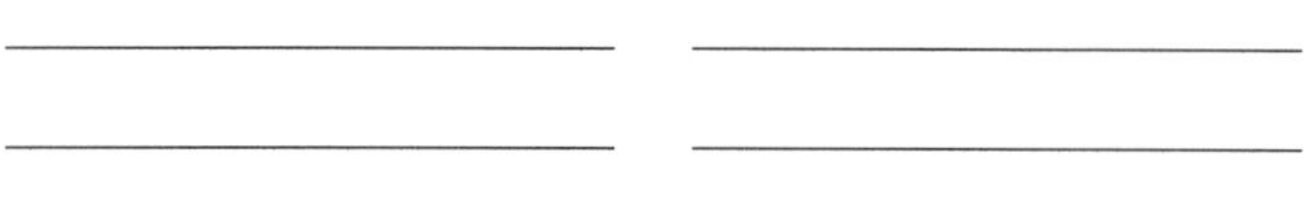

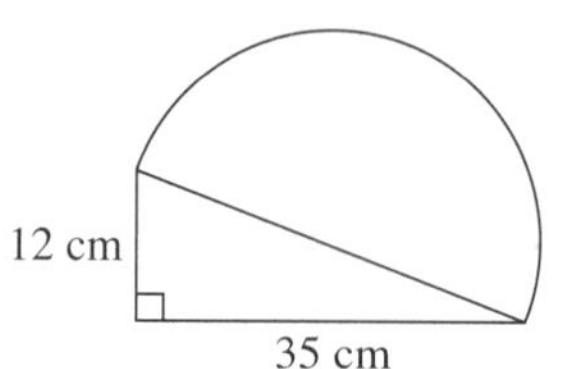

1
1

10 This dot plot shows the scores in a quiz.

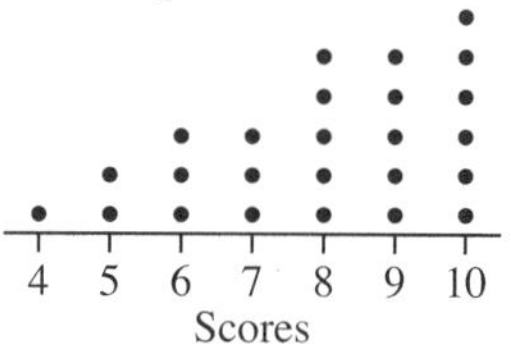

a Briefly describe the shape of the data display. 1

b What is the mean of the scores? 1

c What is the relative frequency, as a decimal, of a score of 5? 1

11 Find.

a the length of BD. **b** the size of $\angle CBA$.

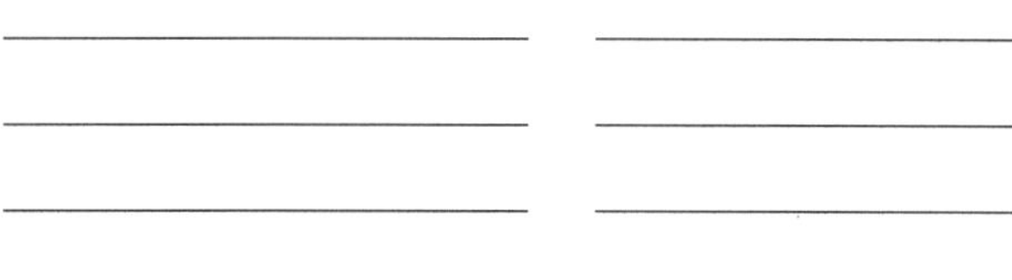

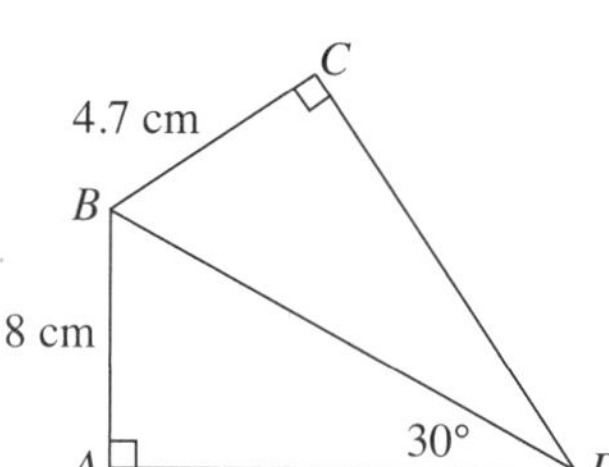

1
1

12 Consider the equation of the line $3x - 4y = 12$

a What is the x-intercept? **b** What is the y-intercept? **c** Graph the line.

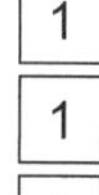

1
1
1

13 Consider the curve $y = 16 - x^2$

a For what values of x does $y = 0$? **b** Sketch the curve.

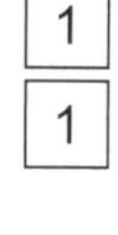

1
1

Total marks achieved for PART C /25

Exam Paper 2

Instructions for all parts
Time allowed: 1 hour

- Attempt all questions.
- Calculators are allowed.

Part A: Allow about 10 minutes for this part.
Part B: Allow about 20 minutes for this part.
Part C: Allow about 30 minutes for this part.

Total marks: 50

EXAM PAPER 2 — PART A

Fill in only one circle for each question.

Marks

1 $3m^2 \times 2m^3$ equals

Ⓐ $6m^6$ Ⓑ $5m^6$ Ⓒ $6m^5$ Ⓓ $5m^5$ [1]

2 $6a^2c$ is equal to

Ⓐ $6 \times 6 \times a \times a \times c \times c$ Ⓑ $6 \times a \times a \times c$
Ⓒ $6 \times 6 \times a \times a \times c$ Ⓓ $6 \times a \times a \times c \times c$ [1]

3 Paul works 35 hours at \$28.80 per hour, 5 hours overtime at time-and-a-half and 3 hours overtime at double time. His pay is

Ⓐ \$1185.60 Ⓑ \$1259.70 Ⓒ \$1383.20 Ⓓ \$1396.80 [1]

4 Which shows p decreasing at an increasing rate?

Ⓐ Ⓑ Ⓒ Ⓓ [1]

5 $(2x^2y)^3$ equals

Ⓐ $2x^6y$ Ⓑ $6x^2y^3$ Ⓒ $8x^6y^3$ Ⓓ $2x^5y$ [1]

6 The distance between the points $A(-2, 0)$ and $B(8, 0)$ is

Ⓐ 10 units Ⓑ 6 units Ⓒ $\sqrt{60}$ units Ⓓ 2 units [1]

7 The median of the scores 4, 2, 7, 3, 8, 2, 9 is

Ⓐ 3 Ⓑ 2 Ⓒ 5 Ⓓ 4 [1]

8 0.0002 equals

Ⓐ 2×10^{-4} Ⓑ 2×10^4 Ⓒ 2×10^{-3} Ⓓ 2×10^3 [1]

9 $8a^0$ equals

Ⓐ 8a Ⓑ 0 Ⓒ 1 Ⓓ 8 [1]

10 If $3.4 - x = 6$ then x equals

Ⓐ –2.6 Ⓑ 2.6 Ⓒ –9.4 Ⓓ 9.4 [1]

Total marks achieved for PART A /10

EXAM PAPER 2 PART B

Write only the answer in the answer column. For any working use the question column.

Questions | **Answers** | **Marks**

1 Simplify: $\dfrac{3p + 2p - p}{2 \times 2p}$ ______ [1]

2 Evaluate $12.56^2 - 7.15^2$ correct to three significant figures. ______ [1]

3 Write 21 600 in standard form. ______ [1]

4 Evaluate $a^2 - 7a + 5$ if $a = -1$ ______ [1]

5 What is the equation of the y axis? ______ [1]

6 Simplify $8x^0 \times (8x)^0$ ______ [1]

7 Factorise $2p - 4q$ ______ [1]

8 Simplify $(2a^3)^2$ ______ [1]

9 Calculate the volume of a cube of side 3 cm. ______ [1]

10 Find the average speed of a train which travels 60 km in 30 minutes. ______ [1]

11 Convert 2.5 mL to litres. ______ [1]

12 The mean of 7, 4 and x is 5. Find the value of x. ______ [1]

13 Solve the equation $2x - 5 = 13$. ______ [1]

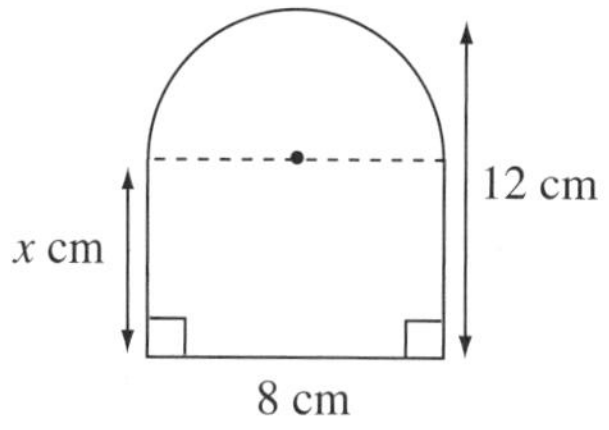

14 The diagram shows a rectangle and a semi-circle. Find x. ______ [1]

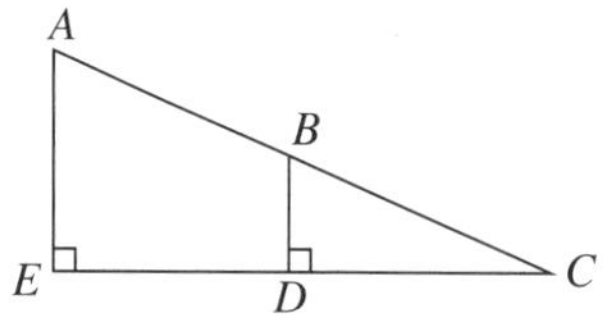

15 Complete $\Delta AEC \,|||\, \Delta$ ______ [1]

Total marks achieved for PART B ___ /15

EXAM PAPER 2 PART C

Show all working for each question.

Marks

1 Expand and simplify $10x - 4(5x + 2) + 7 =$ ______________________________

______________________________ 1

2 Solve these equations.

a $3x - 1 = 7 + 2x$ 1

b $4(x - 3) = 5 - 2(x + 1)$ 1

3 Find θ to the nearest degree. 1

50 m

19 m

θ

4 A pair of dice is thrown. What is the probability of getting:

a a double number? 1

b a total of 6? 1

c a total of less than 6? 1

5 a Show that ΔABC is isosceles. 1

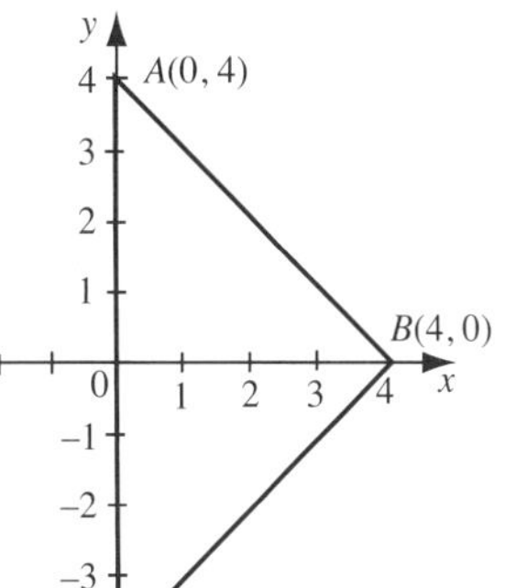

b Show that ΔABC is right-angled. 1

c Find the mid-point of AC. 1

d Find the mid-point M of AB. 1

e Find the gradient of OM. 1

Continued on the next page

EXAM PAPER 2 PART C

Show all working for each question.

Marks

6 A cylindrical water tank has a diameter of 3 m. It holds water to a height of 1.6 m.

a What is the volume of water in the tank in cubic metres (to 1 decimal place). 1

b How many litres of water does the tank hold? 1

c If the water is used at the rate of 720 litres per day, how long will the water last if there is no rain to replenish the supply? 1

7 Find the shaded area. 1

28 cm

8 Simone buys a car priced at $32 000. She pays 15% deposit and borrows the balance at a simple interest rate of 7% pa over 5 years.

a How much is the deposit? 1

b What is the total amount of interest? 1

c If Simone repays the loan in equal monthly instalments, how much is each instalment? 1

Continued on the next page

EXAM PAPER 2 PART C

Show all working for each question.

Marks

9 This back-to-back stem-and-leaf plot shows the scores by students in two separate classes in an exam.

9R		9Y
	4	7
6 3	5	2 5 8 8
4 4 1	6	0 2 3 6 8
9 5 2 2 2	7	3 4 5 7 9
9 7 6 5 5 3 0	8	1 1 4 6
8 7 5 2	9	0 3

a What is the mode for class 9R? 1

b Which class had the greater range and by how much? 1

c Briefly comment on any similarities or differences between the results for the two classes. 1

10 Is this triangle right-angled (diagram not to scale)? 1

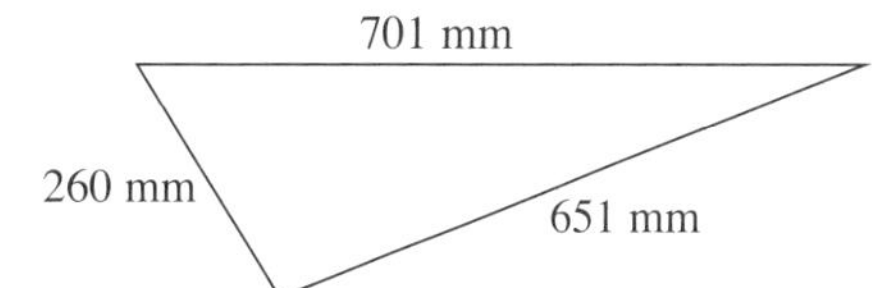

11 The diagram shows a sketch of the parabola $y = ax^2 + c$.

a What is the value of c? 1

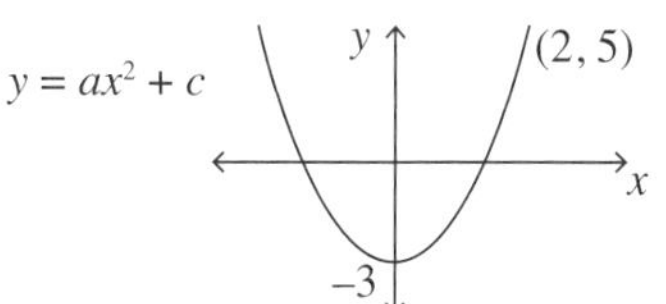

b Find the value of a. 1

Total marks achieved for PART C /25

Exam Paper 3

Instructions for all parts
Time allowed: 1 hour

- Attempt all questions.
- Calculators are allowed.

Part A: Allow about 10 minutes for this part.
Part B: Allow about 20 minutes for this part.
Part C: Allow about 30 minutes for this part.

Total marks: 50

EXAM PAPER 3 — PART A

Fill in only one circle for each question.

Marks

1 The point (3, 6) lies on which of these lines. [1]

Ⓐ $x + 2y + 12 = 0$ Ⓑ $x + 2y - 12 = 0$ Ⓒ $2x + y + 12 = 0$ Ⓓ $2x + y - 12 = 0$

2 7.06×10^{-6} equals [1]

Ⓐ 0.000 706 Ⓑ 0.000 000 706 Ⓒ 0.000 007 06 Ⓓ 0.000 070 6

3 4 km + 58 m + 19 cm equals [1]

Ⓐ 4.058 19 m Ⓑ 458.19 m Ⓒ 4058.19 m Ⓓ 45 819 m

4 Which type of data is categorical? [1]

Ⓐ heights of seedlings Ⓑ colour of hair
Ⓒ weights of oranges Ⓓ number of siblings

5 38 581 written to three significant figures is equal to [1]

Ⓐ 386 Ⓑ 396 Ⓒ 38 500 Ⓓ 38 600

6 Approximately how many kilobytes are in a terabyte? [1]

Ⓐ 1000 Ⓑ 1 000 000 Ⓒ 1 000 000 000 Ⓓ 1 000 000 000 000

7 Which could be the equation of this curve? [1]

Ⓐ $y = 3^x$ Ⓑ $y = 5^{-x}$
Ⓒ $y = -7^x$ Ⓓ $y = -10^{-x}$

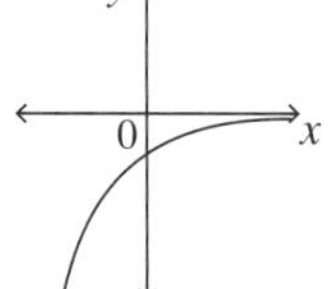

8 Which shows data that is both negatively skewed and bimodal? [1]

Ⓐ 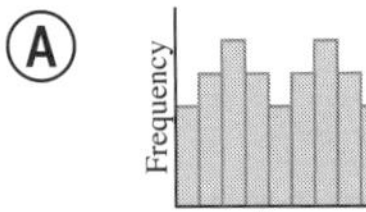 

Ⓑ 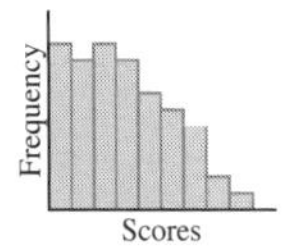

Ⓒ 

Ⓓ 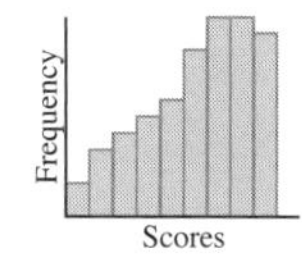

9 Which is closest to the curved surface area of a cylinder of diameter 18 cm and height 11cm? [1]

Ⓐ 622 cm² Ⓑ 1018 cm² Ⓒ 1244 cm² Ⓓ 2799 cm²

10 Water is poured into this container at a constant rate. [1]
Which diagram shows the height of the water as time passes?

Ⓐ 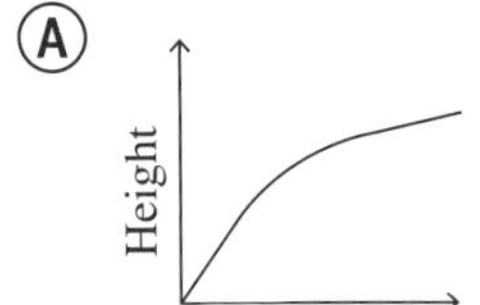

Ⓑ 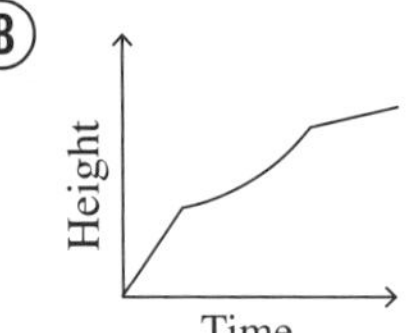

Ⓒ 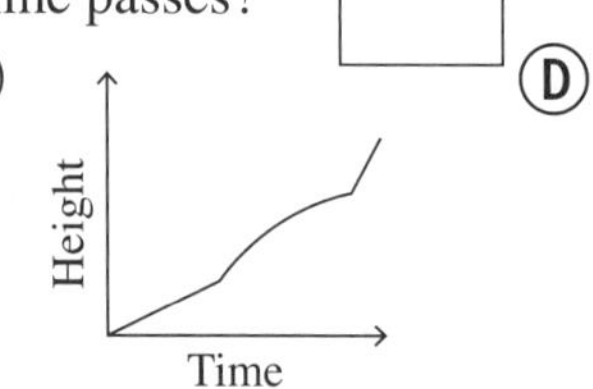

Ⓓ 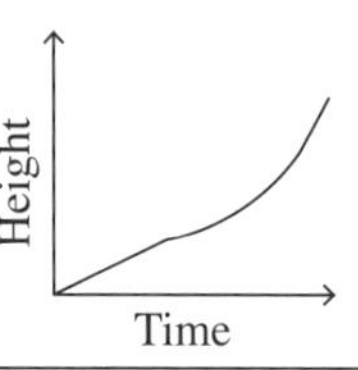

Total marks achieved for PART A /10

EXAM PAPER 3 — PART B

Write only the answer in the answer column. For any working use the question column.

Questions	Answers	Marks
1 Calculate $\sqrt{304.1} + (3.041)^2$ to five significant figures. ______	______	1
2 Simplify $(3a^3)^3$ ______	______	1
3 Factorise $a^2b - ab^2$ ______	______	1
4 What is the median of these scores? 12, 10, 1, 9, 10, 4, 5, 10, 2 ______	______	1
5 Two dice are thrown simultaneously. Find the probability that they both show a 5. ______	______	1
6 Find the gradient of the line $2x + 3y - 3 = 0$ ______	______	1
7 Jenny earns \$15.50 an hour. Calculate her week's wages if she worked 40 hours at the normal rate and 6 hours overtime at time-and-a-half. ______	______	1
8 Fertiliser was spread on a paddock of 20 hectares at the rate of 275 kilograms per hectare. How many tonnes of fertiliser were used for the paddock? ______	______	1
9 Write 2^{-3} without indices. ______	______	1
10 Find the simple interest on \$12 000 at 7.5% pa for 18 months. ______	______	1
11 A square photo of side length 6 cm was enlarged. Each side is now 2 cm longer. What was the enlargement factor? ______	______	1
12 Of 30 students, 19 have a dog, 17 have a cat and 8 have both. What is the probability that a student, selected at random from the group, has neither? ______	______	1
13 Expand $2x^2y(4xy - 5y^2)$ ______	______	1
14 Referring to the table, what is the relative frequency, as a fraction in simplest form, of a score of 4? ______	______	1
15 What is the surface area of a cube of side length 7 cm? ______	______	1

Table for question 14:

Result	1	2	3	4	5	6
Frequency	10	12	8	15	9	16

Total marks achieved for PART B ___ / 15

EXAM PAPER 3 PART C

Show all working for each question.

Marks

1 Solve the following equations.

a $\frac{3p-7}{2} = 7$

b $8(x-4) - 2(x+1) = 30$

1

1

2 Find the area of this figure to two decimal places.

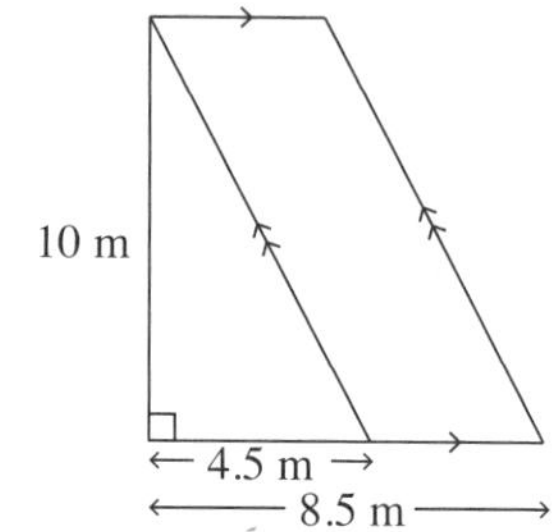

1

3 In the diagram opposite:

a find the value of x.

b find the area.

1

1

4 In one week Julie worked 40 hours plus 12 hours overtime at time-and-a-half. Altogether she earned $835.20. What is her hourly rate?

1

5 A computer is advertised for a cash price of $1860 or 20% deposit then $95.50 per month for 2 years. How much extra is paid by buying over the two years?

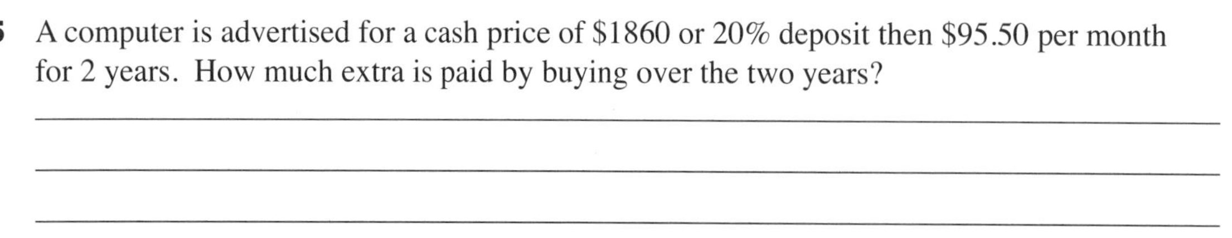

1

6 A cylinder of height 18.5 cm has a volume of 1500 cm³. What is the length of the radius of the cylinder?

1

Continued on the next page

Show all working for each question.

Marks

7 **a** Briefly explain why these two triangles are similar. 1

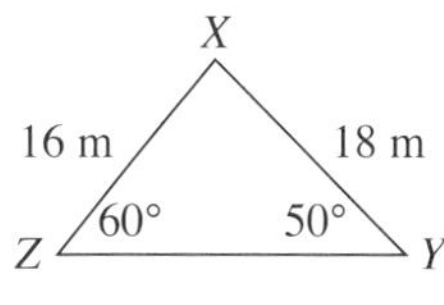

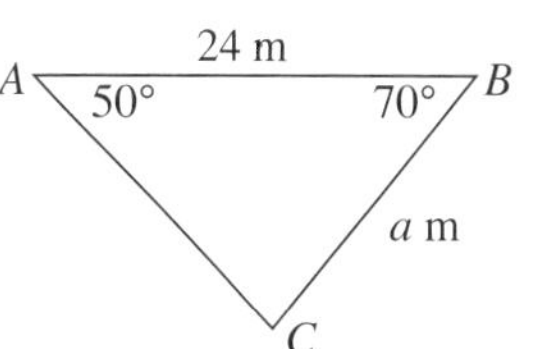

b Complete $\Delta XYZ \parallel\!| \, \Delta$________ 1

c Find the value of a. 1

8 The diagram shows that a 5 m high building casts a shadow, on level ground, that is 3.5 m long.

a Find the value of θ to the nearest degree. 1

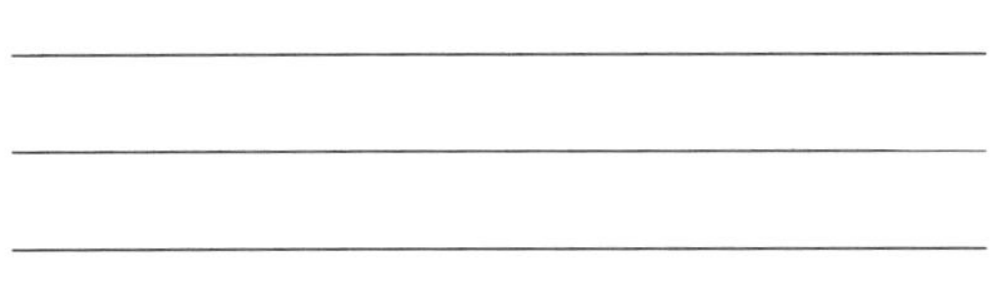

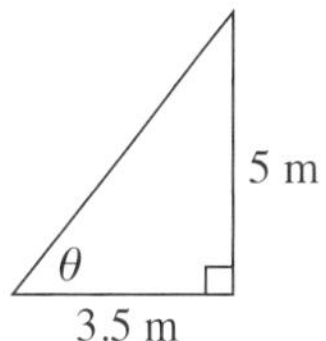

b At the same time, a tree casts a shadow of 23.8 m. Find the height of the tree to the nearest metre. 1

9 A bag holds 2 red and 1 black jellybean. Two jellybeans are taken from the bag, one after the other without replacement.

a Draw a tree diagram to show the possible outcomes. 1

b What is the probability that the jellybeans are different colours? 1

Continued on the next page

EXAM PAPER 3 PART C

Show all working for each question.

Marks

10 a What is the equation of a circle, centre $(0, 0)$, radius 8 units? ____________ 1

b Find the distance between the points $O(0, 0)$ and $P(4, -7)$.

____

____

____________ 1

c Does P lie inside, on or outside the circle? ____________ 1

11 a What is the gradient of the line shown?

____

____

____

____________ 1

b What is the y-intercept of the line?

____

____________ 1

c What is the equation of the line?

____

____________ 1

12 A class of students was given a test. Later, the same test was given again to the class. These two dot plots show the results.

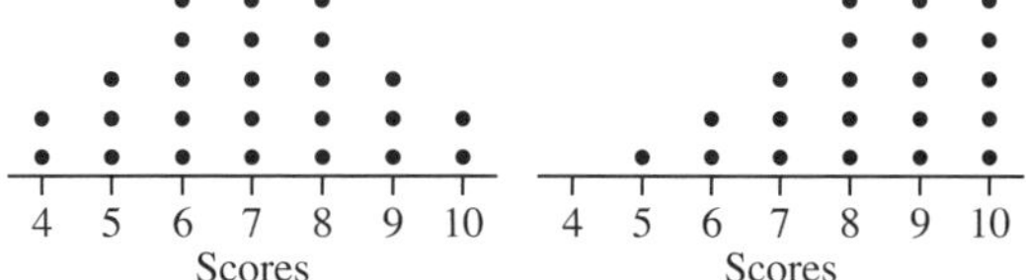

By how much did:

a the mean increase? **b** the median increase? **c** the range decrease?

____________ ____________ ____________ 1

____________ ____________ ____________ 1

____________ ____________ ____________ 1

d Comment on any similarities or differences between the two data sets.

____

____

____________ 1

Total marks achieved for PART C /25

Answers

Chapter 1 – Rational numbers, rates and measurements

Page 1 1 a 2513680 b 2514000 c 3000000 d 2510000 e 2513700 f 2500000 2 a 5.8 b 6.7 c 8.6 d 4.3 e 3.0 f 0.7 g 0.6 h 0.8 i 14.0 j 11.7 k 0.5 l 11.0 3 a 6.745 b 8.236 c 5.739 d 0.218 e 0.007 f 0.090 g 15.019 h 86.008 i 153.541 4 a 7480 b 19700 c 63600 d 106000 e 3980000 f 2030000 g 0.168 h 13.3 i 7.25 j 0.00653 k 0.00702 l 0.0000862 5 a 19630 b 35670000 c 857000 d 30030 e 5913000 f 160700 g 12.23 h 3.621 i 2.908 j 0.0002792 k 0.06071 l 0.04260 6 a 9 b 3 c 0 d 0 e 7 f 8

Page 2 1 a 32 b 81 c 125 d 10000000 e 1 f 256 2 a 7^7 b 5^7 c 2^{10} d 3^3 e 11^7 f 5^8 3 a $\frac{1}{2}$ b $\frac{1}{3}$ c $\frac{1}{7}$ d $\frac{1}{5}$ e $\frac{1}{10}$ f $\frac{1}{8}$ g $\frac{1}{36}$ h $\frac{1}{9}$ i $\frac{1}{78\,125}$ 3 a $\frac{1}{2}$ b $\frac{1}{3}$ c $\frac{1}{7}$ d $\frac{1}{5}$ e $\frac{1}{10}$ f $\frac{1}{8}$ g $\frac{1}{36}$ h $\frac{1}{9}$ i $\frac{1}{78\,125}$ 4 a 2^{-1} b 3^{-1} c 7^{-1} d 5^{-1} e 10^{-1} f 2^{-3} g 6^{-2} h 3^{-2} i 5^{-4} 5 a $\frac{1}{2}$ b $\frac{1}{3}$ c $\frac{1}{7}$ d $\frac{1}{5}$ e $\frac{1}{10}$ f $\frac{1}{8}$ g $\frac{1}{36}$ h $\frac{1}{9}$ i $\frac{1}{625}$ j $\frac{1}{49}$ k $\frac{1}{32}$ l $\frac{1}{1\,000\,000}$ 6 a 6^{-1} b 11^{-1} c 13^{-1} d 2^{-2} e 3^{-2} f 2^{-3} g 10^{-3} h 7^{-2} i 10^{-4}

Page 3 1 a 1000 b 10000 c 1000000 d 1000000000 e 10000000 f 100000000 g 100000 h 100 i 10000000000 j 1000000000000 k 100000000000 l 1 2 a 10^1 b 10^4 c 10^5 d 10^2 e 10^0 f 3×10^7 g 4×10^3 h 9×10^8 i 10^0 j 8×10^9 k 7×10^4 l 5×10^5 3 a 30000 b 5000000 c 800 d 9000 e 600000 f 7000000000 g 200000 h 90000 i 400000 j 60000000 k 500000000 l 3000 4 a 3×10^3 b 2×10^4 c 5×10^4 d 6×10^5 e 6×10^2 f 4×10^7 g 8×10^5 h 9×10^4 i 6×10^5 j 7×10^5 k 1×10^6 l 3×10^4 5 a $\frac{1}{10}$ b $\frac{1}{10\,000}$ c $\frac{1}{100}$ d $\frac{1}{1000}$ e $\frac{1}{100\,000}$ f $\frac{1}{1\,000\,000}$ g $\frac{1}{10\,000\,000}$ h $\frac{1}{10\,000\,000\,000}$ i $\frac{1}{100\,000\,000}$ 6 a 0.1 b 0.000001 c 0.000000001 d 0.00001 e 0.01 f 0.0001 g 0.001 h 0.0000001 i 0.00000001

Page 4 1 a 7.35×10^3 b 5.25×10^4 c 8.15×10^5 d 9.386×10^6 e 3.5×10^6 f 6.856×10^3 g 6.95×10^4 h 4.3687×10^4 i 7.8643×10^6 j 8.5363×10^5 k 1.9643×10^4 l 9.83×10^5 2 a 7.5×10^{-3} b 9.82×10^{-4} c 5.4×10^{-2} d 9.5×10^{-5} e 5.283×10^{-1} f 6.813×10^{-4} g 9.8×10^{-3} h 6.54×10^{-1} i 6.325×10^{-1} j 1.7×10^{-3} k 7.18×10^{-6} l 8.352×10^{-4} 3 a 870 b 80000 c 4900 d 780000 e 250000 f 17000 g 0.037 h 0.0046 i 0.00093 j 0.023 4 a 2.76×10^{12} b 5.00×10^7 c 8.82×10^8 d 2.57×10^3 e 5.36×10^{-1} f 5.10×10^7 g 1.44×10^{10} h 2.00×10^4

Page 5 1 a 6.3×10^9 b 1.785×10^{10} c 6.75×10^2 d 3.9936×10^6 e 3.25×10^9 f 7.56×10^6 g 2.496×10^6 h 1.512×10^6 2 a 3×10^4 b 2×10^3 c 2×10^3 d 6×10^4 e 4.083×10^1 f 7×10^{-8} g 2×10^3 h 6.631×10^7 3 a 2×10^2 b 9×10^5 c 5×10^4 d 1.575×10^{11} e 5.76×10^{10} f 3.264×10^2 g 8.16×10^8 h 4.2225×10^9 i 7.989×10^{10} j 1.440×10^4 4 a 3.869×10^{10} b 4.527×10^3 c 1.250×10^{10} d 5.146×10^7 e 1.777×10^{10} f 1.066×10^{-3} g 2.638×10^6 h 3.764×10^{31} i 8.900×10^{35} j 7.192×10^2 5 a 2.7×10^{19} b 2.5×10^{-5} c 2.646×10^7 d 1.62×10^{18} e 2×10^{10} f 4×10^{18} g 4.8×10^{12} h 1.009×10^{-12} i 4.628×10^{-4} j 2×10^1 k 2×10^0 l 7.521×10^{12} 6 a 2.688×10^{11} b 1.35×10^2 c 5.510×10^4 d 2.704×10^{-3} e 4.930×10^{17} f 4×10^1 g 5.12×10^3 h 2.521×10^4 i 4.589×10^{-4} j 1.382×10^{-8}

Page 6 1 a 5×10^3 b 8×10^9 c 1.5×10^6 d 2.1×10^6 e 3×10^{-2} f 6.3×10^2 g 6.3×10^{-5} h 8.3×10^{-3} 2 a $3.5 \times 10^{-3}, 3.5 \times 10^5, 3.5 \times 10^8$ b $8 \times 10^{-9}, 8 \times 10^{-6}, 8 \times 10^{-4}$ c $2.5 \times 10^{-5}, 1.86 \times 10^3, 3.1 \times 10^4$ d $8 \times 10^2, 8 \times 10^4, 8 \times 10^6$ e $2.1 \times 10^7, 3.8 \times 10^7, 5.4 \times 10^7$ f $6 \times 10^{-5}, 6 \times 10^{-4}, 6 \times 10^{-2}$ g $3.5 \times 10^{-3}, 3.9 \times 10^{-3}, 5.6 \times 10^{-3}$ h $5.7 \times 10^{-2}, 8.9 \times 10^0, 3.6 \times 10^5$ 3 a $3.2 \times 10^7, 2.8 \times 10^7, 1.5 \times 10^7$ b $9 \times 10^3, 8 \times 10^3, 5 \times 10^3$ c $3.5 \times 10^9, 3 \times 10^9, 2.5 \times 10^9$ d $4 \times 10^{-3}, 4 \times 10^{-5}, 4 \times 10^{-6}$ e $5.1 \times 10^{-6}, 3.7 \times 10^{-6}, 2.5 \times 10^{-6}$ f $4.6 \times 10^3, 3.8 \times 10^2, 3.9 \times 10^{-4}$ g $4.9 \times 10^{-1}, 3.6 \times 10^{-2}, 2.5 \times 10^{-7}$ h $8.3 \times 10^6, 5.4 \times 10^4, 3.5 \times 10^3$

4 a $7 \times 10^3, 5 \times 10^4, 8 \times 10^5$ b $6.7 \times 10^5, 5.3 \times 10^5, 3.2 \times 10^5$ c $2.5 \times 10^{-4}, 8.5 \times 10^{-3}, 3.7 \times 10^{-2}$ d $5.4 \times 10^{-1}, 6.4 \times 10^{-2}, 6.2 \times 10^{-3}$ e $9.6 \times 10^2, 8.35 \times 10^5, 7.69 \times 10^6$ f $9.2 \times 10^3, 8.5 \times 10^3, 7.9 \times 10^3$ g $3.5 \times 10^{-6}, 6.2 \times 10^{-5}, 5.4 \times 10^{-3}$ h $3.17 \times 10^{-3}, 5.17 \times 10^{-4}, 8.15 \times 10^{-6}$ 5 a 8.6×10^5 b 5.04×10^{-2} c 8.6×10^3 d 3.2×10^{-6} e 5.79×10^{-6} f 5×10^{-7} g 3.71×10^{-7} h 8.6×10^{-6} 6 a $8 \times 10^2, 8 \times 10^3, 8 \times 10^4, 8 \times 10^5$ b $3.8 \times 10^5, 5.2 \times 10^5, 7.6 \times 10^5, 8.2 \times 10^5$ c $6.3 \times 10^{-5}, 5.4 \times 10^{-4}, 3.8 \times 10^{-3}, 9.1 \times 10^{-2}$ d $4.8 \times 10^{-3}, 7 \times 10^{-3}, 8.1 \times 10^{-3}, 9.2 \times 10^{-3}$ e $4.3 \times 10^{-6}, 4.3 \times 10^{-5}, 4.3 \times 10^{-3}, 4.3 \times 10^{-2}$ f $3.6 \times 10^{-4}, 4.9 \times 10^{-2}, 3.7 \times 10^0, 5.7 \times 10^4$ g $5.9 \times 10^2, 9.2 \times 10^2, 6.8 \times 10^3, 8.6 \times 10^4$ h $8.31 \times 10^{-3}, 4.56 \times 10^{-2}, 5.12 \times 10^2, 3.42 \times 10^3$

Page 7 1 4.9×10^6 cm 2 4.0075×10^4 km 3 152100000 km 4 9.5×10^{12} km 5 5.28×10^{13} km 6 4.16×10^{13} km 7 2.773×10^9 km 8 0.0007 mm 9 0.0000000297 cm 10 1.3×10^9 cm 11 1.5×10^{11} m 12 1.4×10^{-7} mm 13 a 5×10^6 cm b 6×10^6 g c 1.2×10^6 m^2 d 3.8×10^8 mL 14 a 9.15×10^{-4} b 2.21×10^3 c 1.54624×10^{11} d 2.665625×10^5 15 2.6559×10^{-20}

Page 8 1 a 5000 b 6200 c 370 d 70 e 0.75 f 87.25 g 290 h 16000 i 0.003 j 500 k 60 l 2 m 100000 n 0.6 o 67.5 p 0.7 q 12.5 r 95000 s 180 t 480 u 4 h 12 min v 10 w 100 x 1000

2

Prefix	nano	micro	milli	(unit)	kilo	mega	giga	tera
	n	μ	m	-----	k	M	G	T
Meaning	$\frac{1}{1\,000\,000\,000}$	$\frac{1}{1\,000\,000}$	$\frac{1}{1000}$	1	1000	1000000	1000000000	1000000000000
	10^{-9}	10^{-6}	10^{-3}	-----	10^3	10^6	10^9	10^{12}

3 a 3000000 b 0.000007 c 0.018 d 60 e 7 f 8000000000 g 5000000000 h 23000000 i 0.00000006 j 0.0002 k 50000 l 8
4 Not exactly correct. There are 1024 bytes in a kB because a number must be a power of two. But, there are approximately 1000 bytes in a kilobyte 5 a 6000000 b 92 c 4000000 d 5000000 e 0.35 f 45 6 a $\frac{1}{1\,000\,000\,000}$ b $\frac{1}{1\,000\,000}$ 7 a 3600000000 b 86400000000000

Answers

Page 9 **1** He could say 'the table is two metres long' but measurements are never exact. **2 a** 10 m **b** 6 m **c** 60 m^2 **d** 63.318024 m^2 **e** 63 m^2 **f** 63 m^2 is more accurate. The errors in the rounded measurements are multiplied to produce a less accurate result. **3** It is not a sensible answer. The radius is given to nearest hundred kilometre so an answer expressed to the nearest hundred metres makes no sense. He should say the answer is about 40 200 km. **4 a** 35 m and 45 m **b** 355 m and 365 m **c** 1495 m and 1505 m **d** 2.295 km and 2.305 km **5 a** 7.75 m and 7.85 m **b** 3.35 cm and 3.45 cm **c** 21.45 km and 21.55 km **d** 156.65 m and 156.75 m **6** to the nearest 5 mm **7** to the nearest 100 g

Page 10 **1 a** 72 km/h **b** $6\frac{1}{4}$ h **2 a** 2.6 s **b** 5.7 km **3 a** 5 min **b** 4 min 10 s **4 a** 33.6 L **b** 8 cents **5 a** direct **b** indirect **c** direct **d** indirect **6 a** 0.32 **b** 112 **c** 450 m

Page 11 **1 a** 6720 **b** 300 **c** 76.8 **d** 5 **e** 780 **f** 1.2 **2 a** 1200 **b** 72 000 **c** 72 **3 a** 90 000 **b** 1500 **c** 25 **4 a** 54 km/h **b** 35 m/s **c** 21.6 L/h **5 a** 26 ha **b** 32 ha **c** 84 acres **d** 17 acres **e** 2020 ha **f** 325 kg **g** 13

6 a

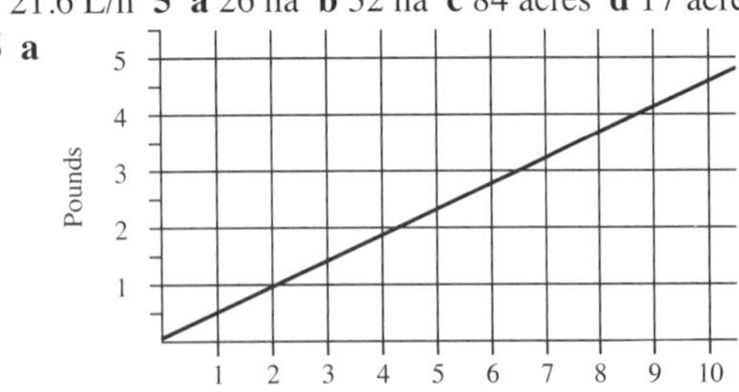

b 180 pounds **c** \$8000

Page 12 **1 a** 8 am **b** 8 h **c** 50 km **d** 11.15 am **e** car stationary **f** Between 1.30 pm and 4 pm; the line is steepest then. **g** 80 km/h **2** It is impossible to be in two different places at the same time. **3 a** C **b** B **c** D **d** E **e** F **f** A **4 a** B **b** D **c** C **d** increasing **e**

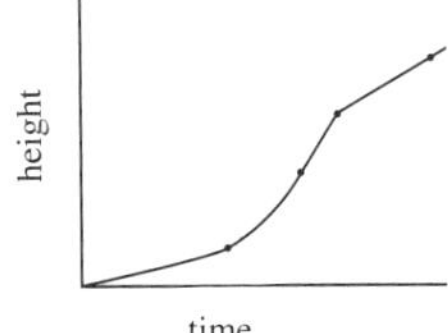

Page 13 **1** D **2** B **3** D **4** A **5** C **6** C **7** A **8** D **9** C **10** B

Page 14 **1 a** 0.0031 **b** 0.003 08 **c** 3.075×10^{-3} **2 a** 675 km **b** $8\frac{1}{4}$ h **c** 250 m/s **3 a** 0.0007 **b** 2 500 000 **4 a** 540 **b** 6 days 22 h 40 mins **d** 22h 40 mins **5 a** 3.6 **b** 86.4 cm **c** 35 cm **6 a** middle **b**

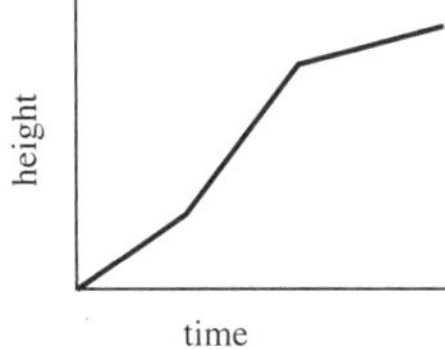

Chapter 2 – Algebraic techniques

Page 15 **1 a** $a + b$ **b** xy **c** m^2 **d** $\sqrt{p}$ **e** $7x + 2y$ **f** k^3 **g** $25x^2$ **h** $8p - 3q$ **i** $\frac{3x}{7}$ **j** $9a^2$ **2 a** \$$dm$ **b** $60T$ min **c** $x + 2$ **d** hk kilometres **e** $4l$ cm **3 a** $\frac{2k}{15}$ **b** $27ab + 7$ **c** $8(5x + 11y)$ **d** $9(3x + 14)$ **e** $9x - a(2b + 3c)$ **4 a** $\frac{2x}{3y} + z$ **b** $1000k$ **c** $\frac{M}{1000}$ **d** $1000Y$ **e** $1000x$ **f** $\frac{s}{3600}$ **5 a** 9 and 6 **b** 27 and 9 **c** 18 and 36 **d** $\frac{1}{3}$ and $\frac{1}{27}$ **e** 45 and 75 **f** 34 and 64

Page 16 **1 a** $16x$ **b** $18x$ **c** $17x$ **d** $37x$ **e** $43x$ **f** $21a$ **g** $17ab$ **h** $10mn$ **i** $18p$ **j** $20x^2$ **k** $23a^2$ **l** $24n$ **2 a** $15a$ **b** x **c** $5y$ **d** $6m$ **e** $6x$ **f** $4xy$ **g** $5x^2$ **h** $9n$ **i** $6p$ **j** $3a^2$ **k** $3y$ **l** $-4x$ **3 a** $11a$ **b** $6x$ **c** $16a$ **d** $7mn$ **e** $9p^2$ **f** $16ab$ **g** $7t$ **h** $19a$ **i** $8m^2$ **j** $16t$ **k** $7x$ **l** $9mn$ **4 a** $3a + 4b$ **b** $20x + 2y$ **c** $27a^2 - 6b^2$ **d** $11m + 3n$ **e** $a + 12b$ **f** $2m + n$ **g** $19x - 9y$ **h** $8p + 6q$ **i** $15ab^2 + a^2b$ **j** $12x - 8y$ **5 a** $18 - 4x$ **b** $15x^2 - 9y^2$ **c** $7a + 7b$ **d** $3m + 3n$ **e** $14x + 2y$ **f** $20xy - 8yz$ **g** $23p - 9q$ **h** $16ab$ **i** $8t - 2$ **j** $xy + yz$

Page 17 **1 a** $15a$ **b** $35y$ **c** $72x$ **d** $30b$ **e** $3x$ **f** $5y$ **g** $10a$ **h** $2xy$ **i** $42ab$ **j** $135xy$ **2 a** $32xy$ **b** $48ab$ **c** $-80a$ **d** $-40a$ **e** $-30x$ **f** $-45y$ **g** $-18xy$ **h** $-21ab$ **i** $-90abc$ **j** $-36ab$ **3 a** $40m^2$ **b** $63a^2b$ **c** $72b$ **d** $18x^2y^2$ **e** $30a^2$ **f** $6a^2b^2$ **g** $5mn^2$ **h** $28a$ **i** $-21a^2bc$ **j** $12ab$ **4 a** $-6a^2$ **b** $10a^3$ **c** $30x^3$ **d** $-24m^2n$ **e** $-6a^3$ **f** $120xy$ **g** $30x^3y^2$ **h** $15a^2$ **i** $-16a^3$ **j** $-6x^3$ **5 a** $12xy$ **b** 0 **c** $-30ay$ **d** $-15x^2y$ **e** $-35m^2n$ **f** $-30xy$ **g** $54a^2$ **h** $56a^2b$

Page 18 **1 a** $3ab$ **b** $8x$ **c** $5q$ **d** $\frac{4b}{a}$ **e** $8x$ **f** $2x$ **g** $12x$ **h** $5bc$ **i** $3a^2$ **j** $6a$ **2 a** $-\frac{12b}{a}$ **b** $2a$ **c** $-36y$ **d** $3n$ **e** -5 **f** $-\frac{a}{2}$ **g** $3y$ **h** $10n$ **i** $3b$ **j** $9y$ **3 a** $4y$ **b** $4x$ **c** $-5b$ **d** $2y$ **e** 6 **f** $-3x$ **g** $-2a$ **h** $-\frac{7xy^2}{2}$ **i** $9y$ **j** $3yz$ **4 a** 3 **b** -1 **c** $-\frac{24z}{5}$ **d** 1 **e** $-2m$ **f** a **g** c **h** 1 **i** $\frac{5abc}{2}$ **j** $2x$ **5 a** 1 **b** c **c** 1 **d** b **e** 2 **f** y **g** 1 **h** $\frac{56a}{-27}$

Page 19 **1 a** $20x$ **b** $24xy$ **c** $20x^2y$ **d** $9y^2$ **e** $-18mn$ **f** $6pq$ **g** $-30a^3$ **h** $-18x^4$ **i** $-20a^2b^2$ **j** $24xy$ **k** xy^2z **l** $4x^2y^2$ **2 a** $2x$ **b** 3 **c** x^2 **d** $4n$ **e** 12 **f** $\frac{3}{x}$ **g** $-3y$ **h** -8 **i** $2z$ **j** $-5y$ **k** $4b$ **l** $-4ac$ **3 a** $-9b$ **b** -1 **c** n **d** $9y$ **e** $\frac{2ab}{c}$ **f** $\frac{3}{x}$ **g** $2x$ **h** b **i** $\frac{3}{2ab}$ **j** $-3a$ **4 a** $30k^2y$ **b** $24xyz$ **c** $84x^2$ **d** $6x^2$ **e** 1 **f** $21m$ **g** $24y$ **h** $3abc$ **i** $\frac{a^2b^2}{3}$ **j** $10x^2$ **5 a** $2y^2$ **b** 1 **c** $3x$ **d** $28y^2$ **e** $6a^2$ **f** 24 **g** 2 **h** $\frac{10}{a}$ **i** $10ab$ **j** $\frac{a}{b}$

Answers

Page 20 **1 a** x^5 **b** y^6 **c** a^7 **d** m^7 **e** p^{10} **f** n^{13} **g** a^9 **h** x^8 **i** y^{14} **2 a** x^6 **b** x^8 **c** x^6 **d** y^3 **e** y^7 **f** a^8 **g** m^{11} **h** m^2 **i** m^4 **3 a** $5x^9$ **b** $9x^7$ **c** $3a^{16}$ **d** $5m^9$ **e** $56k^8$ **f** $40a^{13}$ **g** m^8n^9 **h** x^7y^{11} **i** x^6y^4 **4 a** x^3 **b** y^3 **c** a^2 **d** $3m^2$ **e** $2n^3$ **f** $4a^2$ **g** $3y^5$ **h** x^2y^2 **i** a^3b^3 **5 a** $6a^5$ **b** $6p^5$ **c** $81y^{11}$ **d** $35m^7$ **e** $32a^8$ **f** $2x^{15}$ **g** $24x^{12}$ **h** $72a^9$ **i** $80p^{11}$ **j** $24x^{20}$ **k** a^7b^7 **l** $36x^7y^5$ **6 a** $2a^4$ **b** $4m^2$ **c** $4ab$ **d** $3a^{10}$ **e** $4k^5$ **f** $2a^2b^5$ **g** $4a^{12}$ **h** $4a^2$ **i** $-9x^{10}y$ **j** $4m^7$ **k** $2m^2n^2$ **l** $8a^2b^2$

Page 21 **1 a** a^6 **b** b^{20} **c** a^{30} **d** x^{21} **e** b^{14} **f** x^{56} **g** $9x^6$ **h** $81x^8$ **i** $27b^{12}$ **2 a** $2a^6$ **b** $3y^{25}$ **c** $64x^4y^6$ **d** $5m^{10}$ **e** $6x^{21}$ **f** $729a^3b^6$ **g** ax^{24} **h** $64a^6$ **i** $125m^{12}$ **j** x^2y^{42} **k** $81p^4$ **l** $1000a^9b^9$ **3 a** 1 **b** 1 **c** 1 **d** 1 **e** 1 **f** 1 **g** 9 **h** 2 **i** 1 **4 a** 4 **b** 1 **c** 8 **d** 5 **e** a^6 **f** 1 **g** 1 **h** 1 **i** 1 **j** 2 **k** 5 **l** 8 **m** 0 **n** 4 **o** 4 **5 a** 3^6 **b** 2^6 **c** x^6 **d** m^{12} **e** $4x^{30}$ **f** $16y^{12}$ **g** a^{10} **h** y^8 **i** a^{22}

Page 22 **1 a** $\frac{x^6}{y^2}$ **b** $\frac{x^6}{y^9}$ **c** $\frac{a^{20}}{b^{15}}$ **d** m^{10} **e** $\frac{a^{10}}{b^{14}}$ **f** $\frac{m^{16}}{n^{14}}$ **g** $\frac{m^{24}}{16}$ **h** $\frac{x^8}{y^{12}}$ **i** $\frac{a^{27}}{b^6}$ **2 a** $16a^{11}$ **b** $8x^{15}$ **c** $4a^{22}$ **d** $x^{13}y^{15}$ **e** $p^{16}q^{13}$ **f** $144a^4b^2$ **g** $400x^8y^{10}$ **h** $\frac{1}{2}p^4q^{19}$ **i** $9x^5y^4$ **j** $m^{10}n^9p^4$ **k** $2m^6$ **l** $3x$ **m** $216a^6$ **n** $20x^2$ **o** $125x^2$ **3 a** 5 **b** 8 **c** $\frac{1}{9}$ **d** 1 **e** 1 **f** $\frac{1}{6}$ **g** 40 **h** 8 **i** 63 **4 a** 7 **b** 6 **c** $512a^5$ **d** $\frac{64c}{3}$ **e** $3m^7$ **f** $\frac{a^{13}}{9}$ **g** $\frac{y^{17}}{64}$ **h** $\frac{27y^5}{4}$ **i** $6x^4y^2$ **j** m^8n^7 **k** $\frac{1}{36}$ **l** $3x$ **m** $8x^4$ **n** $432k^7$ **o** y^{11}

Page 23 **1 a** $\frac{1}{x^2}$ **b** $\frac{1}{a}$ **c** $\frac{1}{e^3}$ **d** $\frac{1}{p^7}$ **e** $\frac{2}{m^2}$ **f** $\frac{5}{n^4}$ **g** $\frac{6}{x^6}$ **h** $\frac{1}{4a^3}$ **i** $\frac{1}{27x^3}$ **j** $\frac{1}{49y^2}$ **k** $\frac{1}{32a^5}$ **l** $\frac{1}{x^4y^4}$ **2 a** x^{-4} **b** a^{-6} **c** e^{-10} **d** x^{-9} **e** $3n^{-3}$ **f** $4m^{-8}$ **g** ab^{-5} **h** $7a^{-12}$ **i** $\frac{1}{4}a^{-2}$ or $(2a)^{-2}$ **j** $\frac{1}{8}x^{-3}$ or $(2x)^{-3}$ **k** $\frac{1}{2}x^{-7}$ **l** $\frac{1}{81}a^{-4}$ or $(3a)^{-4}$ **3 a** x^5 **b** a^3 **c** m^{-5} **d** $10p^3$ **e** $24h^{-7}$ **f** $16x^{-3}$ **g** x^{10} **h** a^{12} **i** b^7 **j** $3m^{-7}$ **k** $5n^{-10}$ **l** $6a^7$ **m** $18a^{-4}$ **n** $7x^{10}$ **o** 8 **p** b **q** pq^2 **r** $m^{-3}n$ **4 a** $\frac{1}{x^4}$ **b** $\frac{1}{a^6}$ **c** $\frac{1}{x^2}$ **d** $\frac{10}{p}$ **e** $\frac{4}{a^6}$ **f** $\frac{x^5}{y^5}$

Page 24 **1 a** $3x + 6$ **b** $2a + 10$ **c** $8y - 4$ **d** $18a + 21$ **e** $40 - 5a$ **f** $12k - 18$ **g** $5n^2 - 5n$ **h** $12 - 9a$ **i** $14n + 49$ **j** $2y^2 + 7y$ **k** $m^2 + 10m$ **l** $6a^2 - 14a$ **2 a** $-4a - 6$ **b** $-15n + 12$ **c** $-y - 8$ **d** $-35 - 10t$ **e** $-15x - 54$ **f** $-12x + 8$ **g** $-6x - 11$ **h** $-8x + 18$ **i** $-20x + 25$ **j** $-3a + 42$ **k** $-8x + 80$ **l** $-2 + 5x$ **3 a** $3x - 5$ **b** $4x - 2$ **c** $-6y + 2$ **d** $2a^4 + 3a^3$ **e** $3a^3 + 4a^2b$ **f** $-6y^2 - 14y$ **g** $-3y^3 + 6y^2$ **h** $20t^3 - 32t^2$ **i** $-3m^3 - 5m^2$ **j** $-18p^3 - 30p$ **k** $-32x^2 + 4x$ **l** $24n^3 + 21n^2$ **4 a** $-10x - 2y + 2z$ **b** $-6a - 9b + 12c$ **c** $4a^2 - 12a + 28$ **d** $-5t^2 + 3t - 4$ **e** $6xy + 9xy^2 - 24x$ **f** $8a^3b^2 - 12a^2b^2 + 6a^2b^3$ **g** $-15a^2 + 10ab - 20ac$ **h** $24p^2 - 6pq + 9pr$ **i** $4a^3 + 8a^2b - 12a^2c$ **j** $-2a^2 - 3ab + 9ac$ **k** $-2t^4 - 3t^3 + 5t^2$ **l** $72x - 56y + 16z$ **5 a** $3t^5 - 15t^4 + 6t^3 - 24t^2 - 21t$ **b** $5m^5 - 3m^4 + 2m^3 - m^2 - m$ **c** $4x^2y^2 - 3x^3y + 4x^3 - 7x^2y$ **d** $a^5b - a^4b + 4a^2b^2 - 2a^3b + 3a^2b^3$ **e** $-20a^4 + 16a^3 - 12a^2 + 8a$ **f** $-16y^3 - 14y^2 + 2xy^2 - 12y$ **g** $-a^4b - ab^3 + 2a^2b^2 - abc$ **h** $-4x^4 - 4xy^2 + 8x^2y + 4x^2$

Page 25 **1 a** $7x + 10$ **b** $5a - 1$ **c** $24m - 10$ **d** $2a - 1$ **e** $6y - 13$ **f** $15x + 4$ **g** $3t^2 + 5t + 21$ **h** $6x + 1$ **i** $2x + 22$ **j** $10m - 34$ **k** 9 **l** $5x + 17$ **2 a** $10a + 23$ **b** $19t - 49$ **c** $12m - 15$ **d** $3p + 16$ **e** $34y - 3$ **f** $23x - 60$ **g** $21 - 16n$ **h** $9y^2 + 27y - 4$ **i** $-2a + 12$ **j** $35 - 8x$ **k** $6x + 2y - z$ **l** $19 - t$ **3 a** $6x + 2$ **b** $5a^3 - 11a^2 - 24a$ **c** $3x^3y - 3xy^2 - 21xy - x^3 - 3x^2$ **d** $-3m + 33n - 16$ **e** $-13t^3 + 4t^2 + 11t$ **f** $7a^4 - 5a^3 - 4a^2 + 7a - 20$ **4 a** $9a - 2b$ **b** $13x + 2y - 2z$ **c** $-2m + 2$ **d** $13a - 3$ **e** $3y^2 - 6y + 3$ **f** $3t^2 - 9t - 6$

Page 26 **1 a** $8x + 30$ **b** $13a + 34$ **c** $5m - 23$ **d** $-2n - 16$ **e** $14x + 14$ **f** $14x - 3$ **g** 3 **h** $-21x - 26$ **i** $-17x - 2$ **j** $x^2 - x - 6$ **k** $3x^2 + 3$ **l** $3a^2 - 7a + 2$ **m** $14x^2 - 41x + 15$ **n** $7x^2 - 9x$ **o** $-3x$ **p** $6b$ **q** $18m^3 - 25m^2 + 12m$ **r** $14a^3 - 3a^2b + 2ab$ **2 a** $2x^3y + 5x^2y^2 - 3xy^2$ **b** $6a^2b - 6ab^2 - 6a - 3b$ **c** $25x^2y + 5xy^2 - 24x^2 + 6y$ **d** $8 - 12x - 14y$ **e** $-2m^2n + 3mn^2 - 16mn$ **f** $-x^2 + 5x - 5$ **g** $13xy - 4xz - 3yz$ **h** $21x^4y - 5x^2y^2 - 2xy^4$ **i** $4p^3q - 4pq^2 - p^4 + p^2q$ **j** $12x^{10} - 8x^6y^2 - 16xy^3 - 40y^2$ **k** $-3a^5 + 12a^2 - 2a^6 - 6a$ **l** $2xy - 4x + 5y$

Page 27 **1 a** 1 **b** 5 **c** 7 **d** 2 **e** -3 **f** -1 **g** 0 **h** -8 **i** 16 **j** 13 **k** -6 **l** $-\frac{1}{6}$ **2 a** 36 **b** 144 **c** 15 **d** 121 **e** -10 **f** 2 **g** 1 **h** 540 **i** 5 **j** 11 **k** 21 **l** 15 **3 a** $\frac{5}{6}$ **b** $\frac{1}{6}$ **c** 5 **d** $\frac{1}{5}$ **e** $5\frac{1}{5}$ **f** $\frac{13}{36}$ **g** $\frac{5}{36}$ **h** $\frac{1}{5}$ **i** 1 **j** $\frac{25}{36}$ **k** $\frac{1}{36}$ **l** $\frac{13}{6}$ **4 a** 229.8 **b** 375.7 **c** 77.5 **d** 187.7 **e** 45.2 **f** 4.5 **g** 94.3 **h** 1.2 **i** 2.4 **j** 1062.8

Page 28 **1 a** $4(x + 4)$ **b** $9(a - 3)$ **c** $5x(1 - 5x)$ **d** $7a(1 + 3ab)$ **e** $5ab(1 + 5ab)$ **f** $7m(1 - 3mn)$ **g** $a^2b^2(a - b)$ **h** $14x^2y^2(xy - 2)$ **i** $5b(3a - 5c)$ **j** $3a(4b + 5a)$ **k** $xy(xy - 7)$ **l** $bc(a - 6d)$ **2 a** $-4(a + 7)$ **b** $-3(a + 5)$ **c** $-8(x + 4)$ **d** $-5y(2x + 3)$ **e** $-8(y - 5)$ **f** $-m^2(m + 1)$ **g** $-x^2(x + 10y^2)$ **h** $-6x(x - 2)$ **i** $-2y(5y - 6)$ **j** $-x(5 + 9x)$ **k** $-3m(1 + 6m^2)$ **l** $-9m(1 - 4m^3)$ **3 a** $(a + 2)(a + b)$ **b** $(x + y)(3 - a)$ **c** $(x - y)(9 + 2a)$ **d** $(2a + 3b)(5 - c)$ **e** $(5 - y)(x^2 - 3)$ **f** $(2x - 9)(x + 5)$ **g** $(a - b)(m - n)$ **h** $(x^2 + 7)(12 - y)$ **i** $(x + 8)(5 + y)$ **j** $(3b - 5c)(4a + 2)$ **k** $(2n - p)(m - q)$ **l** $(2a - 5b)(3x^2 + y^2)$ **4 a** $m(x + y + z)$ **b** $c(a + b + d)$ **c** $m(5 - n + 6p)$ **d** $5(2a + 5b + 7c)$ **e** $4(5xy - 2x^2 + 9)$ **f** $n(n - 8m + 10)$ **g** $5a(a + 3bc - 2)$ **h** $xy(y - 2 + x)$ **i** $3a(1 - 3b - 5a)$ **j** $5m(1 - 2n + 4mn)$ **k** $x^2y^2(x - 2 + 3y)$ **l** $x^2y^2(12z^2 - x + y)$

Page 29 **1** C **2** A **3** D **4** D **5** B **6** B **7** B **8** B **9** C **10** C

Page 30 **1 a** $25x^6$ **b** $4m^2$ **c** $10x^4y^5$ **d** 10 **e** $p + 5q$ **f** $12x^2 - 7x$ **g** $12x^8y^9$ **h** $2a^{-2}$ **i** $15a^{-6}$ **j** $49a^{13}b^7$ **k** 1 **2 a** $12xy + 24x^2 - 9y^2$ **b** $-6x^4 - 8x^3y^2$ **3 a** $-4c$ **b** $14x - 28$ **c** $47a - 43$ **d** $-5x^3y + 6x^2 + 12y^3$ **4** a^2 **5** $9a - 3b - 12$ **5** $\frac{13}{18}$ **6** $8a(a + 3b - 2)$

Chapter 3 – Pythagoras' theorem

Page 31 **1 a** c **b** f **c** h **d** JL **e** MN **f** PR **2 a** AB **b** EF **c** JL **d** PQ **e** AC **f** VZ **3 a** c **b** a **c** b **d** c **e** a^2 **f** b^2 **g** c^2 **h** hypotenuse

Page 32 **1 a** 16, 12, 20, 256, 144, 400, 400 **b** 9, 12, 15, 81, 144, 225, 225 **c** 24, 10, 26, 576, 100, 676, 676 **d** 30, 16, 34, 900, 256, 1156, 1156 **e** 4, 3, 5, 16, 9, 25, 25 **f** 15, 20, 25, 225, 400, 625, 625 **g** 5, 12, 13, 25, 144, 169, 169 **h** 8, 6, 10, 64, 36, 100, 100 **i** 8, 15, 17, 64, 225, 289, 289 **j** 40, 9, 41, 1600, 81, 1681, 1681 **k** 24, 18, 30, 576, 324, 900, 900 **l** 80, 18, 82, 6400, 324, 6724, 6724

Page 33 **1 a** C **b** C **c** B **d** C **e** C **f** C **g** C **h** B **i** C **j** C **k** C **l** B **m** C **n** A **o** C **p** B

Answers

PAGE 34 1 a 25 b 225 c 784 d 961 e 8464 f 81 g 3136 h 49 i 3721 j 1024 k 7225 l 6084 2 a 2 b 1 c 3 d 4 e 7 f 8 g 5 h 9 i 10 j 12 k 6 l 11 3 a 28 b 17 c 37 d 13 e 14 f 49 g 21 h 34 i 18 j 16 k 15 l 63 4 a 1.69 b 31.36 c 62.41 d 27.04 e 44.89 f 69.7225 g 68.89 h 69.2224 i 126.5625 j 94.09 k 29.2681 l 492.84 5 a 31.4721 b 10.24 c 39.8161 d 60.84 e 28.09 f 182.25 g 34.81 h 46.24 i 231.04 j 44.89 k 84.64 l 80.1025 6 a 2.3 b 2.6 c 7.3 d 2.8 e 1.8 f 9.7 g 2.8 h 2.6 i 7.9 j 2.9 k 2.9 l 8.6

PAGE 35 All answers are in cm. 1 a 5 b 13 c 10 d 26 e 17 f 25 2 a 9.8 b 7.1 c 14.0 d 8.7 e 5.9 f 18.8 g 10.8 h 8.5 i 7.2

PAGE 36 All answers are in cm. 1 a 6 b 8 c 24 d 4 e 9 f 15 2 a 9.90 b 12.39 c 13.89 d 17.35 e 8.39 f 10.40 g 12.39 h 20.03 i 6.62

PAGE 37 1 a 5 b 37 cm c 19.7 m 2 a 12 b 24 m c 8 cm 3 a 10.9 cm b 12.0 m c 10.6 km d 14.5 e 14.1 f 14.6

PAGE 38 All answers are in cm. 1 a 5 b 5 c 8 d 7 e 9 f 10 2 a 15.0 b $x = 10.0$, $y = 10.4$ c 9.9 d 9.0 e 9.0 f 7.8

PAGE 39 1 e, f, g, i , j, k, l 2 a $8^2 + 15^2 = 289, \sqrt{289} = 17$ b $4^2 + 3^2 = 25, \sqrt{25} = 5$ c $9^2 + 40^2 = 1681, \sqrt{1681} = 41$ d $7^2 + 24^2 = 625, \sqrt{625} = 25$ e $5^2 + 12^2 = 169, \sqrt{169} = 13$ f $11^2 + 60^2 = 3721, \sqrt{3721} = 61$ 3 a right-angled b not right-angled c right-angled

PAGE 40 1 a 7.1 cm b 29.4 cm 2 6.8 m 3 a 47.5 b 42.1 cm 4 10.39 cm 5 328 m

PAGE 41 1 4.58 m 2 17 km 3 5.20 m 4 Carlo will need 6.04 m, 6 m is not enough 5 24.25 m

PAGE 42 1 D 2 D 3 A 4 C 5 D 6 B 7 A 8 D 9 B 10 C

PAGE 43 1 69 2 yes 3 a 6 cm b 12 cm c 50 cm d 8.7 m e 5.0 cm f 12.5 m g 21.8 cm h 20.8 m i 604 mm 4 a 5.6 m b 11.9 m 5 a 15.7 m b 9.1 m

CHAPTER 4 – Financial mathematics

PAGE 44 1 \$500 2 \$21.60 3 \$1666.80 4 \$655.20 5 \$837.50 6 \$2802 7 \$34 268 8 a \$38 880 b \$747.69 9 \$109 10 a \$5620 b \$28.10 11 a \$149 512 b \$21 358.86 per day, headline is correct.

PAGE 45 1 a \$22.05 b \$29.40 2 a \$126 b \$126 3 \$1053.50 4 a \$3740.80 b \$654.64 c \$4395.44 5 a \$21.25 b \$1253.75 6 \$1800 7 \$6349.70 8 a \$16.80 b \$201.60 c 6 hours

PAGE 46 1 \$550 2 a \$2100 b \$2350 3 \$586.50 4 \$190.90 5 a \$1700 b \$2450 6 \$9250 7 \$301.50 8 \$1400

PAGE 47 1 \$1642.18 2 \$703.85 3 a \$74 040 b \$20 510.80 c \$4709.20 4 \$7950

PAGE 48 1 a \$720 b \$3360 c \$14 400 d \$2808 e \$520 f \$200 g \$14 760 h \$21 125 i \$13 530 j \$354 k \$937.50 l \$1213.33 m \$384.38 n \$205.48 2 a 4.63 years b 4.43 years 3 a 5.56% b 16.67%

PAGE 49 1 a \$4500 b \$2777.78 2 a \$600 b \$3600 3 a 12.5% b \$12 500 c 3 years d \$4500

PAGE 50 1 a 5.926 years b 3.968 years 2 a 6.52% p.a. b 10% p.a. 3 a \$4444.44 b \$7625 4 a \$4200 b \$300 c \$1125

PAGE 51 1 a \$8000 b \$466.67 2 a \$600 b \$2400 c \$1080 d \$3480 e \$96.67 3 a \$13 560 b \$3060 c \$ 8.5% 4 a \$14 780 b \$246.33

PAGE 52 1 a \$60 000 b \$32 400 c \$2400 2 a \$305 000 b \$4600 3 a \$162 000 b \$196 345 c 30.9%

PAGE 53 1 B 2 C 3 B 4 B 5 B 6 B 7 C 8 C 9 A 10 B

PAGE 54 1 a \$87.50 b \$14.50 2 a \$2573.08 b \$128.65 3 a \$1248.30 b \$1691.78 c \$5867.01 4 a \$900 b 10% 5 a \$1600 b \$14 400 c \$3888 d \$18 288 e \$508 f \$19 888

CHAPTER 5 – Linear and non-linear relationships

PAGE 55 1 a 3 units b 2 units c 4 units d 1 unit e 5 units f 4 units g 4 units h 4 units 2 a 2 units b 3 units c 5 units d 6 units e 3 units f 6 units g 4 units h 7 units i 6 units 3 a 4 units b 5 units c 5 units d 4 units e 5 units f 7 units g 5 units h 5 units i 6 units

PAGE 56 1 a $\sqrt{45}$ b $\sqrt{41}$ c $\sqrt{73}$ d $\sqrt{65}$ e 10 f $\sqrt{52}$ 2 a $\sqrt{85}$ b $\sqrt{50}$ c $\sqrt{41}$ d $\sqrt{52}$ e $\sqrt{65}$ f $\sqrt{53}$

PAGE 57 1 a $\sqrt{89}$ units b $\sqrt{32}$ units c $\sqrt{34}$ units d 5 units e 5 units f $\sqrt{13}$ units g $\sqrt{2}$ units h $\sqrt{85}$ units i $\sqrt{13}$ units j $\sqrt{80}$ units k $\sqrt{8}$ units l $\sqrt{20}$ units 2 $(1 + \sqrt{130} + \sqrt{149})$ units 3 34

PAGE 58 1 a 8 b 8 c 8 d 5 e 5 f 5 2 a 3 b 1 c 5 d 3 3 a 8 b 8 c 6 d 9 e 7 f 3 g 2 h 0 i 10 j 5 k 10 l 10 4 a 5 b 4 c (5, 4) 5 a 12 b 3 c 9 d 7 e 5 f 4 g 2 h 2 i 1 j 7 k –4 l –5

PAGE 59 1 a (1, 5) b (5, 9) c (–4, 1) d (5, 5) e (6, 0) f (3, 7) g $(1\frac{1}{2}, -3\frac{1}{2})$ h (9, 5) i (5, 9) j (5, 7) k (–1, –1) l (–2, –2) 2 midpoint of $AB = (4, 10)$, midpoint of $BC = (1\frac{1}{2}, 6)$, midpoint of $AC = (-4\frac{1}{2}, 5)$ 3 $\frac{7 + -7}{2} = 0, \frac{3 + -3}{2} = 0$

PAGE 60 1 a (2, 5) b (–3, 6) 2 a (7, 8) b (9, 11) c (8, –7) d (–4, 6) e (9, 11) f (8, 6) g (7, 13) h (10, 13) i (–1, 7) j (8, 16) k (11, 6) l (9, –3) 3 a (4, 7) b (4, 8) c (2, 0) d (2, 6) e (4, 6) f (8, 12) g (2, 13) h (1, 2) 4 a (6, 10) b $a = 3, b = -5$ c $p = 10$ d $x = -7, y = 7$

PAGE 61 1 a positive b negative c positive d negative e positive f positive g negative h negative 2 a 2 b 1 c $\frac{2}{11}$ d $-\frac{7}{2}$ e –1 f $-2\frac{1}{2}$ 3 a *DF* b *CD* c *AF* d *FC* or *AB* e *OF* f *ED*

PAGE 62 1 a –12 b $1\frac{1}{2}$ c $-\frac{2}{3}$ d $\frac{1}{3}$ e $-\frac{1}{6}$ f $-\frac{1}{2}$ g $3\frac{1}{2}$ h 3 i $\frac{2}{5}$ j –1 k $-\frac{2}{5}$ l $1\frac{1}{2}$ 2 $\frac{5+1}{-1-1} = \frac{-7-5}{3+1} = \frac{-7+1}{3-1} = -3$

Answers

3 m of $AB = m$ of $CD = \frac{-4}{3}$ and m of BC $= m$ of $DA = \frac{3}{4}$

PAGE 63 **1** **a** (5, 1.5) **b** (9, 4.5) **c** $\frac{3}{4}$ **d** $\frac{3}{4}$ **e** same **f** 10 units **g** 5 units **h** 2 **2** **a** 5 units **b** $\sqrt{50}$ units **c** 5 units **d** yes **e** isosceles **3** (7, 11) **4** $m = \frac{4}{3}$

PAGE 64 **1** **a**

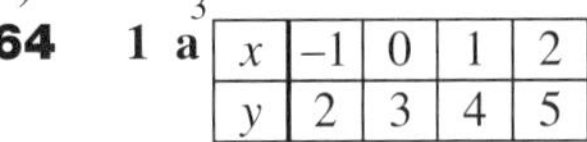

x	–1	0	1	2
y	2	3	4	5

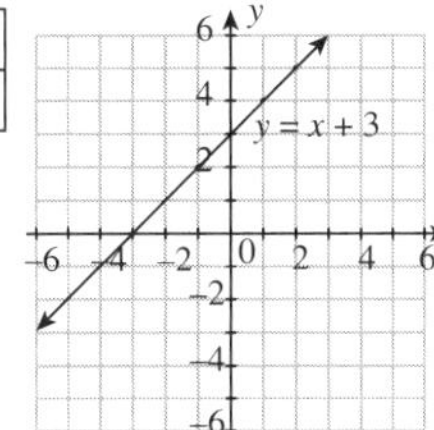

b

x	–1	0	1	2
y	–3	–1	1	3

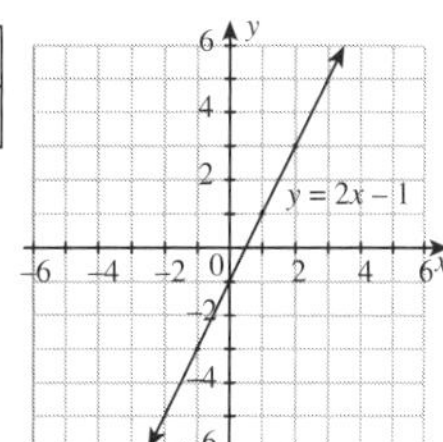

c

x	–1	0	1	2
y	2	1	0	–1

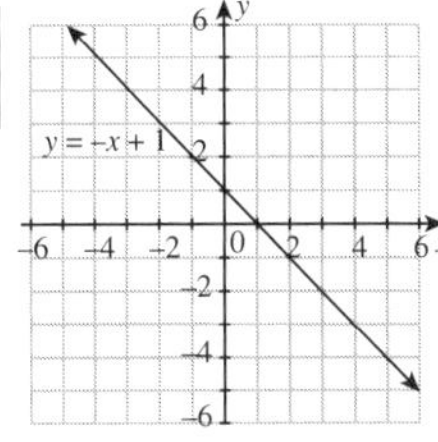

d

x	0	1	2	3
y	–1	0	1	2

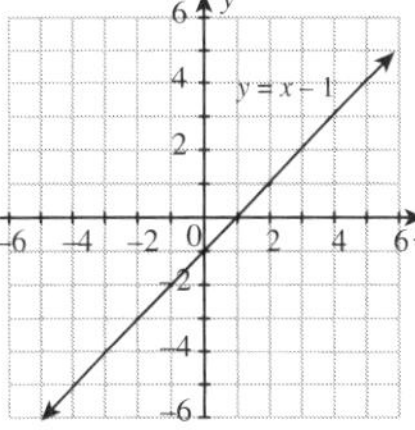

e

x	0	1	2	3
y	0	3	6	9

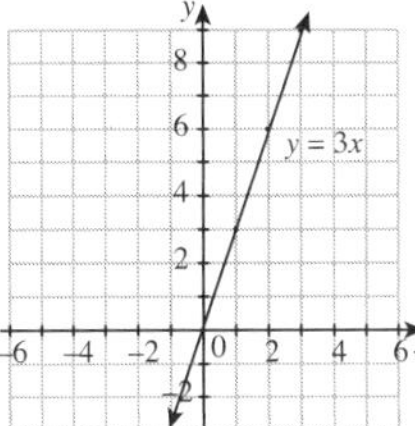

f

x	0	1	2	3
y	2	4	6	8

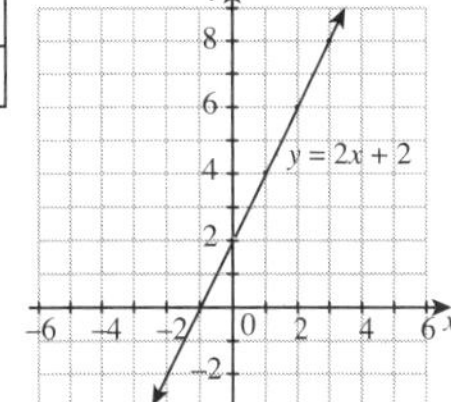

2

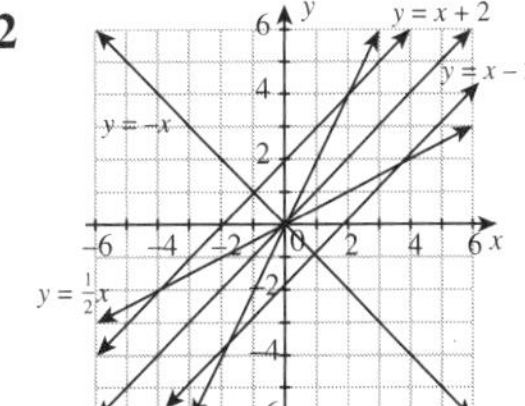

3

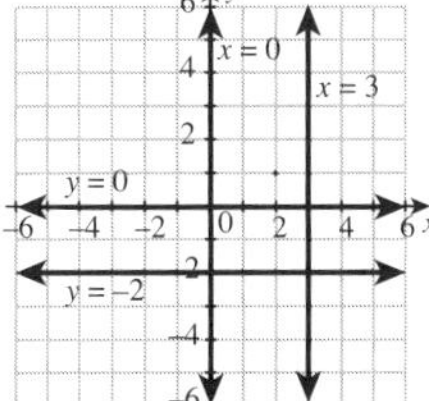

PAGE 65 **1** **a** $x = 2, y = 2$ **b** $x = 4, y = -4$ **c** $x = 3, y = 6$ **d** $x = 6, y = -2$ **e** $x = 4, y = -2$ **f** $x = 1\frac{1}{2}, y = -3$ **g** $x = 4, y = -3$ **h** $x = 1, y = -3$ **2** **a**

b

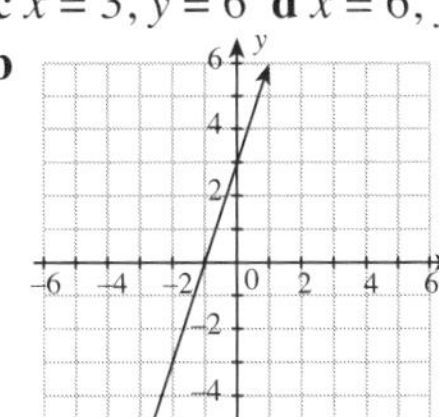

c

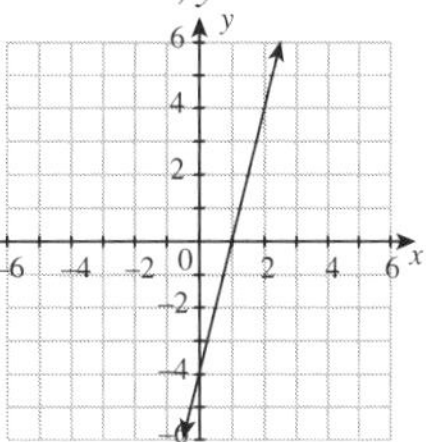

3 **a**

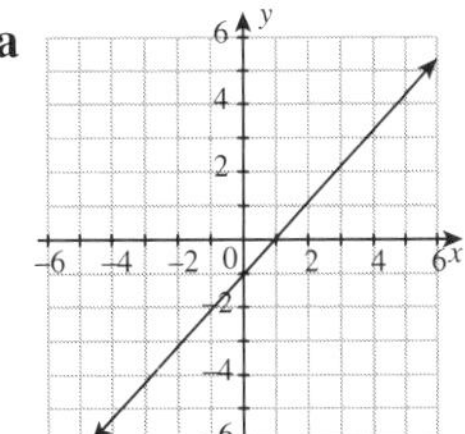

b

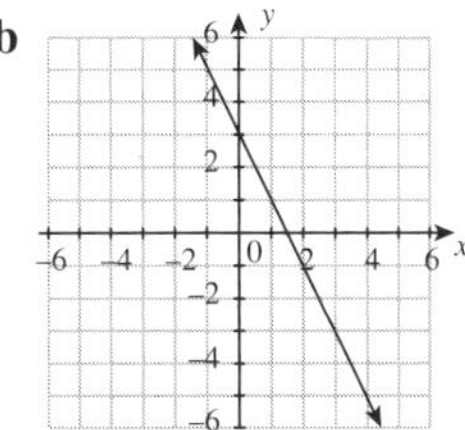

c

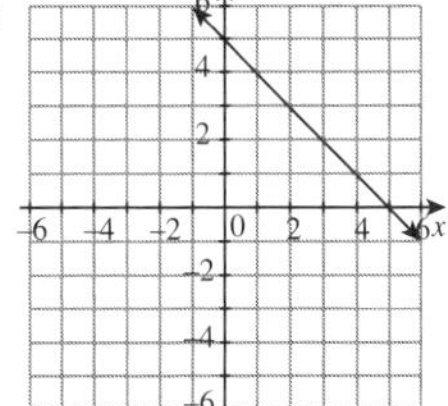

d

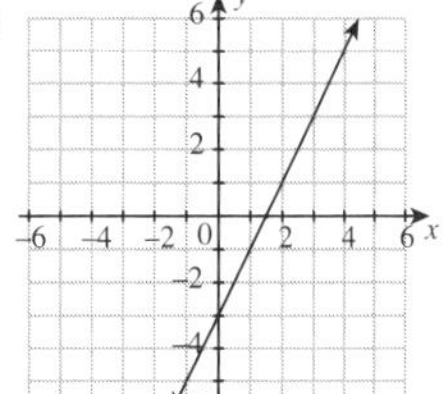

e

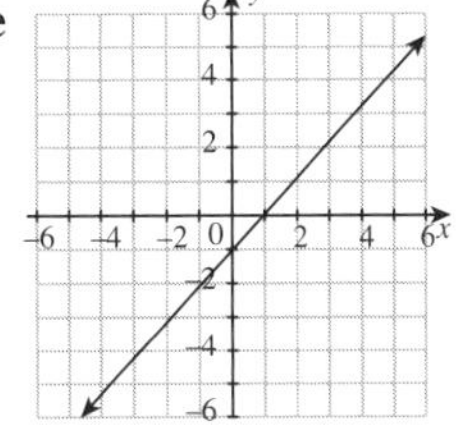

f

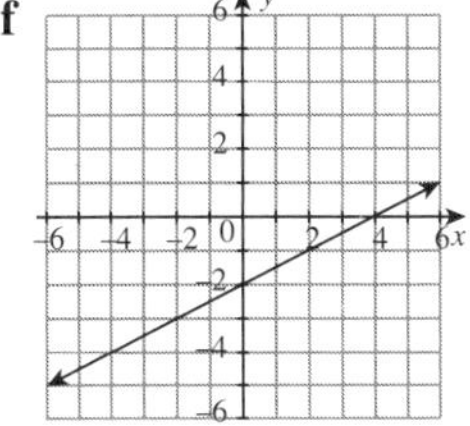

Answers

PAGE 66 **1 a** 2, 3, 4 **b**

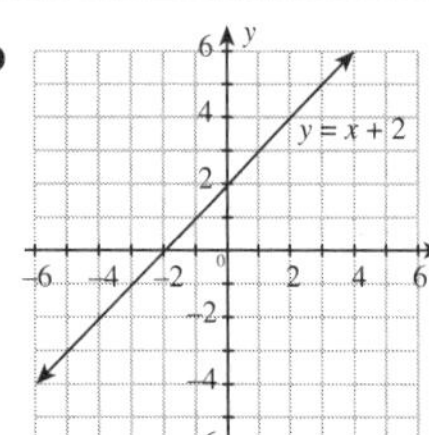

c 1 **d** positive **e** right **f** 1 **g** yes **h** 2 **i** yes

2 a 1, –1, –3 **b**

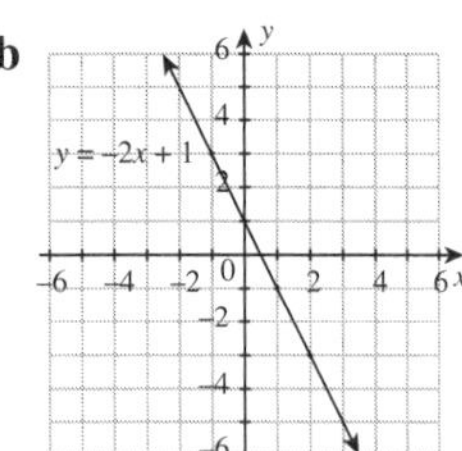

c –2 **d** negative **e** left **f** –2 **g** yes **h** 1 **i** yes **3** gradient, y intercept **4 a** 3, –8 **b** 4, 7 **c** –2, 5

PAGE 67 **1 a** 2, 7 **b** 3, 1 **c** 7, 0 **d** 4, –3 **e** $\frac{1}{2}$, 6 **f** 1, 4 **g** –3, 8 **h** –1, –5 **i** –2, 11

2 a 2, 3

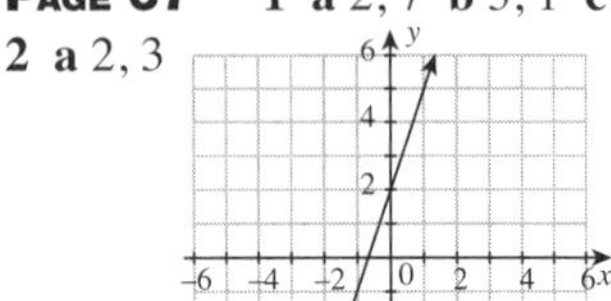

b –1, 2

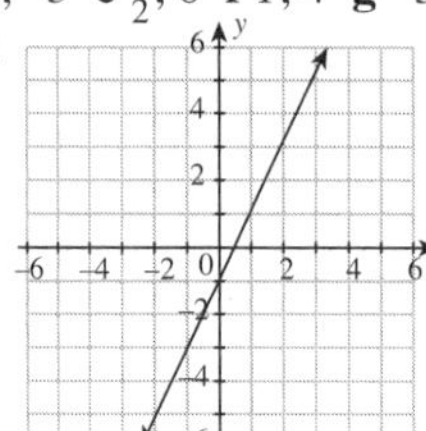

c –5, 3

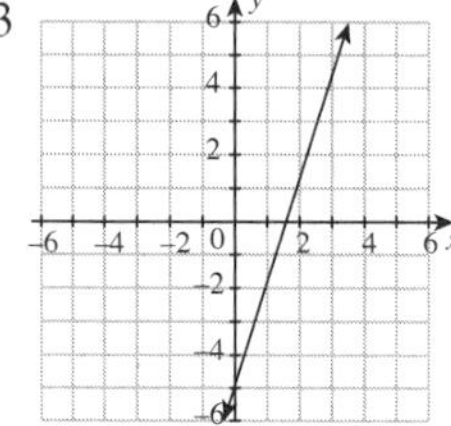

d 0, 1

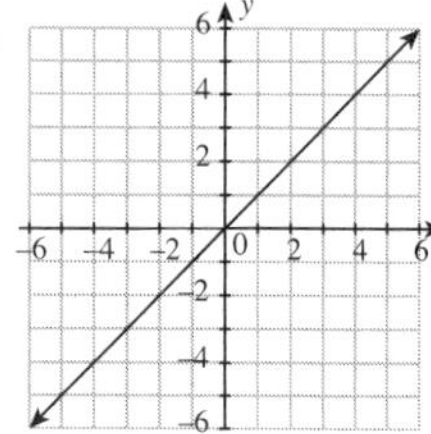

e 1, –2

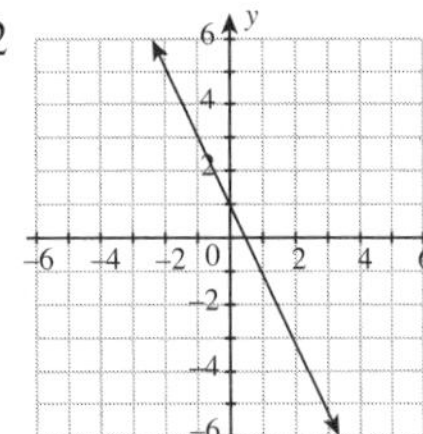

f 4, $\frac{1}{2}$

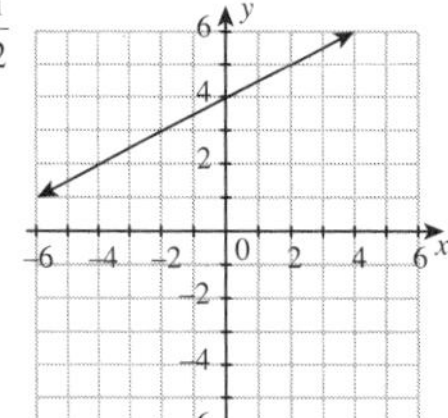

PAGE 68 **1 a** $2x - 5y - 9 = 0$ **b** $3x + 4y - 8 = 0$ **c** $5x - 2y - 7 = 0$ **d** $4x - 8y + 3 = 0$ **e** $2x + y - 9 = 0$ **f** $8x - y + 7 = 0$ **g** $2x - 3y + 6 = 0$ **h** $8x - 9y + 12 = 0$ **i** $x - 6y + 3 = 0$ **2 a** $y = -\frac{2}{3}x + \frac{8}{3}$ **b** $y = -\frac{1}{5}x + \frac{7}{5}$ **c** $y = \frac{3}{2}x - \frac{3}{2}$ **d** $y = x + 7$ **e** $y = -2x + 9$ **f** $y = \frac{5}{6}x + \frac{11}{6}$ **g** $y = \frac{3}{2}x - 3$ **h** $y = -\frac{4x}{5} - \frac{3}{5}$ **i** $y = 2x + 6$ **3 a** $y = 4x + 3$; $4x - y + 3 = 0$ **b** $y = 2x - 5$; $2x - y - 5 = 0$ **c** $y = 3x + 7$; $3x - y + 7 = 0$ **d** $y = \frac{1}{2}x + 4$; $x - 2y + 8 = 0$ **e** $y = \frac{2}{3}x + 6$; $2x - 3y + 18 = 0$ **f** $y = -\frac{5}{6}x + 3$; $5x + 6y - 18 = 0$

PAGE 69 **1 a** (0, 3) **c** (–4, 6) **e** (4, 0) **f** (8, –3) **2 b** $2y = 3x$ **c** $x - 5y = 0$ **e** $y = -2x$ **3 a** yes **b** yes **c** no **d** yes **e** yes **f** no **4** $m = 3$ **5** $a = 5$ **6 a** (0, –2) **b** (2, 4) **c** (1, 1) **d** (5, 13) **e** (–1, –5) **f** (–2, –8)

PAGE 70 **1 a** $y = 2x + 3$ **b** $y = 3x + 13$ **c** $y = -x + 8$ **d** $y = \frac{1}{2}x + 4$ **e** $y = -\frac{2}{3}x - 4$ **f** $y = \frac{4}{3}x - \frac{8}{3}$ **2 a** $2x + y + 1 = 0$ **b** $x - 4y + 18 = 0$ **c** $3x + 2y + 17 = 0$ **3 a** $y = \frac{1}{2}x + \frac{7}{2}$ **b** $y = -\frac{2}{7}x + \frac{17}{7}$ **c** $y = \frac{2}{7}x - \frac{8}{7}$ **4 a** $x - 2y - 8 = 0$ **b** $9x - 5y - 1 = 0$ **c** $x + y - 1 = 0$

PAGE 71 **1 a i** –6 **ii** 0 **iii** 4 **b i** 0 **ii** 1 **iii** 3 **c** Because $y = 2x - 4$, so finding the x-value when $y = -4$, for example, is the same as finding the x-value for which $2x - 4 = -4$ **2 a** $x = 1$ **b** $x = -1$ **c** $x = 0$ **d** $x = -2$ **e** $x = 2$ **f** $x = \frac{2}{3}$ **g** $x = \frac{4}{3}$ **h** $x = -\frac{4}{3}$ **i** $x = -\frac{5}{3}$

3 a

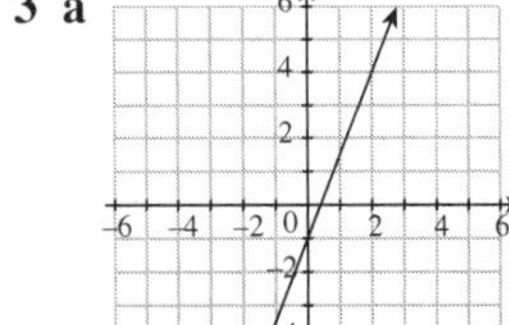

b i $x = \frac{1}{2}$ **ii** $x = 0$ **iii** $x = 2$ **iv** $x = -2$ **4 a**

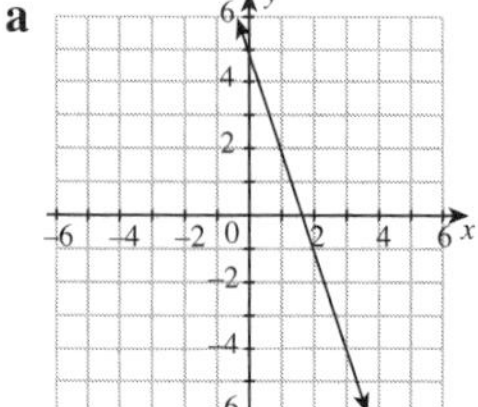

b i $x = 1$ **ii** $x = 3$ **iii** $x = 0$ **iv** $x = 2$

Answers

PAGE 72 1

x	–3	–2	–1	0	1	2	3
$y = x^2$	9	4	1	0	1	4	9
$y = x^2 + 2$	11	6	3	2	3	6	11
$y = x^2 - 2$	7	2	–1	–2	–1	2	7

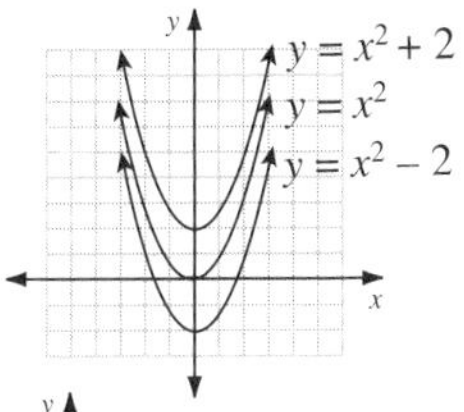

2 a

x	–3	–2	–1	0	1	2	3
$y = 2x^2$	18	8	2	0	2	8	18
$y = \frac{1}{2}x^2$	$4\frac{1}{2}$	2	$\frac{1}{2}$	0	$\frac{1}{2}$	2	$4\frac{1}{2}$

b

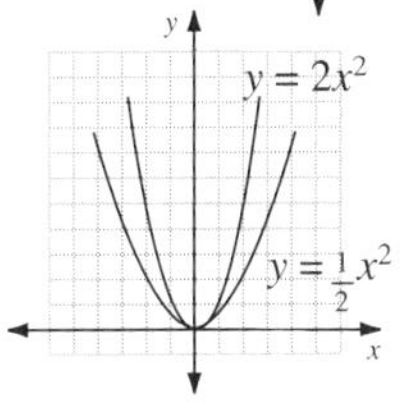

3 a

x	–3	–2	–1	0	1	2	3
$y = -x^2$	–9	–4	–1	0	–1	–4	–9
$y = -x^2 + 4$	–5	0	3	4	3	0	–5
$y = 9 - x^2$	0	5	8	9	8	5	0

b

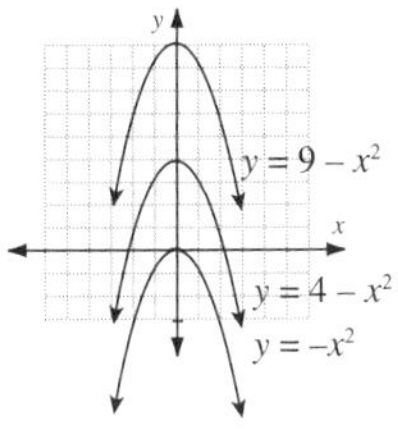

4 a concave down **b** $x = 0$ **c** 8 **d** –4 and 4 **e** $y = -\frac{1}{2}x^2 + 8$

PAGE 73 1 **a** $y = 2^x$

x	–3	–2	–1	0	1	2	3
y	$\frac{1}{8}$	$\frac{1}{4}$	$\frac{1}{2}$	1	2	4	8

b

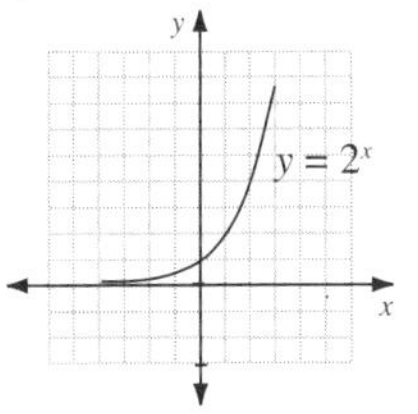

c i $\frac{1}{16}$ **ii** 16 **d** no
e y becomes very large **f** $y = 1$

2 a $y = 3^x$

x	–3	–2	–1	0	1	2	3
y	$\frac{1}{27}$	$\frac{1}{9}$	$\frac{1}{3}$	1	3	9	27

b

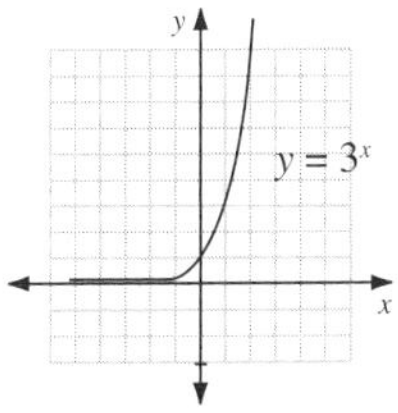

c 81 **d** no **e** (0, 1) **3 a**

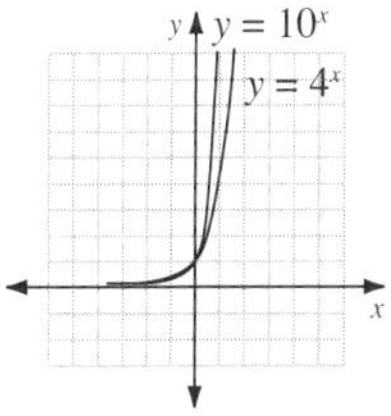

b As x gets large y becomes very large for both curves, more so for $y = 10^x$. As x gets smaller y approaches O for both curves. Both intersect the y-axis when $x = 0$ at $y = 1$ **c** 81 **4 a**

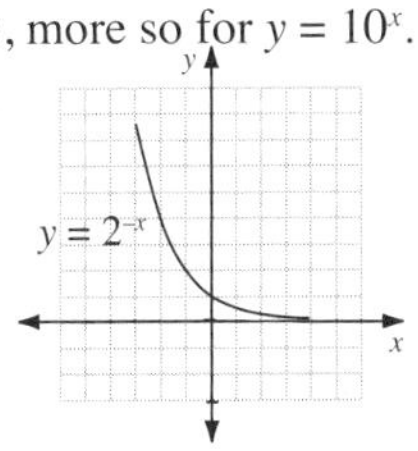

b

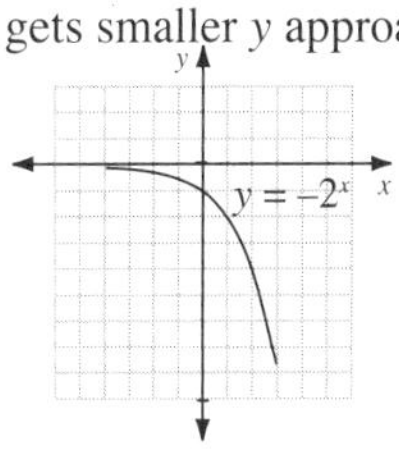

c

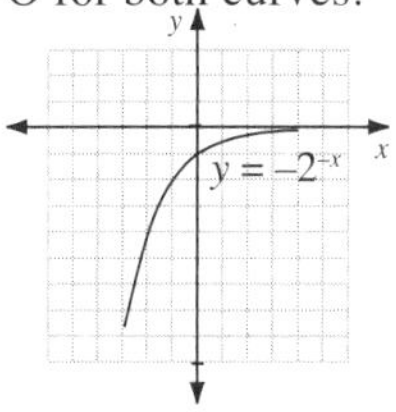

PAGE 74 1 **a** 3 units **b** 7 units **c** 12 units **d** 1 unit **2 a** $x^2 + y^2 = 100$ **b** $x^2 + y^2 = 36$ **c** $x^2 + y^2 = 169$ **d** $x^2 + y^2 = 289$
3 a $x^2 + y^2 = 16$ **b** $x^2 + y^2 = 4$ **c** $x^2 + y^2 = 25$ **d** $x^2 + y^2 = 12\frac{1}{4}$
4 a

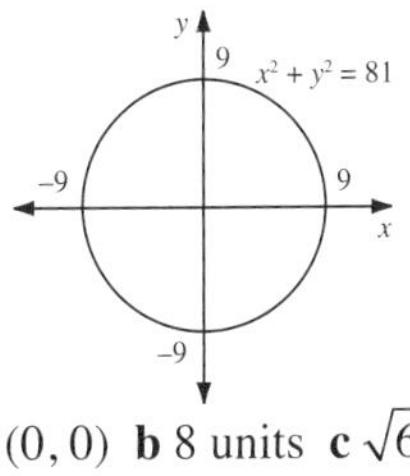

b

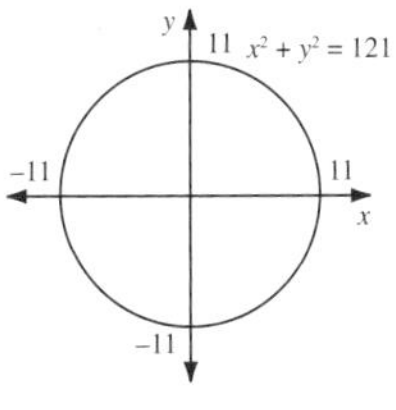

c

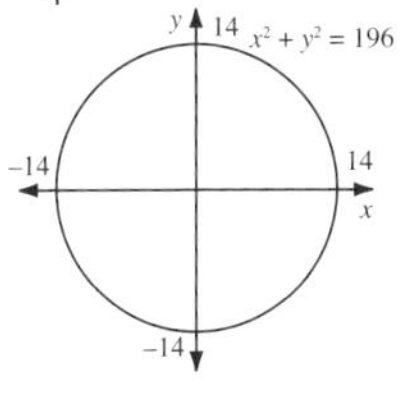

d

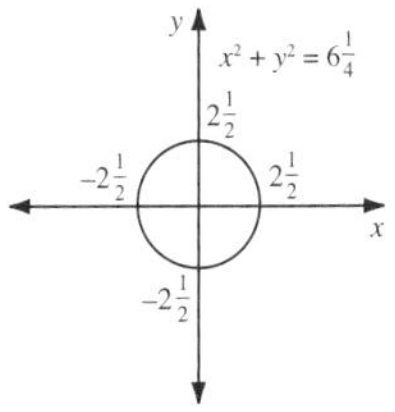

5 a (0, 0) **b** 8 units **c** $\sqrt{61}$ units **d** inside
PAGE 75 1 A **2** C **3** C **4** D **5** B **6** D **7** B **8** B **9** C **10** B
PAGE 76 1 **a** $-\frac{4}{3}$ **b** (0, 2) **c** 10 units **d**

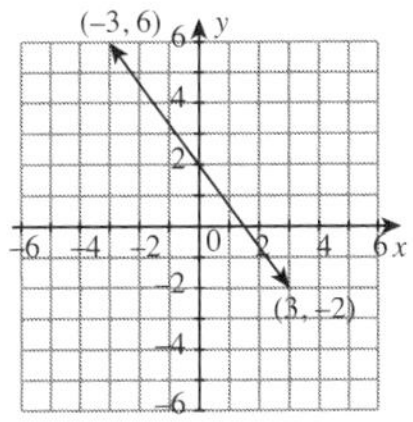

e 2 **d** $y = -\frac{4}{3}x + 2$

Answers

2 a

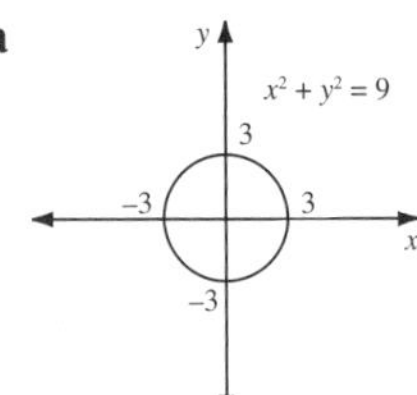

b

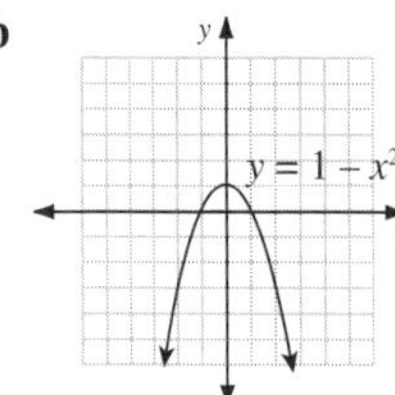

c

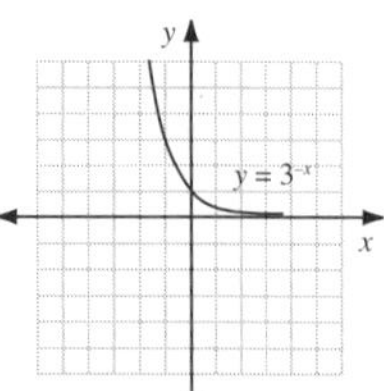

3 a 2 **b** 8 **c**

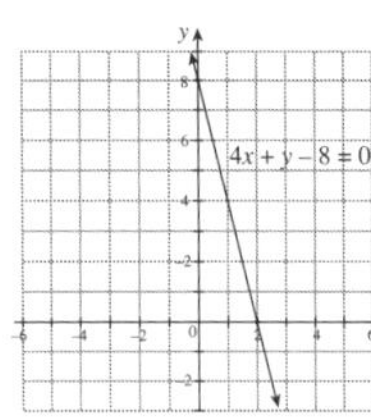

d yes **4 a** $3x - y + 6 = 0$ **b** $y = 3x + 6$

Chapter 6 – Equations

Page 77 **1 a** $x = 4$ **b** $x = 10$ **c** $x = 14$ **d** $x = 4$ **e** $x = 9$ **f** $x = -9$ **g** $x = 11$ **h** $x = 2$ **i** $x = 4$ **j** $y = 54$ **k** $m = 40$ **l** $a = -17$ **m** $a = -40$ **n** $p = -6$ **o** $a = 56$ **p** $a = 5.6$ **q** $m = -22$ **r** $x = 8$ **s** $y = 20$ **t** $y = -9$ **2 a** $x = 4$ **b** $x = 4$ **c** $x = 7$ **d** $x = 5$ **e** $x = 18$ **f** $x = 45$ **g** $x = 45$ **h** $x = -15$ **i** $x = 16$ **j** $x = 10$ **k** $x = 15$ **l** $x = 24$ **m** $x = 14$ **n** $m = 20$ **o** $x = 30$ **3 a** $x = 10$ **b** $x = 10$ **c** $x = 15$ **d** $x = 5$ **e** $x = 2$ **f** $x = 6$ **g** $x = 7$ **h** $x = 10$ **i** $m = -4$ **j** $y = 2\frac{1}{3}$ **k** $y = 1\frac{4}{9}$ **l** $p = 6$ **m** $x = 15$ **n** $x = 10\frac{2}{3}$ **o** $x = 37\frac{1}{2}$

Page 78 **1 a** $x = 2$ **b** $y = 3$ **c** $x = 3$ **d** $x = 20$ **e** $m = 4$ **f** $x = 9$ **g** $k = 10$ **h** $x = 10$ **i** $x = 3$ **j** $m = 4$ **k** $x = 20$ **l** $x = 7$ **m** $x = 3$ **n** $a = 5$ **o** $n = 4$ **2 a** $x = 6$ **b** $y = 27$ **c** $x = 19$ **d** $x = -1$ **e** $m = 10$ **f** $m = 6$ **g** $y = -1$ **h** $a = 3$ **i** $b = 0.9$

Page 79 **1 a** $x = 12$ **b** $y = 10$ **c** $m = 9$ **d** $x = -21$ **e** $x = 34$ **f** $m = -3$ **g** $t = 11$ **h** $y = 17$ **i** $x = 3$ **j** $m = 6$ **k** $a = 7$ **l** $x = -12$ **m** $a = 4$ **n** $x = 5$ **o** $x = -\frac{4}{3}$ **2 a** $x = 4$ **b** $a = 12$ **c** $y = 6$ **d** $m = 5$ **e** $p = 5$ **f** $x = 2$ **g** $x = -\frac{14}{9}$ **h** $y = 5$ **i** $m = 3$

Page 80 **1 a** $x = 10$ **b** $x = 17$ **c** $x = 16$ **d** $x = 6$ **e** $x = 15$ **f** $x = 4$ **g** $m = 3$ **h** $x = 7$ **i** $x = 2$ **j** $x = 2$ **k** $x = 9$ **l** $a = 46$ **2 a** $x = -7$ **b** $x = 6$ **c** $m = -2\frac{1}{4}$ **d** $t = 1$ **e** $y = 27$ **f** $y = 8$ **g** $a = 6$ **h** $x = 4$ **i** $m = -9$ **j** $n = 4$ **k** $x = 3$ **l** $x = 2$ **m** $m = 3$ **n** $a = 11$ **o** $x = 8$

Page 81 **1 a** $m = 5$ **b** $a = 6$ **c** $x = 4\frac{2}{5}$ **d** $a = -32$ **e** $a = 0$ **f** $a = 21$ **g** $m = 1\frac{1}{2}$ **h** $x = 23$ **i** $a = -57$ **j** $x = 68$ **k** $x = -1\frac{1}{2}$ **l** $x = 5$ **2 a** $n = 3$ **b** $n = 12$ **c** $x = 2$ **d** $p = 4$ **e** $x = 4$ **f** $x = 2$ **g** $x = 2$ **h** $x = 3$ **i** $y = 17$

Page 82 **1 a** $x = 1\frac{1}{2}$ **b** $x = 8$ **c** $x = 10$ **d** $x = 6$ **e** $x = -4$ **f** $m = -6$ **g** $x = 29$ **h** $x = 0$ **i** $x = 38$ **j** $x = 17$ **k** $x = -3\frac{1}{2}$ **l** $x = -8$ **2 a** $a = -6$ **b** $x = 6$ **c** $m = 23$ **d** $y = -8$ **e** $t = -2$ **f** $x = 1$ **g** $a = 4\frac{1}{8}$ **h** $m = -5\frac{1}{2}$ **i** $a = \frac{-7}{11}$ **j** $a = 1$ **k** $x = 0$ **l** $m = -4$

Page 83 **1 a** $x = 1\frac{1}{3}$ **b** $a = 1\frac{2}{5}$ **c** $y = 3\frac{3}{4}$ **d** $y = 9$ **e** $p = -24$ **f** $m = 21$ **g** $a = 30$ **h** $m = 10\frac{1}{2}$ **i** $x = 10\frac{1}{2}$ **j** $x = 3\frac{3}{4}$ **k** $m = 2$ **l** $m = 7$ **2 a** $x = -4\frac{1}{7}$ **b** $x = -6$ **c** $m = 3$ **d** $x = 9$ **e** $x = 9\frac{3}{5}$ **f** $x = 2$ **g** $x = \frac{1}{10}$ **h** $x = -2\frac{1}{2}$ **i** $x = 2$ **j** $a = 11\frac{1}{5}$ **k** $k = 51$ **l** $m = 30$

Page 84 **1 a** $x = 8$ **b** $p = 27$ **c** $y = 45$ **d** $a = 19$ **e** $m = 8$ **f** $x = 3$ **g** $x = -5\frac{2}{7}$ **h** $x = -2$ **i** $x = 9$ **j** $x = 7$ **k** $x = 60$ **l** $x = 20$ **2 a** $x = 22\frac{1}{2}$ **b** $x = 25\frac{2}{3}$ **c** $x = 16\frac{1}{4}$ **d** $x = \frac{1}{2}$ **e** $m = 2$ **f** $x = \frac{1}{8}$ **g** $x = -2\frac{4}{13}$ **h** $x = 1\frac{5}{7}$ **i** $y = -4$ **j** $y = 1$ **k** $x = -1\frac{1}{3}$ **l** $x = \frac{1}{2}$

Page 85 **1 a** $a = 10\frac{2}{7}$ **b** $x = -30$ **c** $x = 40$ **d** $a = 128$ **e** $y = 27$ **f** $t = 5\frac{1}{7}$ **g** $p = 7\frac{1}{2}$ **h** $x = \frac{10}{19}$ **2 a** $x = \frac{3}{31}$ **b** $x = -12\frac{1}{2}$ **c** $m = -13$ **d** $t = -1\frac{1}{4}$ **e** $x = 2$ **f** $m = 5\frac{2}{5}$ **g** $m = 3\frac{22}{23}$ **h** $a = 4\frac{2}{5}$ **i** $x = 35$ **j** $x = 560$ **k** $x = \frac{1}{5}$ **l** $x = \frac{11}{12}$

Page 86 **1 a** 5 **b** 20 **c** 17, 19, 21 **d** 10 **e** 6 cm, 18 cm **f** 18 **2 a** 30°, 60°, 90° **b** 18 **c** 42 years **d** Melissa is 24 years, Steven is 12 years **e** $288

Page 87 **1** C **2** C **3** D **4** C **5** B **6** C **7** B **8** B **9** C **10** B

Page 88 **1 a** $x = 35$ **b** $y = -72$ **c** $m = 48$ **2** 10 **3 a** $x = 5\frac{1}{2}$ **b** $x = 3$ **c** $x = 2$ **d** $a = 3\frac{1}{2}$ **e** $x = 1\frac{1}{6}$ **f** $y = \frac{3}{10}$ **g** $m = 8$ **h** $a = 3$ **i** $x = 1\frac{2}{7}$ **j** $m = 2$ **k** $x = 1\frac{2}{7}$

Chapter 7 – Area and volume

Page 89 **1 a** 48 cm^2 **b** 126 cm^2 **c** 24 cm^2 **2 a** 64 cm^2 **b** 72 m^2 **c** 518 cm^2 **d** 112 m^2 **e** 25 cm^2 **f** 90 m^2 **3 a** 8 m^2 **b** 198 km^2 **c** 60 m^2

Page 90 **1 a** $A = s^2$ **b** $A = lb$ **c** $A = \frac{1}{2}bh$ **d** $A = bh$ **e** $A = \frac{1}{2}xy$ **f** $A = \frac{1}{2}h(a + b)$ **g** $A = \frac{1}{2}ab$ **h** $A = \pi r^2$ **i** $A = \frac{\theta}{360°} \times \pi r^2$ **2 a** 68.04 cm^2 **b** 68.89 cm^2 **c** 25.44 cm^2 **d** 170.52 cm^2 **e** 188.34 cm^2 **f** 69.3 cm^2 **3 a** 60.0 cm^2 **b** 452.4 cm^2 **c** 200.0 cm^2 **d** 615.8 cm^2 **e** 80.0 cm^2 **f** 42.3 cm^2

Page 91 **1 a** 50.3 cm^2 **b** 346.4 cm^2 **2 a** $\frac{1}{4}$ **b** $\frac{3}{4}$ **c** $\frac{1}{8}$ **d** $\frac{1}{3}$ **3 a** 63.6 cm^2 **b** 537.6 cm^2 **c** 33.5 cm^2 **d** 472.5 m^2 **e** 2832.9 km^2 **f** 16 890.2 mm^2

Page 92 **1 a** 103.7 m^2 **b** 125.7 cm^2 **2 a** 68 m^2 **b** 42 cm^2 **3** 163.4 m^2 **4** 163 cm^2

Answers

Page 93 **1 a** 198 cm^2 **b** 864 cm^2 **c** 130 cm^2 **d** 240 cm^2 **e** 144 cm^2 **f** 1200 cm^2 **g** 5700 cm^2 **h** 624 cm^2 **i** 175 cm^2 **j** 54 cm^2 **k** 160 cm^2 **l** 117 cm^2 **m** 380 cm^2 **n** 340 cm^2 **o** 300 cm^2

Page 94 **1 a** 108 cm^2 **b** 44 cm^2 **c** 56 cm^2 **d** 292 cm^2 **e** 6096 cm^2 **f** 29 cm^2 **2 a** 10 cm **b** 30.5 cm^2

Page 95 **1 a** 144 cm^2 **b** 350 cm^2 **c** 220 cm^2 **d** 361 cm^2 **e** 90 cm^2 **f** 130 cm^2 **2 a** 195.0 cm^2 **b** 30.9 cm^2 **c** 182.8 cm^2

Page 96 **1 a** 710 cm^2 **b** 926 cm^2 **c** 2047.6 cm^2 **d** 298.45 cm^2 **e** 322 cm^2 **f** 102.73 cm^2 **3 a** 292.25 cm^2 **b** 201.06 cm^2 **c** 35.16 cm^2 **d** 181.44 cm^2 **e** 20.62 cm^2 **f** 678.58 cm^2

Page 97 **1 a** 294.0 m^2 **b** 433.5 cm^2 **c** 541.5 m^2 **2 a** 472.0 cm^2 **b** 632.2 cm^2 **c** 1925.7 cm^2 **d** 788.0 cm^2 **e** 1861.6 cm^2 **f** 1249.1 cm^2 **3 a** 894 cm^2 **b** 23 774 cm^2 **c** 4984 cm^2

Page 98 **1 a** 736 cm^2 **b** 768.0 cm^2 **c** 1432.0 m^2 **d** 524.0 cm^2 **e** 862.0 cm^2 **f** 627.0 cm^2 **2 a** 2320 cm^2 **b** 14 976 cm^2

Page 99 **1 a** 201.06 cm^2 **b** 1005.31 cm^2 **2 a** 170 cm^2 **b** 470 cm^2 **c** 450 cm^2 **d** 310 cm^2 **e** 1700 cm^2 **f** 430 cm^2 **3 a** 1385.4 cm^2 **b** 1407.4 cm^2 **c** 754.0 cm^2

Page 100 **1 a** 340 cm^2 **b** 5792 cm^2 **2 a** 1950 cm^2 **b** 108 cm^2 **c** 3780 cm^2 **3 a** 576 cm^2 **b** 287.9 cm^2 **c** 2537.6 cm^2 **b** 1407.4 cm^2 **c** 754.0 cm^2

Page 101 **1 a** 27 m^3 **b** 125 cm^3 **c** 592.704 cm^3 **2 a** 192 cm^3 **b** 350 cm^3 **c** 834.8 cm^3 **3 a** 280 cm^3 **b** 4200 m^3 **c** 1794 m^3 **4 a** 17.5 m^2 **b** 70 m^3

Page 102 **1 a** 195.1 cm^3 **b** 188.2 cm^3 **c** 213.6 cm^3 **d** 1235 cm^3 **e** 223.2 cm^3 **f** 21.06 cm^3 **g** 6118 cm^3 **h** 351.0 cm^3 **i** 1717 cm^3 **2 a** 1072.5 m^3 **b** 1968 m^3 **c** 520 m^3

Page 103 **1 a** 2300 cm^3 **b** 5200 cm^3 **c** 41 000 cm^3 **2 a** 612 cm^3 **b** 688 cm^3 **c** 905 cm^3 **d** 1924 cm^3 **e** 3848 cm^3 **f** 296 cm^3 **3** 1.77 m^3

Page 104 **1** 325.2 cm^3 **2** B The radius is squared in the volume formula. As the radius of *B* is twice that of *A*, volume will be four times larger; but, as *B* is half the height, its volume is now twice that of *A*. **3 a** 664.2 cm^3 **b** 7754.7 cm^3 **c** 1012.1 cm^3 **d** 695.2 cm^3

Page 105 **1 a** 1 mL **b** 1 L **c** 1000 L **2** 15 L **3 a** 1003 cm^3 **b** 1 L **4 a** 6.72 m^3 **b** 6720 L **c** 104 mm **5 a** \$23 998.40 **b** 672 000 L

Page 106 **1 a** 14.14 m^3 **b** 78 days **2 a** 9.425 L **b** 565.5 L **3 a** 89 250 mL = 89.25 L **b** 11 fish **4 a** 1440 cm^3 **b** 27 792 g = 27.79 kg **5 a** 53.4 cm^3 **b** 478.5 g

Page 107 **1** B **2** C **3** B **4** D **5** C **6** B **7** C **8** D **9** D **10** D

Page 108 **1 a** 128.68 cm^2 **b** 916.84 cm^2 **c** 2110.35 cm^3 **2 a** 6.4 m^3 **b** 18.228 m^3 **c** 1224 m^3 **3 a** 25.6 m^2 **b** 46.48 m^2 **c** 850 m^2 **4 a** 1319.7 cm^2 **b** 22.0 m^2 **c** 85.7 cm^2 **5 a** 448.448 m^3 **b** 448.448 tonnes **c** 140 140 L

Chapter 8 – Similarity

Page 109 **1 a i** 17 **ii** 34 **iii** 12.5 **iv** 25 **v** 15 **vi** 30 **b i** 2 **ii** 2 **iii** 2 **iv** 2 **v** 2 **vi** 2 **c** 2 **d** O **2 a i** 30 **ii** 18 **iii** 39 **iv** 21 **v** 20 **vi** 12 **vii** 26 **viii** 14 **b i** $\frac{2}{3}$ **ii** $\frac{2}{3}$ **iii** $\frac{2}{3}$ **iv** $\frac{2}{3}$ **v** $\frac{2}{3}$ **vi** $\frac{2}{3}$ **vii** $\frac{2}{3}$ **c** $\frac{2}{3}$ **3** t he figures are actually reduced in size

4 a

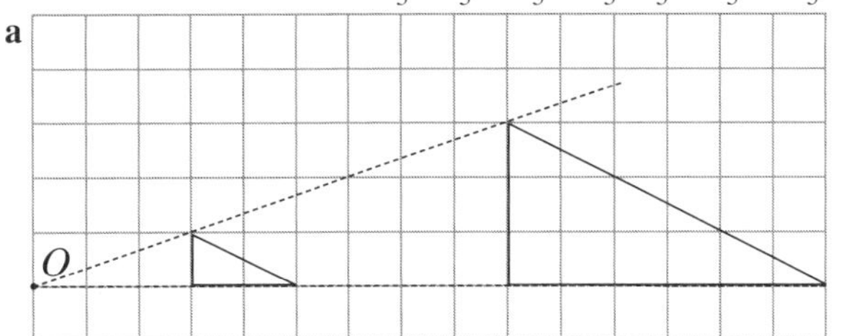

b

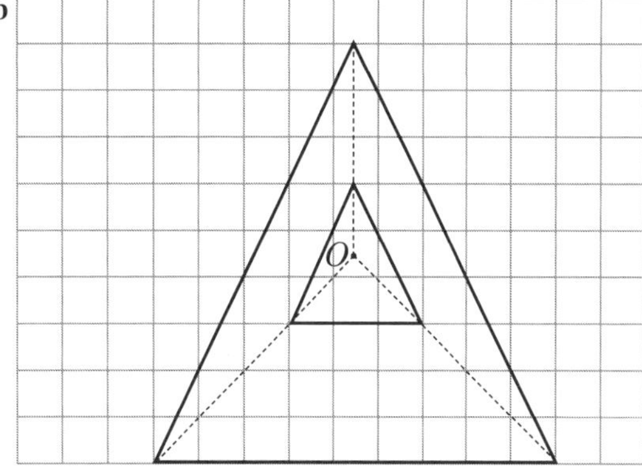

Page 110 **1** 12 cm long and 9 cm wide **2** 6 cm long and 4.4 cm wide **3** 6.5 cm long and 4.5 cm high **4 a** 19 cm **b** 13 cm **5 a** 5 **b** 20 cm **6** 8 cm **7 a** 20 cm **b** 52 cm **8 a** $\frac{4}{3}$ **b** 11.7 cm

Page 111 **1 a** 1.5 **b i** 118° **ii** 62° **iii** 73° **iv** 107° **v** 118° **vi** 62° **vii** 73° **viii** 107° **c** yes **d** yes **e** *EH* **f** $\angle BCD$

2 a

B, 75°, A, 35°, 70°, C

E, 70°, 35°, F, 75°, D

b i $\angle FDE$ **ii** $\angle EFD$ **iii** $\angle DEF$ **d i** *BC* **ii** *BA* **iii** *CA* **3 a** $\angle SPQ = \angle WVU = 136°$; $\angle PQR = \angle VUT = 63°$; $\angle QRS = \angle UTW = 117°$; $\angle RSP = \angle TWV = 44°$ **b i** *VU* **ii** *UT* **iii** *TW* **iv** *WV* **4 a** corresponding, equal **b** corresponding, ratio

Page 112 **1 a** true **b** false **c** true **d** false **e** true **f** false **g** true **h** true **i** true **2** Yes, the diagram is formed with squares of different sizes **3** Similar shapes are the upper window panes, lower window panes and the steps. The upper and lower windows are similar **4** Enlargements of photographs; different sizes of sheets of papers; scale diagrams; models of trains, cars, aeroplanes etc

Page 113 **1 a** *AB* and *DE*, *BC* and *EF*, *AC* and *DF* **b** *AB* and *EF*, *BC* and *FG*, *CD* and *GH*, *DA* and *HE*

Answers

2 a $\frac{AB}{EF} = \frac{BC}{FG} = \frac{CD}{GH} = \frac{DA}{HE}$ **b** $\frac{PQ}{LM} = \frac{QR}{MN} = \frac{RS}{NO} = \frac{SP}{OL}$ **3 a** $\angle PRQ$ **b** $\angle RPQ$ **c** $\angle PQR$ **4 a** *AC* **b** *CB* **c** *AD* **d** *DB* **5 a** proportional **b** shape, size **c** congruent **6 a** Yes. They are all the same shape (but not necessarily the same size) **b** Yes. They are all the same shape (but not necessarily the same size) **c** No, they are not necessarily the same shape (one might be long and thin and another short and wide) **d** Yes, they are all the same shape **e** No, they are not necessarily the same shape.

Page 114 **1 a** $\angle D, \angle E, \angle F$ **b** *DE, EF, DF* **c** *DEF* **2 a** $\frac{1}{2}, \frac{1}{2}, \frac{1}{2}$ **b** *PRQ* **3 a** $\frac{1}{2}$ **b** $\frac{1}{2}$ **c** *EFD* **d** *FED* **4 a** $\frac{2}{3}$ **b** $\frac{2}{3}$ **c** *ZYX*, 90° **d** *ZXY* **5 a** two angles **b** same ratio **c** one angle, same ratio **d** are proportional to the hypotenuse and second side of the other.

Page 115 **1 a** $\Delta ABC \parallel\mid \Delta EFD$, equiangular **b** $\Delta PQR \parallel\mid \Delta TSR$ hypotenuse and side in proportion **c** not necessarily similar **d** $\Delta GHI \parallel\mid \Delta LKJ$, sides in proportion **2 a** equiangular; $\Delta ABC \parallel\mid \Delta DEC$ **b** equiangular; $\Delta ADE \parallel\mid \Delta ABC$ **3 a** $\angle A$ common, as $PQ \parallel BC$ $\angle P = \angle B$ and $\angle Q = \angle C$ **b** $x = 24$

Page 116 **1 a** same ratio **b** one angle; same ratio **c** $\parallel\mid$ or ~ **2 a** *PST* and *PQR* **b** *PQ* **c** 1.5 **d** 9 cm **3 a** equiangular; $\angle B = \angle D$; $\angle ACB = \angle ECD$ being vertically opposite **b** *EDC* **c** 45 **4 a** equiangular **b** $\frac{x}{3} = \frac{6}{4}$ **c** $x = 4\frac{1}{2}$

Page 117 **1 a** equiangular; $x = 5$ **b** equiangular; $y = 4$ **c** equiangular; $m = 20$ **d** two sides and the included angle; $a = 80$, $x = 5$ **2 a** equiangular; $x = 5$ **b** equiangular; $x = 5$ **c** equiangular; $y = 9$ **d** equiangular; $x = 4$

Page 118 **1 a** $x = 8, y = 15$ **b** $x = 3, y = 15$ **c** $x = 6, y = 20$ **d** $x = 9, y = 35$ **e** $x = 25, y = 12$ **f** $x = 12, y = 12$ **2** 55 m

Page 119 **1 a** $\frac{25}{64}$ **b** $\frac{9}{25}$ **c** $\frac{9}{25}$ **2 a** $22\frac{6}{7}$ cm² **b** $3\frac{3}{8}$ cm² **c** $348\frac{4}{9}$ cm² **3 a** 4 times **b** 3:1 **c** Always similar **d** 4:9

Page 120 **1 a** $\frac{8}{27}$ **b** $\frac{1}{8}$ **c** $\frac{8}{125}$ **2 a** 31.89 cm³ **b** 12.98 cm³ **c** 5.1 cm³ **3 a** 125:64 **b i** 8:7 **ii** 512:343 **c** 544 cm³

Page 121 **1** D **2** B **3** A **4** C **5** C **6** D **7** A **8** C **9** B **10** A

Page 122 **1** 18 cm **2 a** All corresponding sides are in proportion **b** *EDC* **c** 37° **3 a** equiangular; $\angle A$ is common, $\angle ADE = \angle ABC$; corresponding $DE \parallel BC$ **b** *ABC* **c** 13.5 cm **4 a** $x = 8$ **b** $y = 10$ **5 a** $x = 4$ **b** $y = 15$ **6 a** 4:25 **b** 8:125 **7 a** 3:7 **b** 27:343

Chapter 9 – Trigonometry

Page 123 **1 a** x = opp., y = adj., z = hyp. **b** x = hyp., y = adj., z = opp. **c** x = opp., y = adj., z = hyp. **d** x = opp., y = adj., z = hyp. **e** x = adj., y = hyp., z = opp. **f** x = hyp., y = opp., z = adj. **2 a** p = opp., q = adj., r = hyp. **b** a = adj., b = opp., c = hyp. **c** d = opp., e = adj., f = hyp. **d** a = opp., b = adj., c = hyp. **e** p = opp., q = hyp., r = adj. **f** l = adj., m = opp., n = hyp. **3 a** *BC* **b** *EF* **c** *PQ*

Page 124 **1 a** $\sin X = \frac{x}{17}, \cos X = \frac{y}{17}, \tan X = \frac{x}{y}$ **b** $\sin\theta = \frac{a}{c}, \cos\theta = \frac{10}{c}, \tan\theta = \frac{a}{10}$ **c** $\sin 30° = \frac{8}{m}, \cos 30° = \frac{p}{m}, \tan 30° = \frac{8}{p}$ **d** $\sin\theta = \frac{a}{c}, \cos\theta = \frac{b}{c}, \tan\theta = \frac{a}{b}$ **e** $\sin\theta = \frac{q}{r}, \cos\theta = \frac{p}{r}, \tan\theta = \frac{q}{p}$ **f** $\sin\theta = \frac{l}{n}, \cos\theta = \frac{m}{n}, \tan\theta = \frac{l}{m}$ **2 a** $\sin\theta = \frac{6}{10}, \cos\theta = \frac{8}{10}, \tan\theta = \frac{6}{8}$ **b** $\sin\theta = \frac{3}{5}, \cos\theta = \frac{4}{5}, \tan\theta = \frac{3}{4}$ **c** $\sin\theta = \frac{12}{13}, \cos\theta = \frac{5}{13}, \tan\theta = \frac{12}{5}$ **d** $\sin\theta = \frac{12}{15}, \cos\theta = \frac{9}{15}, \tan\theta = \frac{12}{9}$ **e** $\sin\theta = \frac{7}{25}, \cos\theta = \frac{24}{25}, \tan\theta = \frac{7}{24}$ **f** $\sin\theta = \frac{15}{17}, \cos\theta = \frac{8}{17}, \tan\theta = \frac{15}{8}$ **3 a** tan **b** sin **c** sin

Page 125 **1 a** 0.934 **b** 0.342 **c** 0.424 **d** 0.122 **e** 0.384 **f** 0.966 **g** 1.111 **h** 0.588 **i** 0.669 **2 a** 3.15 **b** 1.97 **c** 0.686 **d** 7.87 **e** 0.931 **f** 0.414 **g** 0.903 **h** 19.9 **i** 0.461 **3 a** 0.31 **b** 0.04 **c** 22.71 **d** 0.08 **e** 0.14 **f** 28.84 **g** 0.05 **h** 0.15 **i** 65.98 **4 a** 26° **b** 38° **c** 57° **d** 60° **e** 59° **f** 56° **g** 72° **h** 63° **i** 71° **j** 52° **k** 54° **l** 36° **5 a** 36°52' **b** 61°07' **c** 67°0' **d** 66°31' **e** 40°53' **f** 28°50' **g** 52°26' **h** 14°29'

Page 126 **1 a** 6.4 m **b** 3.7 m **c** 12.3 m **d** 7.8 km **2 a** 6.9 cm **b** 13.8 cm **c** 3.4 cm **d** 3.0 cm **3 a** 4.5 cm **b** 10.1 cm **c** 12.4 cm **d** 8.0 cm **4 a** 12.64 cm **b** 22.17 cm **c** 7.45 cm **d** 17.10 cm

Page 127 **1 a** 7.8 cm **b** 3.2 cm **c** 12.2 m **d** 12.2 cm **e** 4.1 cm **f** 11.8 cm **g** 17.6 cm **h** 145.9 mm **2 a** 3.30 cm **b** 16.37 cm **c** 6.38 cm **d** 6.03 km **e** 17.49 m **f** 6.44 m **g** 4.78 m **h** 10.26 km

Page 128 **1 a** 11.8 cm **b** 9.2 cm **c** 15.2 cm **d** 4.7 cm **e** 20.5 cm **f** 11.4 cm **g** 31.3 cm **h** 35.0 cm **i** 15.7 cm **j** 18.2 cm **k** 21.9 cm **l** 50.8 cm **2 a** 9.8 m **b** 2.1 m **c** 61.0 m **d** 93.9 km

Page 129 **1 a** 22° **b** 56° **c** 20° **2 a** 40° **b** 29° **c** 64° **3 a** 22°45' **b** 21°04' **c** 71°34' **d** 66°53' **e** 31°28' **f** 61°31'

Page 130 **1 a** 23°06' **b** 53°08' **c** 23°48' **d** 26°17' **e** 17°43' **f** 64°17' **g** 72°29' **h** 26°42' **i** 48°54' **j** 13°41' **k** 51°45' **l** 63°49' **2** 51° **3** 34°

Page 131 **1 a** 22.6 km **b** 39.4 m **c** 25.6 km **d** 107.2 km **2 a** 46° **b** 37° **c** 32° **d** 58 **3 a** 11.5 cm **b** 13.3 cm **4 a** 117 m **b** 49°

Page 132 **1** 46 m **2** 12.36 cm **3** 30° **4** 6.4 cm **5** 37° **6** 45 m

Page 133 **1** C **2** B **3** D **4** C **5** B **6** D **7** D **8** D **9** D **10** A

Page 134 **1 a** 1.627 **b** 4.096 **c** 0.025 **2 a** 37° **b** 34° **3** 67°23' **4 a** 6.8 m **b** 5.2 m **c** 7.1 m **d** 76.9 m **5 a** 64°47' **b** 23°35' **c** 59°45' **6** 74 m **7** 16.1 cm

Chapter 10 – Probability

Page 135 **1 a** $\frac{1}{6}$ **b** $\frac{1}{6}$ **c** 0 **d** $\frac{1}{2}$ **e** $\frac{1}{2}$ **f** $\frac{1}{3}$ **2 a** $\frac{1}{10}$ **b** $\frac{2}{5}$ **c** $\frac{1}{2}$ **d** $\frac{9}{10}$ **e** 0 **f** $\frac{3}{5}$ **3 a** $\frac{4}{11}$ **b** $\frac{7}{11}$ **c** $\frac{2}{11}$ **d** $\frac{2}{11}$ **e** $\frac{4}{11}$ **f** $\frac{1}{11}$ **4 a** $\frac{3}{5}$ **b** 0 **c** $\frac{2}{5}$ **d** $\frac{1}{5}$ **e** $\frac{1}{5}$ **f** $\frac{3}{5}$ **5 a** $\frac{1}{4}$ **b** $\frac{1}{2}$ **c** $\frac{1}{13}$ **d** $\frac{3}{4}$ **e** $\frac{1}{26}$ **f** $\frac{2}{13}$ **6 a** $\frac{1}{2}$ **b** $\frac{1}{3}$ **c** $\frac{1}{6}$ **d** $\frac{5}{6}$ **e** 0 **f** 1 **6** 0, 1

Answers

PAGE 136 **1 a** $\frac{1}{4}$ **b** $\frac{1}{5}$ **c** $\frac{4}{11}$ **d** $\frac{1}{5}$ **e** $\frac{1}{7}$ **f** $\frac{1}{9}$ **g** $\frac{1}{9}$ **h** $\frac{3}{14}$ **i** $\frac{1}{8}$ **j** $\frac{1}{4}$ **k** $\frac{3}{13}$ **l** $\frac{2}{11}$ **m** $\frac{2}{5}$ **n** $\frac{1}{5}$ **o** $\frac{1}{12}$ **p** $\frac{1}{5}$ **2 a** 0.07, 0.13, 0.10, 0.07, 0.27, 0.13, 0.23 **b** 0.10, 0.17, 0.07, 0.20, 0.10, 0.13, 0.23 **c** 0.15, 0.10, 0.15, 0.20, 0.10, 0.20, 0.10 **d** 0.08, 0.13, 0.10, 0.15, 0.18, 0.25, 0.13 **e** 0.13, 0.10, 0.13, 0.15, 0.20, 0.13, 0.18 **f** 0.12, 0.16, 0.14, 0.08, 0.16, 0.20, 0.14

PAGE 137 **1 a** 100 **b** $\frac{59}{100}$ **c** $\frac{1}{2}$ **d** $\frac{41}{100}$ **e** $\frac{1}{2}$ **f** 1 **g** 50 tails **2 a** $\frac{1}{8}, \frac{5}{24}, \frac{1}{4}, \frac{1}{6}, \frac{1}{9}, \frac{5}{36}$ **b i** $\frac{1}{4}$ **ii** $\frac{35}{72}$ **iii** $\frac{1}{4}$ **iv** $\frac{7}{12}$ **c i** $\frac{1}{6}$ **ii** $\frac{1}{2}$ **iii** $\frac{1}{3}$ **iv** $\frac{1}{2}$

d The answers to parts **ii** and **iv** are fairly close to each other. The answers to parts **i** and **iii** in Lucy's experiment are a bit further from the theoretical probability. Lucy should continue the experiment to see if her answers become closer to the theoretical probability.

PAGE 138 **1 a** 8 **b** 24 **c** 32 **2 a** 5 **b** 15 **c** 20 **d** 5 **e** 10 **f** 35 **3 a** green **b** blue **4** The actual probability of throwing a 5 is 0.17; 0.23 is higher, so Jade's statement is correct **5 a** No. It could be biased because you would expect each number to occur 4 times and 5 has occurred 9 times and 6 only once, but 24 rolls is not enough to be sure. **b** Jimmy should repeat the experiment to see if he gets similar results.

PAGE 139 **1 a i** 147 **ii** 75 **iii** 59 **b i** $\frac{1}{7}$ **ii** $\frac{13}{147}$ **iii** $\frac{38}{147}$ **iv** $\frac{134}{147}$ **v** $\frac{113}{147}$ **2 a**

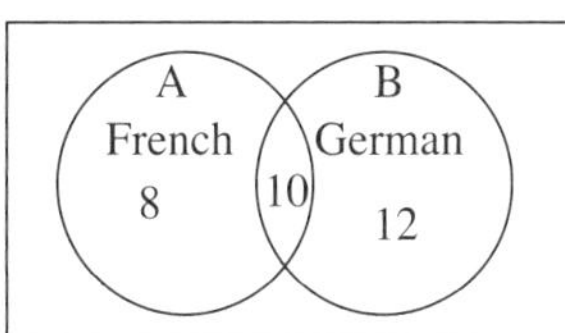

b 10 **c** $\frac{4}{15}$ **3 a** $\frac{1}{11}$ **b** $\frac{3}{55}$ **c** $\frac{1}{5}$ **d** $\frac{31}{55}$ **e** $\frac{4}{11}$ **f** $\frac{13}{55}$ **g** $\frac{41}{55}$ **4 a** $\frac{4}{35}$ **b** $\frac{31}{35}$

PAGE 140 **1 a** 153 **b** 88 **c** 276 **d** $\frac{4}{11}$ **e** 74% **f i** $\frac{51}{92}$ **ii** $\frac{22}{69}$ **2 a** 70% **b** 28% **c i** $\frac{1}{4}$ **ii** $\frac{5}{8}$ **iii** $\frac{7}{40}$

3 a

	Positive	Negative	Total
Male	58	82	140
Female	37	123	160
Total	95	205	300

b i $\frac{41}{60}$ **ii** $\frac{41}{70}$ **iii** $\frac{2}{5}$ **c** Males, more males tested positive and there were fewer males altogether.

PAGE 141 **1 a**

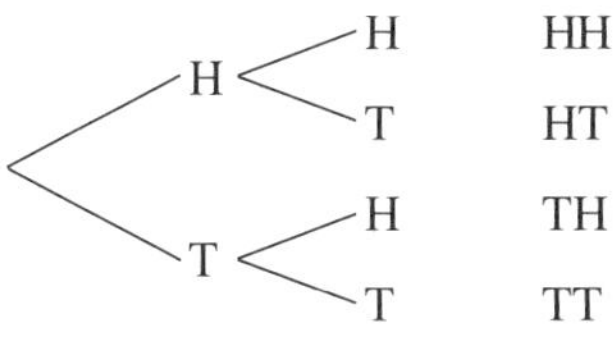

b $\frac{1}{4}$ **c** $\frac{1}{2}$ **d** $\frac{3}{4}$ **e** $\frac{1}{4}$ **2 a**

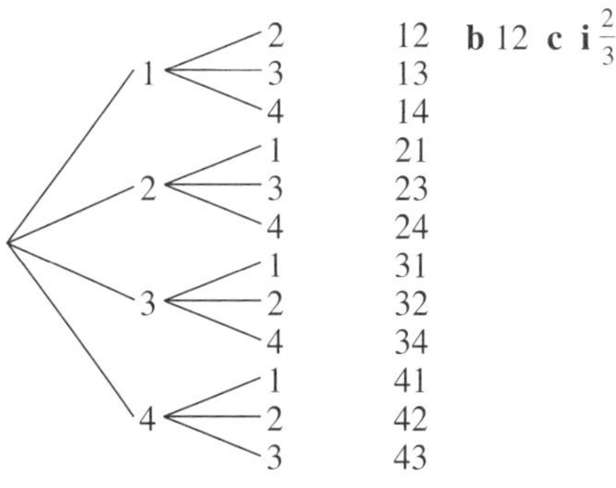

b 12 **c i** $\frac{2}{3}$ **ii** $\frac{1}{3}$ **iii** $\frac{1}{2}$

PAGE 142 **1 a** $\frac{5}{12}$ **b** $\frac{5}{12}$ **c** $\frac{1}{3}$ **d** $\frac{1}{3}$ **e** $\frac{1}{4}$ **f** $\frac{1}{4}$ **2 a** $\frac{5}{12}$ **b** $\frac{4}{11}$ **c** $\frac{1}{3}$ **d** $\frac{3}{11}$ **e** $\frac{1}{4}$ **f** $\frac{2}{11}$ **3 a** 1H 2H 3H 4H 5H 6H 1T 2T 3T 4T 5T 6T **b** $\frac{1}{6}$ **4 a** $\frac{5}{7}$ **b** $\frac{1}{28}$ **5 a** $\frac{4}{9}$ **b** $\frac{2}{3}$

PAGE 143 **1 a** $\frac{5}{14}$ **b** $\frac{9}{14}$ **2 a** 0.8 **b** 0.64 **3 a** $\frac{3}{5}$ **b** $\frac{2}{15}$ **4 a** $\frac{1}{25}$ **b** $\frac{1}{1225}$ **c** $\frac{1128}{1225}$ **5 a** $\frac{1}{12}$ **b** $\frac{1}{4}$ **6 a** $\frac{16}{49}$ **b** $\frac{9}{49}$ **c** $\frac{24}{49}$ **7 a** $\frac{1}{26}$ **b** $\frac{1}{104}$ **8 a** $\frac{1}{26}$ **b** $\frac{1}{24}$

PAGE 144 **1** A **2** D **3** D **4** C **5** A **6** B **7** A **8** D **9** B **10** A

PAGE 145 **1 a** $\frac{1}{2}$ **b** $\frac{1}{2}$ **c** $\frac{1}{3}$ **2 a** $\frac{1}{5}$ **b** 12% **3 a** 200 **b** $\frac{37}{200}$ **c** $\frac{37}{88}$ **d** $\frac{37}{105}$ **4 a**

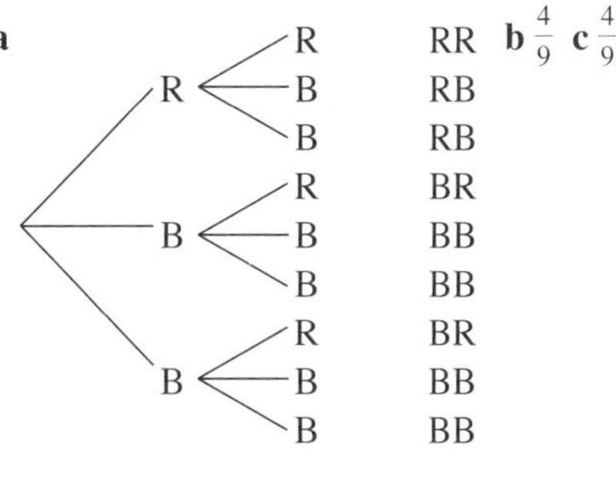

b $\frac{4}{9}$ **c** $\frac{4}{9}$

5 a

R → B, B: RB, RB
B → R, B: BR, BB
B → R, B: BR, BB

b $\frac{1}{3}$ **c** $\frac{2}{3}$

Answers

Chapter 11 – Data representation and analysis

Page 146 **1 a**

Score (x)	Tally	Frequency (f)	$f \times x$	Cumulative frequency
0	𝍸	5	0	5
1	𝍸 𝍸	10	10	15
2	𝍸 𝍸 III	13	26	28
3	𝍸 𝍸 III	13	39	41
4	IIII	4	16	45
5	𝍸	5	25	50
		$\Sigma f = 50$	$\Sigma fx = 116$	

b

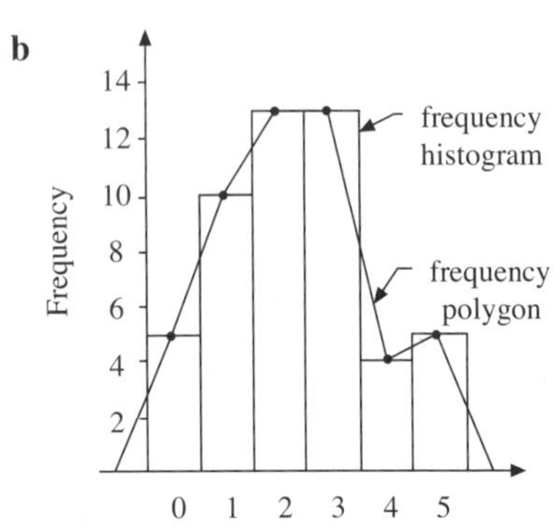

c

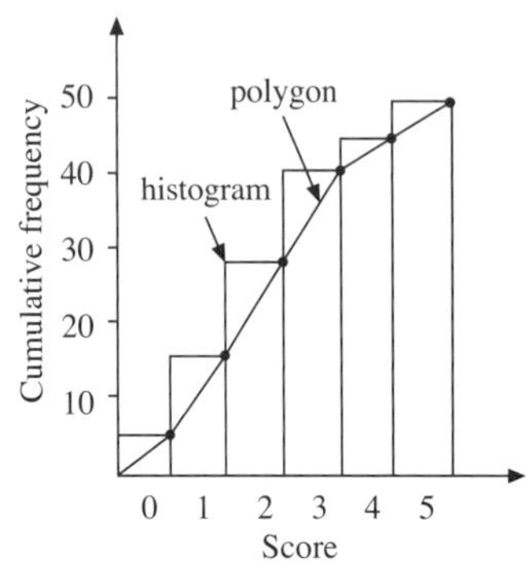

d **i** 2.32 **ii** 2 and 3 **iii** 2 **iv** 5

Page 147 **1 a** mean = 7, mode = 8, median = 7.5, range = 5 **b** mean = 14.9, mode = 15, 18, median = 15, range = 7
c mean = $11\frac{2}{11}$, mode = 8, median = 11, range = 9 **d** mean = 63.625, mode = 56, median = 63, range = 38
e mean = 4.2, mode = 5, median = 4.5, range = 4 **f** mean = 8.125, mode = 9, median = 9, range = 5
g mean = $3\frac{5}{11}$, mode = 3, median = 3, range = 4 **h** mean = 45.5, mode = 52, median = 52, range = 37

2 a

Score (x)	Frequency (f)	$f \times x$	Cumulative frequency
1	3	3	3
2	6	12	9
3	8	24	17
4	7	28	24
5	5	25	29
6	4	24	33

mean = $3.\dot{5}\dot{1}$, mode = 3, median = 3, range = 5

b

Score (x)	Frequency (f)	$f \times x$	Cumulative frequency
5	12	60	12
6	19	114	31
7	18	126	49
8	15	120	64
9	10	90	74
10	13	130	87

mean = 7.4, mode = 6, median = 7, range = 5

c

Score (x)	Frequency (f)	$f \times x$	Cumulative frequency
16	8	128	8
17	6	102	14
18	7	126	21
19	10	190	31
20	5	100	36

mean = $17.9\dot{4}$, mode = 19, median = 18, range = 4

d

Score (x)	Frequency (f)	$f \times x$	Cumulative frequency
16	5	80	5
17	7	119	12
18	8	144	20
19	14	266	34
20	6	120	40

mean = 18.225, mode = 19, median = 18.5, range = 4

Page 148 **1 a** **i** 95.5 **ii** 81 **iii** 96 **b** Barry was above the average but he did not do well, he was second lowest in the class. When data includes atypical scores, the median should be used as the measure of "the middle" **2** Few houses would have identical prices so the mode is not used. If one or several very expensive homes were sold this would significantly increase the mean, the mean would no longer be a good indicator of the price of the majority of houses sold. The median would be unaffected by the few high prices.
3 a 16.7 **b** 14 **c** 16 **d** The shop owner would sell more of this size and so would need to stock more of the modal size.

Answers

PAGE 149 **1 a** $18\frac{1}{9}$ **b** 94% **2 a** \$39 **b** \$35 **c** \$25 **d** \$35 **e** \$39 **f** \$35 **g** \$25 **h** \$35 **i** There are no differences. The new price is the same as the mean so the mean did not change. The other measures of location and spread also did not need to change. **3 a** 58 **b** 6.4 **c** 5 **d** 5 **e** 9 **f** The opposing team, they scored a total of 63 points compared to 58 for the first team. **g** The opposing team, the players on the opposing team were more consistent. **4 a** 15.3 **b** 16 **c** 5 **d** It decreases its value **e** No, the middle value of the ordered numbers does not depend on the smallest size.

PAGE 150 **1 a** symmetrical **b** negatively skewed **c** positively skewed **d** symmetrical **e** none **f** symmetrical **g** positively skewed **h** none **i** negatively skewed **2 a** negatively skewed **b** symmetrical **c** positively skewed

PAGE 151 **1 a** yes **b** 14, 18 **c** yes **d** The mean and median are both 16. Because the graph is completely symmetrical, both the mean and the median must be the same as the middle value. **2 a** The data is negatively skewed. There is a single mode. **b** The data is symmetrical and bi-modal. **c** The data is positively skewed. There is a single mode. **3 a** symmetrical **b** positively skewed **c** negatively skewed **4 a** negatively skewed **b** symmetrical **c** positively skewed

PAGE 152 **1 a** The data has 2 modes, it is not smooth and is positively skewed. **b** The data has 1 mode, it is smooth and symmetrical. **c** data is bi-modal, smooth and symmetrical **2 a**

Stem	Leaf
0	5 7 9
1	2 5 6 8 9
2	0 2 4 5 5 7 8 9
3	0 0 2 6 6 8
4	0 2 5

b 25 **c** No **d** The modal interval was 20 to 29 hours of study, there are no outliers but there is a large spread in the number of hours spent studying.

3 a 2 **b** Negatively skewed. **c** No outliers, the mean would be less than the median.

PAGE 153 **1 a**

Task *A*	Stem	Task *B*
8	0	2 6 7
5 0	1	2 5
	2	1 2 3 3 5 8
8 5 4 2 0	3	1 1 1 1
3	4	8
6 2	5	2 3 6
8 3 1	6	2 9 9 9
9 8 3 1 0	7	
4 4 0	8	1 5 8
9 8 8 2 2 2 2 2	9	3 4 9
	10	0

b 91 **c** 98 **d** 98 **e** 30 **f** 92 **g** 31 **h** 70.5 **i** 39.5 **j** Task A **2 a** 12 **b** 51 **c** 43 **d** The groups both have the same median; 17 **e** 18.3 **f** 21.0 **g** Although both groups have the same median, the mean is higher for females. This reflects the fact that the females have more higher scores and a greater spread. **h** Both groups have positively skewed displays although this is more pronounced for females.

PAGE 154 **1 a**

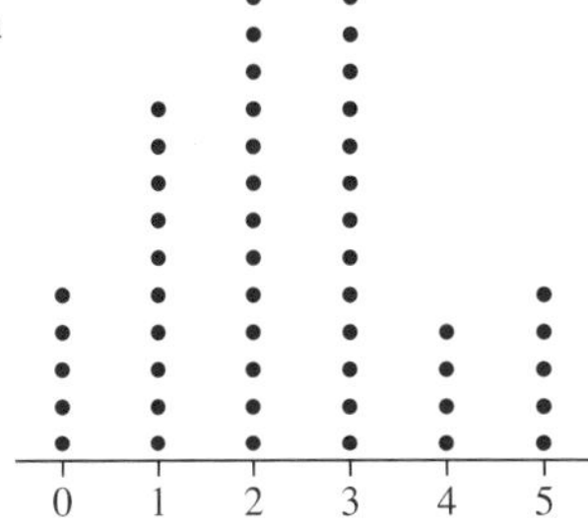

b The data is bi-modal, with modes of 2 and 3. The display is positively skewed.

2 a Class *A*

Class *B*

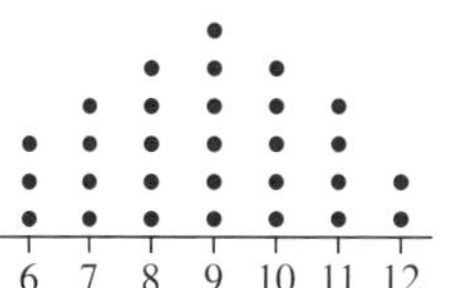

b The data for class *A* is positively skewed while that for class *B* is symmetrical. Class *A* has the greater spread of scores (range of 7 compared to 6 for class *B*). For class *A* the mode is 6, median is 8 and the mean is $8.0\dot{3}$, while for class *B* the mean, mode and median are all 9.

3 a A

B

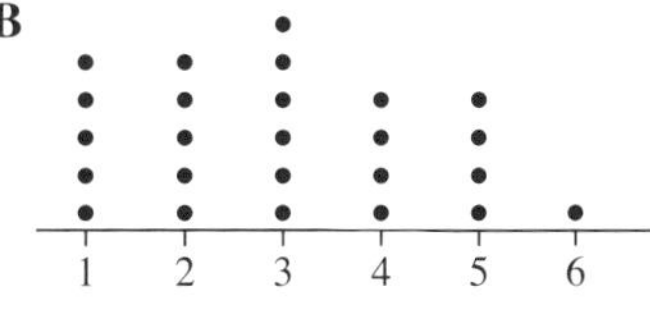

b Both sets are positively skewed and both have the same mode (3). The second set has the higher range (6 compared to 5) but lower median (3 compared to 4) and lower mean ($3.\dot{3}$ compared to 4.2)

PAGE 155 **1** D **2** C **3** C **4** A **5** B **6** B **7** B **8** B **9** C **10** B

Answers

PAGE 156 1 a

Score (x)	Tally	Frequency (f)	$f \times x$	Cumulative frequency
3					3	9	3
4	𝍸	5	20	8
5	𝍸		6	30	14
6	𝍸 ||	7	42	21
7	𝍸 |||	8	56	29
8	𝍸		6	48	35
9					3	27	38
10				2	20	40

b 7 **c** 7 **d** 6 **e** 6.3 **2 a** Females by 13 cm **b** 157.5 cm **c** 13.5 cm **d** 169 cm, 171 cm **e** Both data sets are reasonably symmetrical. The heights for males are higher than for females and the height of the males are also more consistent **3 a** 7 **b** 1.3 **c** 2 **d** Test 2, range is lower (5 compared with 4) **e** The data in Test 1 is symmetrical, that in Test 2 is negatively skewed. The mean, mode and median are all higher than Test 2.

Exam Paper 1

PAGE 157 **1** C **2** C **3** A **4** A **5** C **6** C **7** C **8** A **9** C **10** A

PAGE 158 **1** 4.06×10^{-4} **2** $9a + 3ab$ **3** No **4** \$90 **5** 11.5 **6** $-\frac{2}{3}$ **7** 63.2 **8** $\frac{4}{3}$ **9** 17.5 cm **10** $\frac{1}{4}$ **11** $x^2 + y^2 = 25$ **12** 10 **13** $\frac{1}{25}$ **14** 45 m^3 **15** 525 km

PAGE 159 **1** $p = 3$ **2** $a = -17\frac{1}{2}$ **3 a** $-a + 12$ **b** $3x^2 - 2x - 8$ **4** 4.32 L **5** \$1830 **6 a** 585.4 cm^3 **b** 387.2 cm^2 **7 a** Two pairs of sides are in proportion $(\frac{AB}{AD} = \frac{AE}{AC} = \frac{2}{3})$ and the included angles are equal ($\angle A$ is common) **b** 24 cm **8 a** 15 units **b** $(-1, 2.5)$ **c** $-\frac{3}{4}$

PAGE 160 **9 a** 37 cm **b** 747.6 cm^2 **10 a** negatively skewed **b** 7.92 **c** 0.08 **11 a** 16 cm **b** 133°

12 a 4 **b** –3 **c**

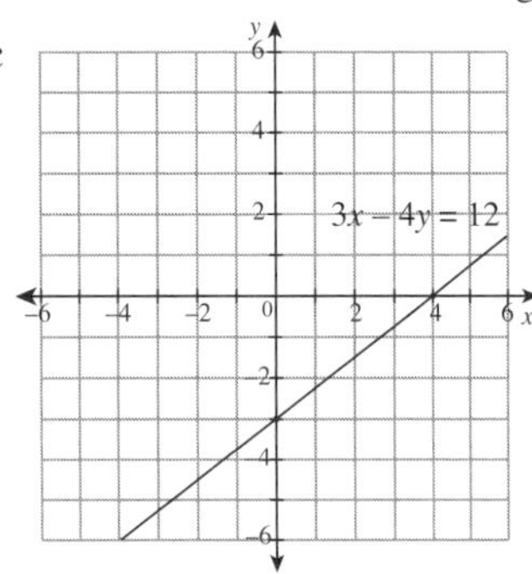

13 a $x = 4$ or -4 **b**

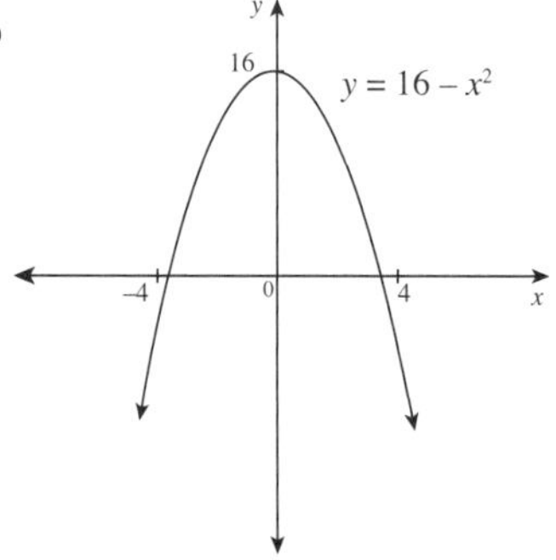

Exam Paper 2

PAGE 161 **1** C **2** B **3** D **4** D **5** C **6** A **7** D **8** A **9** D **10** A

PAGE 162 **1** 1 **2** 107 **3** 2.16×10^4 **4** 13 **5** $x = 0$ **6** 8 **7** $2(p - 2q)$ **8** $4a^6$ **9** 27 cm^3 **10** 120 km/h **11** 0.0025 **12** $x = 4$ **13** $x = 9$ **14** $x = 8$ cm **15** *BDC*

PAGE 163 **1** $-10x - 1$ **2 a** $x = 8$ **b** $x = 2\frac{1}{2}$ **3** 68° **4 a** $\frac{1}{6}$ **b** $\frac{5}{36}$ **c** $\frac{3}{18}$ **5 a** $AB = BC = 4\sqrt{2}$ **b** $\angle OBA = \angle OBC = 45°$ **c** $(0, 0)$ **d** $(2, 2)$ **e** 1

PAGE 164 **6 a** 11.3 m^3 **b** 1130 L **c** 15.7 days **7** 168.3 cm^2 **8 a** \$4800 **b** \$9520 **c** \$612

PAGE 165 **9 a** 72 **b** 9Y by 1 **c** The scores for 9R are slightly negatively skewed while those for 9Y are more symmetrical. 9R performed better than 9Y the median mark in 9R is 80 compared to 73 for 9Y. **10** Yes **11 a** $c = -3$ **b** $a = 2$

Exam Paper 3

PAGE 166 **1** D **2** C **3** C **4** B **5** D **6** C **7** D **8** D **9** A **10** A

PAGE 167 **1** 26.686 **2** $27a^9$ **3** $ab(a - b)$ **4** 9 **5** $\frac{1}{36}$ **6** $-\frac{2}{3}$ **7** \$759.50 **8** 5.5 tonnes **9** $\frac{1}{8}$ **10** \$1350 **11** $1\frac{1}{3}$ **12** $\frac{1}{15}$ **13** $8x^3y^2 - 10x^2y^3$ **14** $\frac{3}{14}$ **15** 294 cm^2

Answers

PAGE 168 **1** **a** $p = 7$ **b** $x = 10\frac{2}{3}$ **2** 62.5 m^2 **3** **a** 8 m **3** 104 m^2 **4** \$14.40 **5** \$804 **6** 5.08 cm

PAGE 169 **7** **a** equiangular **b** *BAC* **c** 21.33 m **8** **a** 55° **b** 34 m **9** **a**

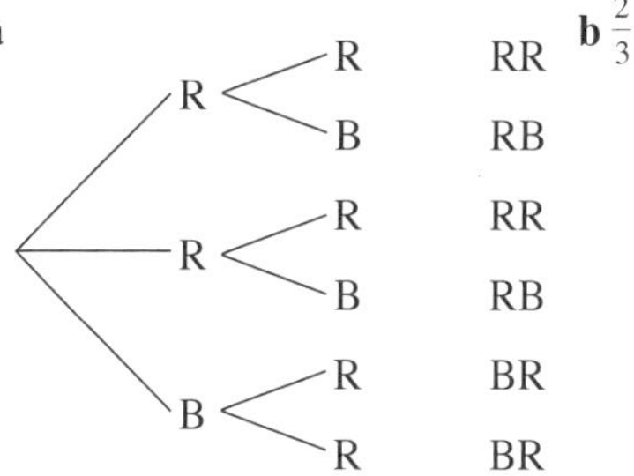

b $\frac{2}{3}$

PAGE 170 **10** **a** $x^2 + y^2 = 64$ **b** $\sqrt{65}$ units **c** outside **11** **a** -2 **b** 1 **c** $y = -2x + 1$ **12** **a** 1.4 **b** 2 **c** 1 **d** The data for test 1 is symmetrical while that for test 2 is negatively skewed. The scores for test 2 are more consistent (range is lower). Also, the mean and median are higher for test 2.

Notes